JOURNAL OF REPRODUCTION AND FERTILITY

SUPPLEMENT No. 41

GENETIC ENGINEERING OF ANIMALS

Proceedings of the Second Symposium

on

Genetic Engineering of Animals

held at

Cornell University, Ithaca, NY, USA

June 1989

EDITED BY

W. Hansel and Barbara J. Weir

Journal of Reproduction & Fertility

1990

First published 1990

ISSN 0449-3087
ISBN 0 906545 19 6

Published by **The Journals of Reproduction and Fertility Ltd.**

Agents for distribution: **The Biochemical Society Book Depot, P.O. Box 32, Commerce Way, Whitehall Industrial Estate, Colchester, CO2 8HP, Essex, U.K.**

Printed in Great Britain by
Henry Ling Ltd., at
The Dorset Press, Dorchester, Dorset

CONTENTS

TRANSGENIC POULTRY

FUTURE AND POTENTIAL

ABSTRACTS OF POSTERS

J. Reprod. Fert., Suppl. **41** (1990), vii

Printed in Great Britain

Preface

Five years have elapsed since the First Symposium on Genetic Engineering in Animals was held at the University of California, Davis (see *Genetic Engineering of Animals: An Agricultural Perspective,* Eds J. W. Evans & A. Hollaender, Plenum Press, New York). At that time the future for the field appeared bright and J. W. Evans and Charles E. Hess, in welcoming the conference participants, pointed out that the field of animal breeding was poised to undergo revolutionary changes as a result of utilization of breakthroughs in molecular genetics, embryo manipulation and gene transfer systems.

It was, perhaps, predictable that this revolution would take some unexpected turns, resulting in disappointments in some areas and unanticipated advances in others. As expected, a number of transgenic pigs, sheep and chickens were produced by random gene insertions by microinjections of gene constructs into pronuclei and by use of replication-defective viral vectors. However, these methods proved to be of limited value in producing transgenic lines of livestock. The efficiency of the microinjection technique in domestic animals turned out to be very low (0·5–1·0%). This fact, coupled with the high cost of research with the large domestic animals made each transgenic animal produced extremely expensive. In addition, our lack of knowledge of promoter and enhancer sequences resulted in production of transgenic animals in which the inserted gene could not be controlled so as to give the desired response (for example, growth) at the desired time.

All of these problems were addressed in the Second Symposium. New techniques described at this symposium may make it possible to carry out targeted gene insertion in domestic animals, as has been done in mice. Perhaps the most promising of these new methods is the use of pluripotent embryo stem cells as vectors for producing desired mutations. Introduction of genetic material into cultured stem cells is clearly more efficient and less costly than microinjection techniques. In addition to making possible targeted gene insertion, this technique can be used to inactivate genes when sequence replacement vectors, rather than insertion vectors are used. This possibility is of particular interest to the livestock industry, because inactivation of genes, such as the genes for inhibin and somatostatin, that have inhibitory physiological effects may result in improvements of a number of economically important productive traits.

The ability to produce multiple copies of an embryo is also of great potential significance to the livestock industry because it enables phenotypic selection to be made, resulting in very rapid changes in selected traits of economic value. Cloning of embryos by nuclear transplantation has now been developed for cattle, sheep, pigs and rabbits and results reported by several workers at this conference indicate that hundreds of pregnancies have already been produced by this technique. Ultimately, it may be possible to combine nuclear transplantation and embryo stem-cell culture techniques to produce transgenic animals without producing germ-line chimaeras. Promising new techniques for transmission of exogenous genes into avian species have also been developed. One of the most promising of these, microinjection of replication-defective reticuloendotheliosis virus vectors beneath the blastoderm of unincubated chick embryos, was reported at the symposium.

These are but a few of the remarkable discoveries in the rapidly advancing field of animal genetic engineering that were reported at the Second Symposium. One can now envision the production of cloned domestic embryos, produced by in-vitro fertilization, nuclear transfer, and embryo stem-cell techniques and possessing highly specific traits for improved growth, lactation, reproduction and disease resistance. The goal of producing significant numbers of marketable genetically engineered animals by the turn of the century may yet be achieved, but the methods used to achieve this goal may be quite different from those envisioned in 1985 at the First Symposium on Genetic Engineering of Animals.

W. Hansel

Sponsors

Major sponsor
Cornell Biotechnology Program, 130 Biotechnology Building, Cornell University, Ithaca, NY 14853, USA

Dr Richard McCarty, Director
Dr Milton Zaitlin, Associate Director
Dr Maureen Hanson, Associate Director
Ms Margaret Arion, Executive Director
Ms Lori Neiderman, Administrative Aide

Academic sponsors
United States Department of Agriculture
United States National Science Foundation

Corporate sponsors
SmithKline Beckman, Animal Health Division, West Chester, PA, USA
Merck Sharp, and Dohme Research Laboratories, Rahway, NJ, USA
Lilly Research Laboratories, Greenfield, IN, USA
Arkzo Pharma, Millsboro, DE, USA
Animal Breeders Service, DeForest, WI, USA
Shaver Poultry Breeding Farms Limited, Cambridge, Ontario, Canada
American Cyanamid Company, Agricultural Research Division, Princeton, NJ, USA
Fisher Scientific, Pittsburgh, PA, USA
Beckman Instruments, Inc., Fullerton, CA, USA

National Organizing Committee

W. Hansel	Cornell University, Ithaca, NY, USA
N. L. First	University of Wisconsin, Madison, WI, USA
J. J. Kopchick	Edison Biotechnology Center, Ohio University, Athens, OH, USA
B. Osburn	University of California-Davis, Davis, CA, USA
C. E. Rexroad, Jr	USDA, Beltsville, MD, USA
J. Rossant	Hospital for Sick Children, Toronto, Canada
R. Silva	USDA, Regional Poultry Research Laboratory, East Lansing, MI, USA
D. Singer	National Cancer Institute, Bethesda, MD, USA
J. Womack	Texas A&M University, College Station, TX, USA

Local Organizing Committee

W. Hansel	Department of Physiology
D. F. Antczak	James A. Baker Institute
R. J. Avery	Department of Microbiology
C. Batt	Department of Food Science
W. R. Butler	Department of Animal Science
R. H. Foote	Department of Animal Science
M. Zaitlin	Department of Plant Pathology

LIST OF PARTICIPANTS

Abplanalp, H.	University of California, Dept. of Avian Science, Davis, CA 95616, USA
Alcivar, A.	Tufts University, Biology Department, Medford, MA 02155, USA
Alestrom, P.	Agriculture University Norway, P.O. Box 36, N-1432 AS-NLH, Norway
Alhonen, L.	University of Kuopio, P.O.B. 6 SF 70211, Kuopio, Finland
Alila, H.	SmithKline Beecham Co., West Chester, PA 19380, USA
Andresen, O.	Norwegian College Vet. Med., Pb 8146 Dep, 0033 Oslo, Norway
Angelos, J.	Cornell University, Baker Inst. Animal Health, Ithaca, NY 14853, USA
Antczak, D.F.	Cornell University, Baker Inst. Animal Health, Ithaca, NY 14853, USA
Appel, M.J.	Cornell University, Vet. Microbiology, Ithaca, NY 14853, USA
Avery, R.J.	Cornell University, Vet. Microbiology, Ithaca, NY 14853, USA
Avis, J.	LaJolla Cancer Research Found., LaJolla, CA 92037, USA
Bachrach, H.	U.S. Congress OTA, P.O. Box 1054, Southold, NY 1197, USA
Baile, C.	Monsanto Company, St. Louis, MO 63198, USA
Baker, J.	Texas A&M University, College Station, TX 77840, USA
Balk, M.W.	Charles River Laboratories, Wilmington, MA 01887, USA
Barr, P.	Cornell University, Vet. Microbiology Immunology, Ithaca, NY 14853, USA
Batt, C.	Cornell University, Food Science, Ithaca, NY 14853, USA
Bean, B.	Eastern A.I. Cooperative, P.O. Box 518, Ithaca, NY 14851, USA
Beerman, D.	Dept. of Animal Science, Cornell University, Ithaca, NY 14853, USA
Bennett, J.	Johns Hopkins University, Dept. of Physiology, Baltimore, MD 21205, USA
Bensadoun, A.	Cornell University, Div. of Nut. Sciences, Ithaca, NY 14853, USA
Bergersen, O.	Dept. of Animal Genetics, P.O. Box 8156, Dep. N-0033 05101, Olso, Norway
Berkowitz, D.	USDA, FSIS/TTA/TS, Washington, D.C. 20250, USA
Betteridge, K.J.	Ontario Veterinary College, University of Guelph, Guelph, Ontario, Canada N1G 2W1
Biggs, P.	Willows, London Rd, St. Ives, Cambs PE17 4ES, UK
Black, D.	University of Massachusetts, Paige Laboratory, Amherst, MA 01003, USA
Bloom, S.E.	Cornell University, Poultry and Avian Sciences, Ithaca, NY 14853, USA
Boisclair, Y.	Cornell University, Dept. of Animal Science, Ithaca, NY 14853, USA
Bosselman, R.A.	1900 Oak Terrace Lane, Thousand Oaks, CA 91360, USA
Bowen, R.	Colorado State University, Dept. of Physiology, Ft. Collins, CO 80523, USA
Brascamp, E.W.	6700 AH Wageningen, The Netherlands
Breitman, M.	Mt. Sinai Research Institute, Toronto, Ontario, Canada M5G 1X5
Brown, G.R.	University of Massachusetts, Amherst, MA 01002, USA
Brumbaugh, J.	University of Nebraska, School of Biological Sciences, Lincoln, NE 68588-0118, USA
Bruns, P.	Dept. of Animal Science, Cornell University, Ithaca, NY 14853, USA
Burfening, P.	Montana State University, Animal and Range Science Dept., Bozeman, MT 59717, USA
Butler, R.	Cornell University, Dept. of Animal Science, Ithaca, NY 14853, USA
Caanitz, H.	Cornell University, Dept. of Physiology, Ithaca, NY 14853, USA
Calnek, B.	Cornell University, Avian & Aquatic Animal Med., Ithaca, NY 14853, USA
Canseco, R.	Virginia Tech., Dairy Science Dept., Blacksburg, VA 24061, USA
Carbone, L.	Cornell University, Center for Research Animal Resource, Ithaca, NY 14853, USA
Carson, C.A.	University of Missouri, Dept. of Vet. Microbiology, Columbia, MO, USA
Carte, I.	Hubbard Farms, Turnpike St. Walpole, NH -3608, USA
Carvallo, D.	Institut Merieux, Lyon 69007, France
Cavalcoli, J. D.	Louisiana State University, Dept. Animal Science, Baton Rouge, LA 70803, USA
Chang, W.-C.	Inst. of Biol. Chem., Academia Sinica, Taipei, Taiwan
Changhsin, W.	Beijing Ag. University, Beijing, China
Chen, N.Y.	Ohio University, Athens, OH 45701, USA
Chen, X.Z.	Ohio University, Edison Animal Biotech. Center, Athens, OH 45701, USA
Chrisman, L.	Purdue University, Dept. of Animal Science, West Lafayette, IN 47906, USA
Cioffi, J.	Ohio University, Edison Animal Biotech. Center, Athens, OH 45701, USA
Cioffi, L.	Embryogen, Athens, OH 45701, USA
Clark, J.R.	Texas Tech University, Dept. of Animal Science, Lubbock, TX 79409-2142, USA
Clement-Sengewald, A.	Dept. of Molecular Animal Breeding, D-8000 Muenchen 22, FRG
Cloud, J.	University of Idaho, Dept. of Biological Sciences, Moscow, ID 83843, USA
Coggins, L.	North Carolina State Univ., College of Veterinary Medicine, Raleigh, NC 27695, USA
Coleman, L.	Cornell University, Baker Inst. Animal Health, Ithaca, NY 14853, USA

Concannon, P.	Cornell University, Physiology, Ithaca, NY 14853, USA
Cook, R.F.	Ohio State University, Dept. Dairy Science, Wooster, OH 44691, USA
Courot, M.	I.N.R.A., Nouzilly, 37380, France
Crittenden, L.B.	USDA, RPRL, East Lansing, MI 48823, USA
Daiss, J.L.	Eastman Kodak Co., Rochester, NY 14650-2119, USA
Daniel, D.	Albright College, Dept. of Biology, Reading, PA 19612, USA
Daniels, L.B.	University of Arkansas, Agri 205, Fayetteville, AR 72701, USA
Davey, H.	University of Guelph, Dept. of Molecular Biology, Guelph, Ontario, Canada
De Buysscher, E.	College of Veterinary Medicine, North Carolina State Univ., Raleigh, NC 27606, USA
Delany, M.E.	Cornell University, Poultry and Avian Sciences, Ithaca, NY 14853, USA
Dickey, J.F.	Clemson University, Dept. Dairy Science, Clemson, SC 29634, USA
Diehl, J.	Clemson University, Dept. of Animal Science, Seneca, SC 29678, USA
Dietert, R.R.	Cornell University, Dept. of Poultry and Avian Sci., Ithaca, NY 14853, USA
Dietz, A.	Texas A&M Dept. of Vet. Pathology, College Station, TX 77843, USA
Donaldson, W.	Cornell University, J. A. Baker Inst. Animal Health, Ithaca, NY 14853, USA
Dougherty, D.	Cornell University, J. A. Baker Inst. Animal Health, Ithaca, NY 14853, USA
Dum, L.V.	Cornell University, Dept. of Avian & Aquatic Sci., Ithaca, NY 14853, USA
Ebert, K.	Tufts University Medical School, North Grafton, MA 01536, USA
Eggleton, K.	Clarion University, Biology, Clarion, PA 16214, USA
Ellendorff, F.	Institut fur Kleintierzucht, Dornbergstr. 25-27, Celle D-3100, FRG
Elliott, J.M.	Cornell University, Department of Animal Science, Ithaca, NY 14853, USA
Engelhardt, D.	Enzo Biochem, Inc., New York, NY 10013, USA
Evans, E.	Cornell University, Vet. Anatomy, Ithaca, NY 14853, USA
Evans, H.	Cornell University, Vet. Anatomy, Ithaca, NY 14853, USA
Evans, M.J.	University of Cambridge, Dept. of Genetics, Cambridge CB2 3EH, UK
Fabricant, J.	KVM, Houston, TX 77099, USA
Farin, C.E.	University of Missouri, 158 Animal Sci. Res. Center, Columbia, MO 65211, USA
Fernandez, E.	Ohio University, Edison Animal Biotechnology, Athens, OH 45701, USA
First, N.L.	University of Wisconsin, Dept. of Meat & Animal Science, Madison, WI 53706, USA
Focht, R.J.	E.I. duPont de Nemours & Co., Wilmington, DE 19880, USA
Foote, R.H.	Cornell University, Animal Science, Ithaca, NY 14853, USA
Fortune, J.	Cornell University, Physiology, Ithaca, NY 14853, USA
Gagne, M.	Labs d'Ontogenie & Repro., CHUL, Quebec, Canada G1V 4G2
Geisow, M.	Biotechnology Interface Ltd, Nottingham NG7 1FD, UK
Geneste, B.	Rhone Poulenc Research Center, Savage, MD 20763, USA
Ghangas, G.S.	Cornell University, Sr. Research Assoc., Ithaca, NY 14853, USA
Giles, J.R.	Cornell University, Animal Science, Ithaca, NY 14853, USA
Godke, R.A.	Louisiana State University, Dept. of Animal Science, Baton Rouge, LA 70803, USA
Gonzalez, I.	Biotrax, Inc., Baltimore, MD 21227, USA
Gordon, K.	394 Marchnost, Boston, MA, USA
Gore-Langton, R.	University of Western Ontario, Dept. of Ob. & Gynaec., London, Ontario, Canada N6A 5A5
Gorman, K.	Eastman Kodak Co., Rochester, NY 14650-02118, USA
Gowe, R.S.	Shaver Poultry Breeding Farm, Cambridge, Ontario, Canada N1R 5V9
Gravance, C.G.	1263 Filbert Ave., Chico, CA 95926, USA
Gray, H.G.	University of RI, Kingston, RI 02881, USA
Grohn, Y.T.	16 Laura Lane, Ithaca, NY 14850, USA
Groschup, M.	Cornell University, Microbio, Immunol. & Parasitology, Ithaca, NY 14853, USA
Haeussler, H.W.	Cornell University, Patents & Technology Marketing, Ithaca, NY 14853, USA
Hagen, D.	Penn State University, Madison, WI 53705, USA
Hammer, R.E.	Univ. of Texas Southwestern Med. Ctr, Dallas, TX 75235, USA
Hansel, W.	Cornell University, Physiology, Ithaca, NY 14853, USA
Hanson, H.	Cornell University, Genetics and Development, Ithaca, NY 14853, USA
Hanson, R.W.	Case Western Reserve Univ. School of Medicine, Cleveland, OH 44106, USA
Harwood, D.	Campbell Institute, Farmington, AR 72730, USA
Hasler, J.F.	Em Tran. Inc., Elizabethtown, PA 17022, USA
Hauge, J.G.	College of Veterinary Medicine, P.O. Box 8146 Dep, 0033 Oslo 1, Norway
Henricks, D.M.	Clemson University, Dept. of Animal Science, Clemson, SC 29634, USA
Highkin, M.	Monsanto Co., Chesterfield, MO 63198, USA

Hines, D.L.	Salsbury Laboratories, Charles City, IA 50616, USA
Hinshelwood, M.M.	Cornell University, Physiology, Ithaca, NY 14853, USA
Hodgetts, R.	University of Alberta, Dept. of Genetics, Edmonton, Alberta, Canada T6G 2E9
Holzschy, D.L.	Eastman Kodak Co., Rochester, NY 14650-2119, USA
Hoshi, M.	University of Wisconsin, Dept. of Meat & Animal Science, Madison, WI 53706, USA
Huang, M.	National Chung-Hsing University, Department of Animal Science, Taichung, Taiwan
Hughes, S.	NCI-Frederick Cancer Research Facility, P.O. Box B, Frederick, MD 21701-1013, USA
Hulsebosch, M.	526 W. Seneca St., Ithaca, NY 14850, USA
Huston, K.	OARDC, Norch Central DAL, Wooster, OH 44691, USA
Hyland, F.	Cornell University, Animal Breeding, Ithaca, NY 14853, USA
Itagaki, Y.	Itoham Foods, Inc., Tokyo 153, Japan
Janne, J.	University of Kuopio, Dept. of Biochemistry, P.O.P. 6 SF-70211, Kuopio, Finland
Jayachandra, S.	Clarion University, Clarion, PA 16214, USA
Jego, Y.	Cornell University, Department of Animal Science, Ithaca, NY 14853, USA
Jochle, W.	Wolfgang Jochle Associates, Denville, NJ 07834, USA
Johnson, P.	Cornell University, Poultry and Avian Sciences, Ithaca, NY 14853, USA
Kamarck, M.	Miles Research Center, Molecular Therapeutics, West Haven, CT 06525, USA
Kamonpatana, M.	Chulalorgkorn University, Bangkok 10330, Thailand
Karaca, K.	Avian Aquatic Animal Medicine, Cornell University, Ithaca, NY 14853, USA
Kent, M.	University of Wisconsin, Dept. of Meat & Animal Science, Madison, WI 53706, USA
Keown, J.	University of Nebraska, A218 Animal Science, Lincoln, NE 68583-0908, USA
King, A. D.	James Baker Institute of Animal Health, Cornell University, Ithaca, NY 14853, USA
Koong, L.J.	Oregon State University, Ag. Expt. Station, Corvallis, OR 97331, USA
Kopchick, J.	Ohio University, Edison Animal Biotech Center, Athens, OH 45701, USA
Kostomakhin, N.	Dept. of Animal Science, Cornell University, Ithaca, NY 14853, USA
Kraemer, D.C.	Texas A & M University, College Station, TX 77843, USA
Kress, D.	Montana State University, ARSD, Bozeman, MT 59717, USA
Krivi, G.	Monsanto Company, St. Louis, MO 63198, USA
Kubisch, M.	University of Guelph, Guelph, Ontario, Canada N1H 3C1
Kulenkamp, A.	Shaver Poultry Breeding Farms, Cambridge, Ontario, Canada N1R 5V9
Kurvink, K.	Moravian College, Biology Dept., Bethlehem, PA 18018, USA
Langston, A.	USDA, APHIS, BBEP, Hyattsville, MD 20782, USA
Lasher, H.N.	P.O. Box 345, Millsboro, DE 19966, USA
Lavoir, M.C.	Cornell University, Veterinary Physiology, Ithaca, NY 14853, USA
Lee, C.N.	University of Hawaii, Dept. of Animal Science, Honolulu, HI 96822, USA
Leichthammer, F.	Dept. of Molecular An. Breeding, D-8000 Muenchen 22, FRG
Leiding, C.	Besamungsverein Neustadt/A, Neustadt 8530, West Germany
Lennon, M.	Clemson University, Clemson, SC 29633-0992, USA
Li, Y.	Edison Animal Biotech. Center, Ohio University, Athens, OH 45701, USA
Lien, S.	Agricultural Univ. of Norway, Dept of Animal Science, N-1432, AS-NLH Oslo, Norway
Linton, A.	Montana State University, Animal/Range Sciences, Bozeman, MT 59717, USA
Liu, C.H.	Chinese Academy of Ag. Science, Institute of Animal Science, Beijing, China
Loftus, R.	Trinity College, Genetics Dept., Dublin 1, Ireland
Lohuis, M.	University of Guelph, Guelph, Ontario, Canada N1H 2P9
Lorton, S.P.	American Breeders Service, DeForest, WI 53532, USA
Lu, J-K.	Purdue University, Dept. of Animal Sciences, West Lafayette IN 47907, USA
Lunt, D.K.	Rt. 1, Box 148, McGregor, TX, USA
Lyuz, M.C.	University of Vermont, Animal Science Dept., Burlington, VT 05468, USA
Ma, X.-D.	80W-115 MSDRL, P.O. Box 2000, Rahway, NJ 07065, USA
Maguire, J.	National Cancer Institute, NIH, Bethesda, MD 20892, USA
Maijala, K.	University of Helsinki, Institute of Biotechnology, 00380 Helsinki, Finland
Mantysaau, E.	Cornell University, Vet. School, Ithaca, NY 14853, USA
Marsh, J.A.	Cornell University, Poultry & Avian Diseases, Ithaca, NY 14853, USA
Marshall, C.E.	Select Sires, 11740 U.S. 42, Plain City, OH 40064, USA
Martin, A.	Praxis Biologics Inc., Rochester, NY 14623, USA
Mascarella, D.	Dept. of Biochem. Cornell University, Ithaca, NY 14853, USA
Massey, J.	2901 Camille, College Station, TX, USA
McCarty, R.E.	Biotechnology Program Director, Cornell University, Ithaca, NY 14853, USA
McDonell, M.	SYNTRO, San Diego, CA 92121, USA

McDonough, P.	Medical College of Georgia, Dept. of Ob/Gyn., Augusta, GA 30912, USA
McGregor, D.	Cornell University, James A. Baker Institute, Ithaca, NY 14853, USA
Meade, H.	Biogen, Cambridge, MA 02142, USA
Medrano, J.F.	University of California, Dept. of Animal Science, Davis, CA 95616, USA
Meltzer, M.	Cornell University, Agricultural Economics, Ithaca, NY 14853, USA
Meyers, N.L.	1 Houghton Grange, Houghton, Cambs PE12 2DA, UK
Middleton, S.	American Cynamid Co., P.O. Box 400, Princeton, NJ 08540, USA
Miller, A.	Amgen, Thousand Oaks, CA 91320, USA
Miller, R.	USDA, ARS, LPSI, BARC-East, Beltsville, MD 20705, USA
Mills, E.O.	Hubbard Farms, Inc., Walpole, NH 03608, USA
Milvae, R.	University of Connecticut, Dept. of Animal Science, Storrs, CT 06269-4040, USA
Mollar, M.	Norsk Hydro Research Center, Porsgrunn 3900, Norway
Montoya-Zavala, M.	Transgenic Sciences, Inc., Worcester, MA 01605, USA
Morrey, J.	Utah State University, Dept. ADVS, Logan UT 84322-5600, USA
Murray, J.	University of California, Dept. of Animal Science, Davis, CA 95616, USA
Murtaugh, M.	University of Minnesota, Vet. Pathobiology, St. Paul, MN 55108, USA
Myers, D.A.	Cornell University, Department of Physiology, Ithaca, NY 14853, USA
Naqi, S.A.	Cornell University, Avian and Aquatic Animal Med., Ithaca, NY 14853, USA
Natuk, R.	Wyeth-Ayerst Research, Biotechnology & Microbiology, Philadelphia, PA 19101, USA
Ninomiya, T.	Snow Brand Milk Products Co., Tochigi, 329-05 Japan
Noble, R.	Tuskegee University, School of Ag. and H. Economics, Tuskegee Inst., AL 36088, USA
Noden, D.M.	Cornell University, Department of Anatomy, Ithaca, NY 14853, USA
Noden, M.L.	Cornell University, Biotechnology Program, Ithaca, NY 14853, USA
Noland, P.	University of Arkansas, Dept. of Animal & Poultry Sci., Fayetteville, AR 72701, USA
Oldham, E.R.	Am. Cyanamid/Animal Ind. Disc., Agricultural Research Division, Princeton, NJ 08540, USA
Oriol, J.G.	Cornell University, Clinical Sciences, Ithaca, NY 14853, USA
Osman, M.	Columbia University, 722 West 168th Street, New York, NY 10032, USA
Paguaga, A.	University of Western Ontario, Dept. of Ob. & Gynaec., London, Ontario, Canada N6A 5A5
Panicali, D.	Applied Biotechnology, Inc., Cambridge, MA 02142, USA
Parks, J.E.	Cornell University, Department of Animal Science, Ithaca, NY 14853, USA
Paul, E.	Cornell University, Physiology, Ithaca, NY 14853, USA
Pauli, B.U.	Cornell University, 215 Veterinary Pathology, Ithaca, NY 14853, USA
Pavasuthipaisit, K.	Mahidol University, Faculty of Science, Bangkok 10400, Thailand
Petters, R.M.	North Carolina State Univ., Dept. of Animal Science, Raleigh, NC 27695, USA
Pevzner, I.Y.	Hubbard Farms Inc., Walpole, NH 03608, USA
Phillips, J.	University of Guelph, Dept. of Biology and Genetics, Guelph, Ontario, Canada N1G 2W1
Pinkert, C.	Embryogen, Athens, OH 45701, USA
Pintado, B.	Dpto. Prod. Animal, CIT-INIA, Carretera de la Corona, Madrid 28040, Spain
Post, L.E.	The Upjohn Company, Kalamazoo, MI 49001, USA
Pothier, F.	CHUL, Molecular Endoc. Lab., SteFoy, Quebec, Canada G1V 4G2
Poulsen, P.H.	Royal Vet. and Ag. University, 1870 Frederiksberg C, Copenhagen, Sweden
Prather, R.S.	University of Missouri, 164 Animal Science Research, Columbia, MO 65211, USA
Pratt, W.	Cornell University, DAAAM, Ithaca, NY 14853, USA
Pursel, V.G.	USDA, Beltsville, MD 20705, USA
Rath, D.	Institut f. Tierzucht u. Tierverhalten, Mariensee, FRG
Reddy, R.	Peterson Farms, Inc., P.O. Box 248, Decatur, AR 72722, USA
Rexroad, C.E.	USDA, Rm. 13, Bldg. 200, BARC-E, Beltsville, MD 20705, USA
Richardson, T.	University of California, Dept. Food Science & Tech., Davis, CA 95616, USA
Rickords, L.	Louisiana State University, Dept. of Animal Science, Baton Rouge, LA 70803, USA
Rigas, C.	Cornell University, Baker Inst. Animal Health, Ithaca, NY 14853, USA
Riggs, P.	Purdue University, Dept. of Animal Science, West Lafayette, IN 47907, USA
Roberts, R.M.	University of Missouri, Animal Science Dept., Columbia, MO 65211, USA
Robertshaw, D.	Cornell University, Physiology, Ithaca, NY 14853, USA
Rocha, J.L.	Texas A&M University, College Station, TX 77840, USA
Roche, J.F.	University College Dublin, Dublin 4, Ireland
Roni, N.	2213 Horniny Branch Court, Columbia, MO 65201, USA
Ryan, A.M.	Texas A&M University, Dept. Veterinary Pathology, College Station, TX 77840, USA

Sabour, M.P.	Agriculture Canada, Animal Research Center, Ottawa, Ontario, Canada K1A 0C6
Saeki, K.	Univ. Wisconsin, Madison, WI 53706, USA
Sakalian, M.	Cornell University, Ithaca, NY 14853, USA
Salter, D.W.	USDA-ARS-RPRL, East Lansing, MI 48823, USA
Sandberg, K.I.	Swedish University of Ag. Sci., Dept. of Animal Breeding & Genet., Uppsala, Sweden S-750 07
Sarkar, S.	Reproductive Tech. Inc., New Haven, CT 06511, USA
Schat, T.	Cornell University, Avian Aquatic An. Medicine, Ithaca, NY 14853, USA
Schatz, G.W.	Dept. Biochem. Molec. Cell Biol., Cornell University, Ithaca, NY 14853, USA
Schlafer, D.H.	Cornell University, Veterinary Pathology, Ithaca, NY 14853, USA
Scott, J.G.	Cornell University, Dept. of Entomology, Ithaca, NY 14853, USA
Sellos-Moros, M.	Cornell University, Ithaca, NY 14853, USA
Sheldon, H.	McGill University, Shelburne, VT 05482, USA
Sherblom, J.	Transgenic Sciences, Inc., Worcester, MA 01605, USA
Shibley, G.	USDA, APHIS, BBEP, VB, Hyattsville, MD 20782, USA
Shire, J.G.M.	The Jackson Laboratory, Bar Harbor, ME 04609, USA
Shore, L.	Kimron Vet. Inst., Bet Dagan, Israel
Shuman, R.M.	North Carolina State Univ., Dept. of Poultry Science, Raleigh, NC 27695-2628, USA
Silva, R.	USDA, Regional Poultry Research Laboratory, East Lansing, MI 48823, USA
Sirois, J.	Cornell University, Physiology, Ithaca, NY 14853, USA
Slanger, W.	National Institutes of Health, Bethesda, MD 20892, USA
Smiley, M.	Institute De Selection Animale, Mauguerand, F 22800 Quintin, France
Smith, C.	University of Guelph, Animal & Poultry Science, Ontario, Canada M1G 2W1
Smith, C.M.	Cornell University, Vet. Pathology, Ithaca, NY 14853, USA
Smith, M.	University of Missouri, 160 Animal Science Center, Columbia, MO 65211, USA
Smith, R.G.	Merck Sharp & Dohme, Rahway, NJ 07065, USA
Southard, L.	Cornell University, Ithaca, NY 14853, USA
Spitsbergen, J.M.	Cornell University, Avian & Aquatic Animal Med., Ithaca, NY 14853, USA
Squires, J.	University of Guelph, Dept. of Animal & Poultry Sci., Guelph, Ontario, Canada
Stewart, L.	Cornell University, Ithaca, NY 14853, USA
Strijker, R.	Genfarm B.V., 2300 RA Leiden, The Netherlands
Sullivan, M.A.	Eastman Kodak Co., Rochester, NY 14650-2119. USA
Sutou, S.	Itoham Foods, Inc., Tokyo 153, Japan
Syed, M.	Norwegian College of Vet. Med., Dept. of Animal Genetics, 0033 Olso 1, Norway
Tai, J.J.	15 Lane 156, Tang-Ann Road, Taiwan
Terrill, C.	ARS-USDA, Beltsville, MD 20705, USA
Thomford, P.	Pitman-Moore, Terre Haute, IN 47802, USA
Tilanus, M.G.J.	Zodiac MHC/EDC, 6709 PG Wageningen, The Netherlands
Tobin, T.	Texas A&M University, Bryan, TX 77801, USA
Tornell, J.	University of Goteborg, Dept. of Physiology, S-400 33 Goteborg, Sweden
Trovo, J.	Cornell University, Animal Science, Ithaca, NY 14853, USA
Tsang, P.	University of New Hampshire, Dept. of Animal & Nut. Science, Durham, NH 03824, USA
Tschunko, A.	Marietta College, Biology Dept., Marietta, OH 45750, USA
Tubbert, M.	Onondaga Community College, Onondaga Hill Campus, Syracuse, NY 13215, USA
Turk, K.L.	Cornell University, Animal Science, Ithaca, NY 14853, USA
Vage, D.	Department of Animal Genetics, P.O. Box 8156, Dep., N-0033 Oslo, Norway
Vitale, J.	Integrated Genetics, Framingham, MA 01701, USA
Voss, A.K.	Cornell University, Physiology, Ithaca, NY 14853, USA
Waldbieser, G.	Purdue University, Dept. of Animal Science, West Lafayette, IN 47907, USA
Walsh, M.	US Congress, Office of Tech. Assessment, Washington, DC 20510-8025, USA
Wang, C.-K.	Dept. of Animal Science, Univ. of Connecticut, Storrs, CT 06268, USA
Wasserman, R.H.	Cornell University, Physiology, Ithaca, NY 14853, USA
Weir, B.J.	J. Reprod. Fert., Cambridge CB5 8DT, UK
Welsh, T.H.	Texas A&M University, Department of Animal Science, College Station, TX 77843, USA
Werner, M.	Dept. of Agriculture, Hyattsville, MD 20782, USA
Wescott, C.R.	8 Hickock Pl., Burlington, VT 05401, USA
Whanger, P.	Oregon State University, Dept. of Ag. Chemistry, Corvallis, OR 97330, USA
White, K.	Louisiana State University, Dept. of Animal Science, Baton Rouge, LA 70803, USA
Wien, J.E.	Cornell University, Animal Science, Ithaca, NY 14853, USA

Wigglesworth, K.	In Vitro Fertilization Lab., Abington Memorial Hospital, Abington, PA 19001, USA
Wight, D.	Ohio University, Edison Animal Biotech. Center, Athens, OH 45701, USA
Wildeman, A.	University of Guelph, Dept. of Molecular Biology, Guelph, Ontario, Canada N1G 2W1
Williams, C.	Tuskegee University, Dept. of Biology, Tuskegee, AL 36088, USA
Williams, J.W.	Tuskegee University, Dept. of Biology, Tuskegee, AL 36088, USA
Wilmut, I.	AFRC, Edinburgh Research Station, Midlothian EH45 8RE, UK
Winkelman, D.	University of Alberta, Dept. of Genetics, Edmonton, Alberta, Canada T6G 2E9
Wolf, E.D.	Cornell University, Electrical Engineering, Ithaca, NY 14853, USA
Womack, J.E.	Texas A&M University, Dept. of Veterinary Pathology, College Station, TX 77843, USA
Woody, C.	University of Connecticut, Dept. of Animal Science, Storrs, CT 06268, USA
Wootton, J.F.	Cornell University, Department of Physiology, Ithaca, NY 14853, USA
Wu, C.	International Living Center, Ithaca, NY 14853, USA
Wu, M.	Taiwan Livestock Res. Inst., Shinhua, Taiwan
Youngs, C.	Iowa State University, Dept. of Animal Science, Ames, IA 50011, USA
Yun, J.	Ohio University, Edison Animal Biotech. Ctr., Athens, OH 45701, USA
Yunghans, W.	SUNY Fredonia, Biology Dept., Fredonia, NY 14063, USA
Zaitlin, M.	Cornell University, Biotechnology, Ithaca, NY 14853, USA
Zhang, C.	Cornell University, Baker Inst. Animal Health, Ithaca, NY 14853, USA
Zhang, X.	University of Western Ontario, Dept. of Ob. & Gynaec., London, Ontario, Canada N6A 5A5
Zinn, S.	Worcester Foundation, Shrewsbury, MA 01545, USA

AN OVERVIEW OF RECENT DEVELOPMENTS

Chairman
W. Hansel

J. Reprod. Fert., Suppl. **41** (1990), 3–14
Printed in Great Britain

New animal breeding techniques and their application

N. L. First

Department of Meat and Animal Science, University of Wisconsin, Madison, Wisconsin 53706, USA

Summary. The new biotechnologies of gene transfer, in-vitro production, cloning and sexing of embryos have been developed and are being refined with efficiencies suitable for use in animal agriculture. Efficient in-vitro systems for maturing oocytes and capacitating spermatozoa, for fertilizing and developing the embryos have resulted in commercial in-vitro production of embryos. Cloning of embryos by nuclear transfer has been accomplished for sheep, cattle, pigs and rabbits with nuclear material supplied by embryos as late as the 120-cell stage in sheep. Embryos have been recloned but much research is needed to increase the efficiency of this procedure. Research is needed to develop the use of cultured cells in embryo cloning so that the number of clones may be increased to thousands or millions. Embryos of most species can be sexed in a non-damaging way with male specific antibodies and a more efficient method, amplified DNA hybridization, is beginning to be tested commercially. Transgenic embryos or offspring have been produced for mice, rats, rabbits, chickens, fish, sheep, pigs and cattle. Genes can be targeted for expression in specific tissues but more efficient methods and a better understanding of the genes to be transferred as well as control by man of the time and tissue of specific gene expression are needed. Before many transgenic animals of value can be made, we must know which genes to introduce. Presently there is a poor understanding of the genes influencing animal growth, efficiency of growth, environmental adaptation, meat, milk or egg composition or animal disease resistance. Their identification will come from badly needed efforts to map the genome of domestic animals. These and other new technologies promise to change livestock breeding drastically in the next decade.

Keywords: embryo cloning; sexing; production; gene transfer

Introduction

Animal agriculture is entering an era of exciting possibilities for rapidly propagating and tailoring animals to meet product and environmental demands. This is made possible by the development of new biotechnologies such as in-vitro embryo production, splitting, cloning, sexing, and genetic transfer, which aim to manipulate embryos of domestic species before implantation. The strengths of these techniques lie in their complementariness and their ability to be readily scaled up for commercial use. These new technologies, along with old techniques such as artificial insemination, embryo transfer and selective breeding, can become part of well organized systems for wide-scale production, sale and transfer of genetically superior animals. Already, in-vitro production, splitting and sexing of embryos are being commercially applied, while cloning and gene transfer are still in the testing and development stages.

In this review I shall look at the status of development of each of these biotechniques and a few of their potential uses. It is important to recognize that there are at least four constraints to the commercial production of transgenic farm animals: (1) how to produce the transgenic efficiently, (2) the appropriate gene construct for expression in the tissue of choice and at the time of choice,

(3) how to propagate efficiently the transgenic or any other animal considered genetically desirable, and (4) identification of the genes controlling the efficiency of growth and lactation, genes controlling product composition and quality and genes controlling animal health.

This review will focus primarily on constraints 3 and 4.

In-vitro production of embryos

The production of embryos *in vitro* from oocytes recovered from the abattoir is best developed for cattle (Lu *et al.*, 1987; Goto *et al.*, 1988; Eyestone & First, 1989; Leibfried-Rutledge *et al.*, 1989), although offspring have been produced from in-vitro fertilization of oocytes completing meiotic maturation *in vitro* in sheep (Crozet *et al.*, 1987) and pigs (Cheng *et al.*, 1986; Nagai *et al.*, 1988). There are at least three reasons for producing embryos of cattle *in vitro*. First, this technique provides large numbers of embryos for commercial transfer and calf production. In Europe and Japan, the value of dairy calves is sufficiently low relative to beef calves that there are economic incentives for transfer of in-vitro produced beef embryos into dairy cow recipients, particularly with the goal of inducing twinning. In Ireland, Scotland and Japan, commercial ventures have been established for in-vitro production of cattle embryos. Second, the economic feasibility of embryo cloning by nuclear transfer requires that the enucleated oocytes be produced *in vitro* from abattoir-recovered ovaries and that the new zygote be developed *in vitro* to a stage suitable for recloning. Third, in-vitro produced embryos are highly valued for research when large numbers or precise timing of fertilization and development are needed.

In-vitro production of embryos requires development of technology in three areas: oocyte maturation, in-vitro fertilization and in-vitro embryo development.

Oocyte maturation

In domestic species, oocytes recovered from follicles matured *in vivo* either with or without superovulation can be fertilized and proceed through embryo development with good success (cattle: Sirard *et al.*, 1985; Leibfried-Rutledge *et al.*, 1987, 1989; pigs: Cheng *et al.*, 1986; Nagai *et al.*, 1988). However, oocytes recovered from small follicles (1–5 mm), many of which have not completed growth and development, produce zygotes which fail to complete embryo development (sheep: Moor & Trounson, 1977; Crosby *et al.*, 1981; cattle: Leibfried-Rutledge *et al.*, 1987, 1989). Embryo development is enhanced when the oocytes from small follicles undergo in-vitro maturation in the presence of hormone-stimulated granulosa cells (Staigmiller & Moor, 1984; Critser *et al.*, 1986a; Lu *et al.*, 1987; Leibfried-Rutledge *et al.*, 1989) and to a lesser extent with cumulus cells confined in a small volume of medium (Critser *et al.*, 1986b; Sirard *et al.*, 1988; Leibfried-Rutledge *et al.*, 1989). Neither the embryo developmental signals developed during cumulus cell co-culture nor the beneficial material from the granulosa cells is identified. Subsequent embryo development is enhanced further if embryos are rigorously screened for normality and follicular source before in-vitro maturation (Leibfried-Rutledge *et al.*, 1989). At present, the frequency of bovine blastocysts developing from in-vitro fertilization of in-vivo matured oocytes is approximately 45% (Brackett *et al.*, 1982; Sirard *et al.*, 1985; Leibfried-Rutledge *et al.*, 1987); from immature oocytes co-cultured with granulosa cells it is approximately 23–63% (Critser *et al.*, 1986a; Lu *et al.*, 1987; Xu *et al.*, 1987; Fukui & Ono, 1988; Leibfried-Rutledge *et al.*, 1989); from culture with significant cumulus contribution per volume of medium it is 20–30% (Critser *et al.*, 1986b; Goto *et al.*, 1988; Sirard *et al.*, 1988); and from immature oocytes cultured under conditions in which the 'helper cell' effect is lost it is ≤20% (unpublished observations; Critser *et al.*, 1986b; Leibfried-Rutledge *et al.*, 1987; Sirard *et al.*, 1988). To increase the efficiency of oocyte maturation, research is needed in two areas. First, a better understanding of how to select oocytes highly competent in embryo development and a better understanding of the mechanisms controlling oocyte maturation are needed. Second, the development of systems for culturing and developing oocytes from the large population of pre-antral follicles is needed to increase the pool of oocytes available.

In-vitro fertilization

The second part of production of embryos *in vitro* is the sperm capacitation and fertilization system. Numerous capacitation systems have been used, including high ionic strength media and glycosaminoglycans such as heparin and fucose sulphate, ageing, pH shift, calcium ionophores and caffeine and oviduct fluid (First & Parrish, 1987, 1988; Parrish *et al.*, 1989). In general, any agent which causes Ca^{2+} entry into the sperm acrosome and causes a pH increase within the spermatozoon causes capacitation (First & Parrish, 1988). With appropriate sperm capacitation, preincubation and incubation in serum-free medium at body temperature, in-vitro fertilization has been successful, resulting in fertilization rates as high as 70–80% in cattle, sheep, pigs and goats.

Development of embryos *in vitro*

Embryos of each domestic species can be developed with good efficiency to the blastocyst stage or later by transfer at the 1-cell or 2-cell stage into the oviduct of the respective species. For cattle, the embryos can also be successfully developed in the oviduct of the rabbit (Boland, 1984; Sirard *et al.*, 1985) or sheep (Eyestone *et al.*, 1987). A single set of culture conditions does not support development *in vitro* to morulae or blastocysts in all species (Wright & Bondioli, 1981). Instead, development is blocked at the transition from maternal to zygotic control of development (Barnes, 1988; First & Barnes, 1989). The embryos remain alive but with cleavage arrested at G_2 of the cell cycle (Eyestone & First, 1988).

Cattle (Lu *et al.*, 1987; Eyestone & First, 1989) and sheep (Gandolfi & Moor, 1987) embryos have been cultured through the period of blocked development and to the blastocyst stage with good efficiency by co-culture with oviduct epithelial cells or media conditioned by cultured oviduct cells (Eyestone & First, 1989). Pig embryos have been successfully cultured from the 1-cell stage to blastocysts in mouse oviducts maintained in culture (Krisher *et al.*, 1989). In cattle the essential oviduct material is in the protein fraction but its identity is unknown. In sheep the essential oviduct component is believed to be a protein of M_r 92 500 either acting alone or in combination with a protein of M_r 46 000. The M_r 92 500, fucose-rich glycoprotein increases greatly in the oviduct just before the block period, translocates to the zona and embryo, and disappears by the blastocyst stage (Gandolfi & Moor, 1988; Gandolfi *et al.*, 1989).

The cattle oviduct factor has not been completely characterized, and its relationship to the sheep oviduct glycoprotein is unknown. The nature of these embryotrophic compounds needs elucidation, as do their modes of action and the way in which the blocked development relates to initiation of embryonic transcription and the transition from short to long cell cycles (Barnes, 1988; First & Barnes, 1989). In spite of these gaps in our knowledge, culture of embryos in fresh or frozen–thawed (W. H. Eyestone, J. M. Vignieri & N. L. First, unpublished) oviduct cell conditioned media has provided an in-vitro method for development of bovine embryos which has resulted in an approximately normal pregnancy rate (50%) after transfer into cows (Eyestone & First, 1989). In pigs, reasonably successful culture systems which do not rely on oviduct cells have been developed (Menino & Wright, 1982; Hagen *et al.*, 1989). It is also an objective for sheep and cattle to develop pure culture systems which do not carry the potential for infection of the embryo by viral or bacterial organisms in the serum or co-culture cells used.

Multiplication of embryos

The ability to produce multiple copies of an individual or embryo is of interest not only to researchers, but also to the livestock industry. Genetically identical individuals provide the perfect control for experimental conditions, thus reducing the genetic variation in experiments to zero. A large number of genetically identical embryos provides a means for embryo phenotypic selection wherein clonal lines descendent from one embryo are selected by progeny test for clonal multiplication to large numbers. This system approaches phenotypic selection and could permit rapid change in selected characteristics such as meat or milk production and, when combined with in-vitro production of embryos, could provide a way to produce large numbers of high quality embryos for frozen storage and commercial transfer. Two methods of embryo multiplication will be discussed here; they are embryo bisection and nuclear transplantation.

Embryo bisection

With embryo bisection an early embryo is bisected to yield either 2 cells as with a 2-cell embryo, or two or more cell masses as with a morula- or blastocyst-stage embryo. This procedure results in identical offspring in sheep (Willadsen, 1982), pigs (Willadsen, 1982; Rorie *et al.*, 1985) and cattle (Ozil *et al.*, 1982; Baker & Shea, 1986; Leibo, 1988). Since this procedure is successful it can be concluded that these cells are totipotent. However, the number of identical individuals produced by this method is limited. If the embryo is divided more than twice, survival to offspring is greatly reduced. This is probably due to the requirement of a minimum cell number at the time of blastulation. In the mouse this minimum is 8–16 cells. If blastulation occurs with fewer cells, a trophoblastic vesicle will form without an inner cell mass (Tarkowski & Wroblewska, 1967). Therefore, the limit to the number of identicals produced by splitting is maximally 4 and efficiently 2 (Robl & First, 1985; Baker & Shea, 1986; Leibo, 1988). This procedure is commonly used in the cattle embryo transfer industry and results in a pregnancy rate nearly equivalent to the whole embryo with the number of offspring nearly doubled (Leibo, 1988).

Nuclear transfer

The second method for producing multiple copies of an embryo is by nuclear transplantation, a procedure that has been shown to be successful in producing viable embryos and offspring in cattle (Prather *et al.*, 1987), sheep (Willadsen, 1986; Smith & Wilmut, 1989), rabbits (Stice & Robl, 1988) and pigs (Prather *et al.*, 1989). This procedure is a modification of a one originally developed and shown to be successful for the frog (Briggs & King, 1952).

As shown by Prather & First (1990) and in Figs 1 and 2, the procedure involves transfer of a blastomere or nucleus from a valuable embryo at a multicellular stage into an enucleated metaphase II oocyte with subsequent development to a multiple cell stage and use as a donor in a serial recloning. This procedure is being developed in private industry as well as in research laboratories. Collectively in the USA and Canada, several hundred pregnancies have been produced in cattle by this procedure and recloning has been performed. So far, the largest number of calves cloned from one embryo has been 8: these were born at Granada Genetics (Marquez, Texas, USA) in 1987.

A system for cloning of embryos useful to the livestock industry depends on the ability to produce offspring from donor embryos of large cell number and the ability to reclone as the clones develop to advanced cell number or to multiply donor cells in culture. Studies with sheep at Edinburgh, Scotland, suggest that this should be possible. The frequency of development to the blastocyst stage after use of donor cells from blastocyst inner cell mass was 56% and pregnancies resulted (Smith & Wilmut, 1989). This is very close to the stage at which embryonic stem cells can be recovered and multiplied in culture in the mouse (see Notarianni *et al.*, 1990). If similar stem cell multiplication can be done in domestic animals and if stem cells should prove useful in cloning by nuclear transfer, then the number of clones possible is unlimited.

Sexing of embryos

Sexing of embryos before transfer is especially sought by the dairy cattle industry in which females are the desired milk-producing unit. To be useful, sexing techniques must be accurate, efficient, rapid and without detrimental effects on the embryos.

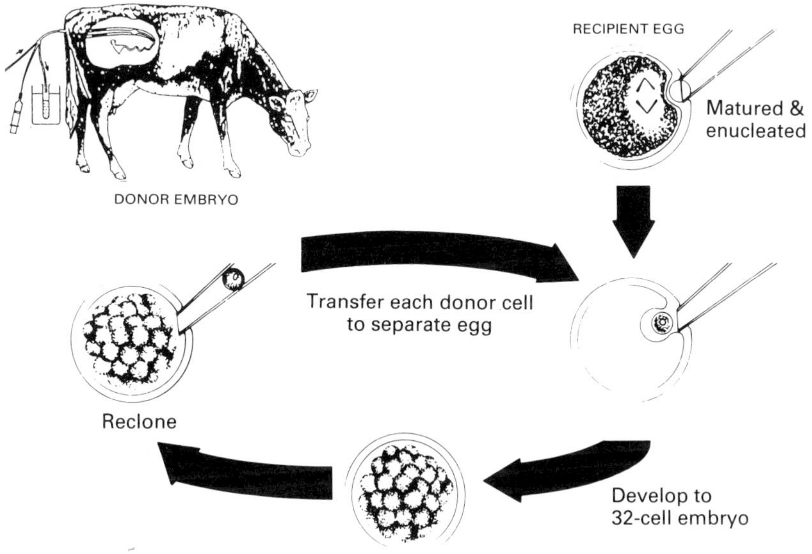

Fig. 1. Procedures used to multiply embryos by nuclear transfer. Valuable morula-stage embryos are recovered by non-surgical flush from the uterus of an inseminated cow, the individual cells or blastomeres are removed from the morula and transferred into enucleated oocytes and the new embryos developed to the morula stage when the process can be repeated. (Presented originally as a report to W. R. Grace and Company by N. L. First.)

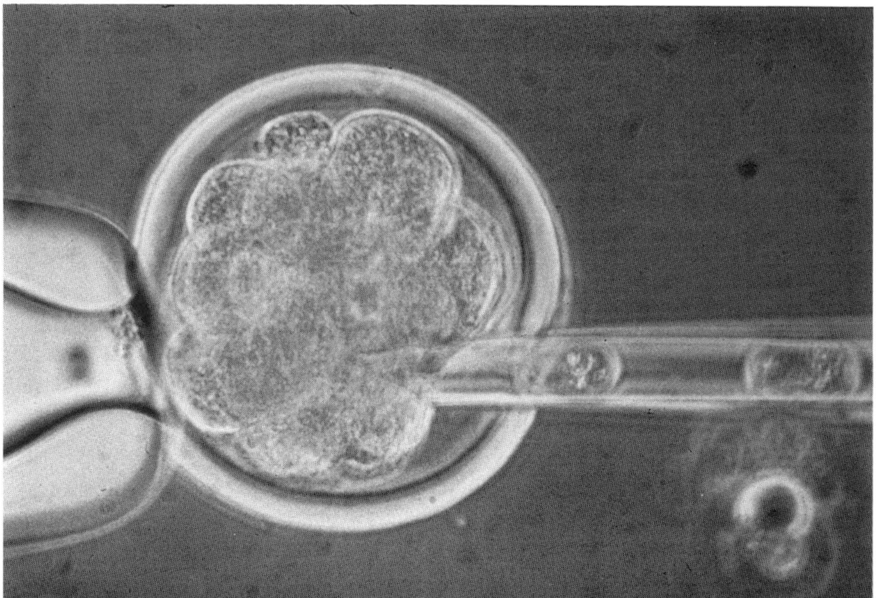

Fig. 2. Removal of blastomeres from a donor bovine embryo for transfer into enucleated oocytes to multiply embryos by nuclear transfer ×400. (From Prather *et al.*, 1987.)

The standard method for sexing embryos has been karyotyping of the cells of the morula or blastocyst. This method is accurate when good metaphase spreads are obtained, about 70% of the time. Unfortunately, to be accurate the method requires nearly half the cells of a morula or blastocyst or all of the cells of earlier stages. This greatly limits use of the method for embryos to be transferred. Embryos can also be sexed by use of antibodies to male-specific antigens such as the H-Y antigen. When a fluorescent second antibody approach is used, this method is not damaging to embryos and is ~85% accurate for cattle embryos (White *et al.*, 1987; Wachtel *et al.*, 1988; White, 1988).

Very accurate methods which use only a few cells and provide a quick answer have recently been developed for sexing embryos. These methods involve the use of Y chromosome-specific DNA hybridization probes, with the DNA identified from specific bands on a gel or from a labelled homologous recognition probe. Y-specific fragments are used as probes to locate homologous sequences present in DNA from blastomeres or trophoblasts. As few as 5 blastomeres can be biopsied from embryos and, using an oligonucleotide polymerase chain reaction for signal amplification, embryonic sex can be determined in 6 h or less. Several Y-specific probes are currently available for sex selection of cattle embryos (Leonard *et al.*, 1987; Ellis *et al.*, 1988; Popescu *et al.*, 1988; Reed *et al.*, 1988; Bondioli *et al.*, 1989; Herr *et al.*, 1989a, b) and one has been developed for pigs.

Gene transfer

The first production of transgenic mice by Gordon & Ruddle (1981) and the evidence that mice transgenic for growth hormone (Palmiter *et al.*, 1982, 1983) grew to nearly twice normal size, greatly excited animal scientists with hopes of doing likewise in their domestic livestock. Especially exciting were the possibilities for targeting gene expression exclusively to skeletal muscle for altering the meat product (Shani, 1985, 1986) or to the mammary gland for altering the composition of the milk to include more or different proteins such as pharmaceuticals (Wilmut *et al.*, 1990). Although in the ensuing years more than 400 strains of transgenic mice have been produced for use in studying problems in biology, medicine and animal agriculture, only a few new transgenic lines of domestic animals have been produced. Microinjection of DNA into a bovine egg is shown in Fig. 3. Slow progress has largely been due to the high economic value of each egg manipulated and to the low efficiency of DNA microinjection into the pronuclei. Even at best, the maximum efficiency obtained in domestic species is 0·5–1% (Hammer *et al.*, 1985; Biery *et al.*, 1988; Murray *et al.*, 1988). Additionally, in many of these cases there has been a failure of the desired response (i.e. growth) at the desired time or in the desired tissues (Rexroad & Pursel, 1988; Pursel *et al.*, 1990). These problems can be and are being resolved as more is learned about promoter and enhancer sequences and the inter-workings of the genes themselves.

Because of low efficiencies with micromanipulation, alternative methods are being sought. Three of these methods include transfer by viral vector, embryonic stem cells and binding onto (Lavitrano *et al.*, 1989) or electroporation into the fertilizing spermatozoon (M. Gagne, F. Pothier & M. A. Sirard, unpublished). The first two approaches have been developed and successfully applied in mice, but not yet in domestic species (see Notarianni *et al.*, 1990; Pursel *et al.*, 1990).

Retroviral vectors have been effective for efficiently transferring genes into mouse embryos at various developmental stages (Eglitis *et al.*, 1985; Soriano *et al.*, 1986; Jaenisch, 1988) to correct genetic defects (Eglitis *et al.*, 1985) or into embryonic stem cells in culture which can be used to produce transgenic mice (Hooper *et al.*, 1987; Kuehn *et al.*, 1987; Thompson *et al.*, 1989). So far, no replication-defective viral vectors of sufficiently high infectivity for efficient transformation have been safety approved for use in domestic species.

Genes introduced by microinjection or viral vector are presently integrated into the genome at random chromosomal sites. In contrast, the stem cell method provides the potential for production of transgenic offspring with site-specific gene insertion. Smithies *et al.* (1985) developed a technique

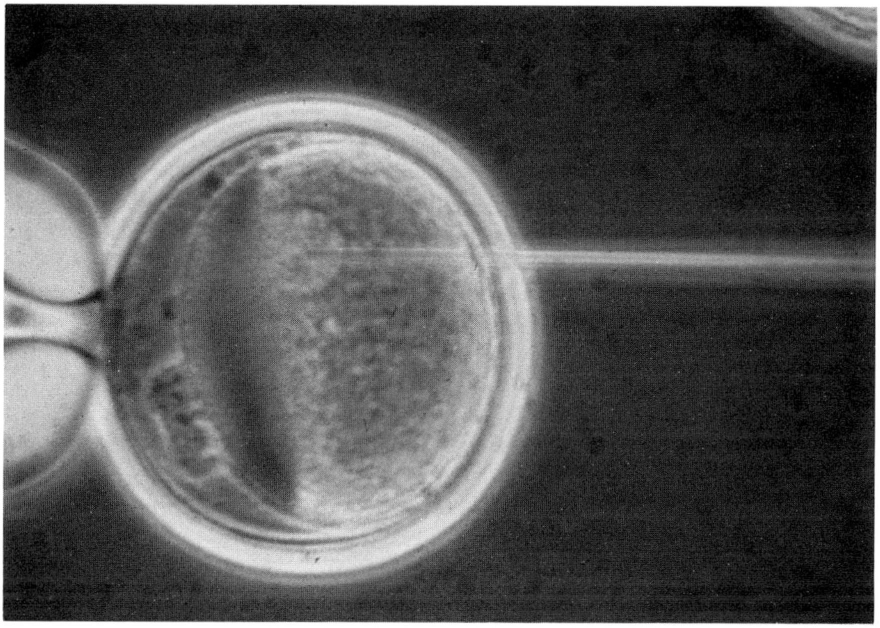

Fig. 3. Microinjection of DNA into a pronucleus of a bovine egg after centrifugation to move the cytoplasm, allowing visualization of the pronuclei. ×400. (Presented originally as a report to W. R. Grace and company by N. L. First.)

for selecting cells after an homologous recombination event between an endogenous gene and a plasmid carrying a genetically engineered copy of the gene. Rather than adding a new copy, this allows replacement of the endogenous copy of the gene by the genetically engineered copy resulting in replacement of the gene (Kuehn *et al.*, 1987). This work was extended by the correction of the HPRT mutation induced in embryonic stem cells (Doetschman *et al.*, 1987). The efficiency of this method has been increased by modifications such as the use of a polymerase chain reaction (Kim & Smithies, 1988) and it has been used to cause gene deletion as well as introduction (Thomas & Capecchi, 1987). Mice expressing a site-specific transgene have been produced and germ-line transmission has occurred (Thompson *et al.*, 1989).

This development could provide the potential for immediate specific gene deletion or replacement in livestock production. At present, this process requires generations of selection and even then is possible only for alternative alleles or rare mutants.

It is expected that gene transfer will become common practice in production of new strains of livestock when more efficient and site-specific transfer systems such as introduction into spermatozoa, viral vectors and embryonic stem cells or their combination are applied in domestic animals. It is likely that males of these new strains will be progeny tested to determine the productivity advantage of the new gene and screened for any defects which may be present in the strain.

To utilize gene transfer fully, much must be learned concerning the genes controlling physiological traits we wish to manipulate and the promoter-enhancer gene sequences required to express the product of a given gene in the tissue desired and at the time and level desired.

Advancements in this area have been rapid and genes can now be targeted to many tissues of the body by use of tissue-specific promoters. For example, targeted expression in the mammary gland is achieved by use of promoter sequences of β-lactoglobulin (Simons *et al.*, 1987), whey acidic proteins (Gordon *et al.*, 1987) or promoter sequences associated with any other gene coding for a protein expressed only in the mammary gland. Similarly, targeted expression in skeletal muscle has been achieved with promoter sequences associated with the skeletal muscle actin gene (Shani, 1986) or myosin light chain kinase (Shani, 1985). So far, tissue-specific expression has been achieved in at least 8 different tissues.

A few promoter sequences have been identified which will turn gene expression on or off in response to endogenous or exogenous signals. For example, expression of the metallothionein I sequence is controlled by heavy metals (Palmiter *et al.*, 1982, 1983), a mouse MHCII antigen by interferon (Pinkert *et al.*, 1985), the insulin gene by glucose (Shelden *et al.*, 1986), pancreatic amylase by insulin (Osborn *et al.*, 1987), whey acidic protein promoter by lactogenic hormones (Andres *et al.*, 1987) and the phosphoenolpyruvate carboxykinase promoter by diet, insulin and cyclic AMP (McGrane *et al.*, 1988). These promoter enhancer sequences are the tools which will allow creation of the desired response from a gene at the desired site and time. This arsenal of tools badly needs expansion, understanding and cataloguing.

Understanding the genes controlling livestock productivity traits

Because gene transfer in domestic animals is expensive, the gene constructs to be introduced must be modelled and understood to perfection. At present the promising genes are few and include growth hormone for promotion of growth (Rexroad & Pursel, 1988; Pursel *et al.*, 1990) and, probably, lactation genes of the MHC complex (Frels *et al.*, 1985; Maguire *et al.*, 1990) and interferon (Chen *et al.*, 1988) for enhancing disease resistance as well as genes coding for proteins enhancing wool production (Ward *et al.*, 1984), genes targeted to the mammary gland to change the composition of milk constituents such as casein, fat or sugar or genes to produce pharmaceutical proteins such as clotting factor IX (Simons *et al.*, 1987) or tissue plasminogen activator (Andres *et al.*, 1987) in milk (Wilmut *et al.*, 1990). Most recently it has become apparent that transgenic plants (Abel *et al.*, 1986; Loesch-Fries *et al.*, 1987), chickens (Salter *et al.*, 1987; Crittenden & Salter, 1990), cultured mammalian cells (Staeheli *et al.*, 1986; Campadelli-Fiume *et al.*, 1988; Petrovskis *et al.*, 1988) and perhaps mammals can be produced which have alterations in their cell membranes preventing recognition of viral organisms. This raises the exciting possibility for permanent immunity to devastating diseases.

Unfortunately, the power of these tools for efficiently propagating and changing the genome of farm animals to meet Man's needs and for their adaptation to unfavourable world environments has little value without an understanding of which genes to introduce. To utilize constructively the above tools it is necessary to identify, isolate and understand the genes controlling animal disease resistance, growth, reproduction, lactation, egg production and meat, milk and egg composition and quality. This could be accomplished by mapping the genome of domestic animals and association of restriction fragments or probed known sequences with production and product quality traits. Much could be learned and borrowed from efforts in the human.

During the past year great strides and commitments have been made in mapping and sequencing the human genome (Donis-Keller *et al.*, 1987; Nakamura *et al.*, 1987; Watkins, 1988). Significant advances have also been made in mapping the genome of mice (O'Brien, 1987). In contrast to the mouse and man, our understanding of the genes of domestic species is very poor, although efforts have begun in cattle (Womack & Moll, 1986; O'Brien, 1987) and to a limited extent in pigs. Fortunately, there is a high degree of conservation of the genome across mammalian species. A large number of conserved linkage groups has been demonstrated in man, mouse and cattle (Womack & Moll, 1986). This means that the DNA hybridization probes and DNA sequence restriction fragments identified and associated with specific physiological functions in the mouse and man can be identified and associated with their functions in domestic animals by borrowing the human and mouse DNA probes, restriction enzymes and information. These tools are now available and ready for use, but require investigations in the target species to confirm similarities and define genetic differences.

From this should come highly effective gene transfer and sophisticated selection and propagation programmes resulting in superior animals and products of high demand on the world food market. Application of these molecular genetic technologies to animal agriculture should bring about exciting changes in animal production and the tailoring of animals to produce products needed by man.

References

Abel, P.P., Nelson, R.S., De, B., Hoffman, N., Rogers, S.G., Fraley, R.T. & Beachy, R.N. (1986) Delay of disease development to transgenic plants that express the tobacco mosaic virus coat protein gene. *Science, NY* **232,** 738–743.

Andres, A.-C., Schonenberger, C.-A., Groner, B., Hennighausen, L., LeMeur, M. & Gerlinger, P. (1987) Ha-ras oncogene expression directly by a milk protein gene promoter: tissue specificity, hormonal regulation, and tumor induction in transgenic mice. *Proc. natn. Acad. Sci. USA* **84,** 1299–1303.

Baker, R.D. & Shea, B.F. (1986) Commercial splitting of bovine oocytes. *Theriogenology* **23,** 3–12.

Barnes, F.L. (1988) *Characterization of the onset of embryonic control and early development in the bovine embryo.* Ph.D. thesis, University of Wisconsin-Madison.

Biery, K.A., Bondioli, K.R. & De Mayo, F.J. (1988) Gene transfer by pronuclear injection in the bovine. *Theriogenology* **29,** 224, abstr.

Boland, M.P. (1984) Use of rabbit oviduct as a screening tool for the viability of mammalian eggs. *Theriogenology* **21,** 126–137.

Bondioli, K.R., Ellis, S.B., Pryor, J.H., Williams, M.W. & Harpold, M.M. (1989) The use of male-specific chromosomal DNA fragments to determine the sex of bovine preimplantation embryos. *Theriogenology* **31,** 95–104.

Brackett, B.G., Bousquet, D., Boice, M.L., Donawick, W.J., Evans, J.F. & Dressel, M.A. (1982) Normal development following in vitro fertilization in the cow. *Biol. Reprod.* **27,** 147–158.

Briggs, R. & King, T.J. (1952) Transplantation of living nuclei from blastula cells into enucleated frogs' eggs. *Zoology* **38,** 455–463.

Campadelli-Fiume, G., Arsenakis, M., Farabegoli, F. & Roizman, B. (1988) Entry of herpes simplex virus in BJ cells that constitutively express viral glycoprotein D is by endocytosis and results in degradation of the virus. *J. Virol.* **62,** 159–167.

Chen, X-Z., Yun, J.S. & Wagner, T.E. (1988) Enhanced viral resistance in transgenic mice expressing the human beta I interferon. *J. Virol.* **62,** 3883–3887.

Cheng, W.T.K., Moor, R.M. & Polge, C. (1986) In vitro fertilization of pig and sheep oocytes matured in vivo and in vitro. *Theriogenology* **25,** 146, abstr.

Critser, E.S., Leibfried-Rutledge, M.L., Eyestone, W.H., Northey, D.L. & First, N.L. (1986a) Acquisition of developmental competence during maturation in vitro. *Theriogenology* **25,** 150, abstr.

Critser, E.S., Leibfried-Rutledge, M.L. & First, N.L. (1986b) Influence of cumulus cell association during in vitro maturation of bovine oocytes on embryonic development. *Biol. Reprod.* **34** (Suppl. 1), 192, abstr.

Crittenden, L.B. & Salter, D.W. (1990) Expression of retroviral genes in transgenic chickens. *J. Reprod. Fert., Suppl.* **41,** 163–171.

Crosby, I.M., Osborn, J.C. & Moor, R.M. (1981) Follicle cell regulation of protein synthesis and developmental competence in sheep oocytes. *J. Reprod. Fert.* **62,** 575–582.

Crozet, N., Huneau, D., Desmedt, V., Theron, M.C., Szollosi, D., Torries, S. & Sevellec, C. (1987) In vitro fertilization of sheep ova. *Gamete Res.* **16,** 159–170.

Doetschman, T., Gregg, R.G., Maeda, N., Hooper, M.L., Melton, W., Thompson, S. & Smithies, O. (1987) Targetted correction of a mutant HPRT gene in mouse embryonic stem cells. *Nature, Lond.* **330,** 576–578.

Donis-Keller, H., Green, P., Helms, C., Cartinhour, S., Weiffenbach, B., Stephens, K., Keith, T.P., Bowden, D.W., Smith, D.R., Lander, E.S., Botstein, D., Akots, G., Rediker, K.S., Gravius, T., Brown, V.A., Rising, M.B., Parker, C., Powers, J.A., Watt, D.E., Kauffman, E.R., Bricker, A., Phipps, P., Muller-Kahle, H., Fulton, T.R., Ng, S., Schumm, J.W., Braman, J.C., Knowlton, R.G., Barker, D.F., Crooks, S.M., Lincoln, S.E., Daly, M.J. & Abrahamson, J. (1987) Genetic linkage map of the human genome. *Cell* **51,** 319–337.

Eglitis, M.A., Kentoff, P., Gilboa, E. & Anderson, W.F. (1985) Gene expression in mice after high efficiency retroviral-medicated gene transfer. *Science, NY* **230,** 1395–1398.

Ellis, S.B., Bondioli, K.R., Williams, M.W., Pryor, J.H. & Harpold, M.M. (1988) Sex determination of bovine embryos using male-specific DNA probes. *Theriogenology* **29,** 242, abstr.

Eyestone, W.H. & First, N.L. (1988) Cell cycle analysis of early bovine embryos. *Theriogenology* **29,** 434, abstr.

Eyestone, W.H. & First, N.L. (1989) Co-culture of early bovine embryos to the blastocyst stage with oviductal tissue or in conditioned medium. *J. Reprod. Fert.* **85,** 715–720.

Eyestone, W.H., Vignieri, J. & First, N.L. (1987) Co-culture of early bovine embryos with oviductal epithelium. *Theriogenology* **27,** 228, abstr.

First, N.L. & Barnes, F.L. (1989) Development of preimplantation mammalian embryos. In *Development of Preimplantation Embryos and Their Environment*, pp. 151–170. Eds K. Yoshinaga & R. Mori. Alan R. Liss, New York.

First, N.L. & Parrish, J.J. (1987) In vitro fertilization of ruminants. *J. Reprod. Fert., Suppl.* **34,** 151–164.

First, N.L. & Parrish, J.J. (1988) Sperm maturation and in vitro fertilization. *Proc. 11th Int. Congr. Anim. Reprod. & AI, Dublin* vol. V, pp. 160–168.

Frels, W.I., Bluestone, J.A., Capecchi, M.R. & Singer, D.S. (1985) Expression of a microinjected porcine class I major histocompatibility complex gene in transgenic mice. *Science, NY* **228,** 577–580.

Fukui, Y. & Ono, H. (1988) In vitro development to blastocyst of in vitro matured and fertilized bovine oocytes. *Vet. Rec.* **122,** 282, abstr.

Gandolfi, F. & Moor, R.M. (1987) Stimulation of early embryonic development in the sheep by co-culture with oviduct epithelial cells. *J. Reprod. Fert.* **81,** 23–28.

Gandolfi, F. & Moor, R.M. (1988) Interactions between somatic and germinal cells during early development. *Proc. 11th Int. Congr. Anim. Reprod. & AI, Dublin,* vol. V, pp. 169–177.

Gandolfi, F., Brevini, T.A.L., Richardson, L., Brown, C.R. & Moor, R.M. (1989) Characterization of proteins secreted by sheep oviduct epithelial cells and their function in embryonic development. *Development* **106,** 303–312.

Gordon, J.W. & Ruddle, F.H. (1981) Integration and stable germline transmission of genes injected into mouse pronuclei. *Science, NY* **214,** 1244–1246.

Gordon, K., Lee, E., Vitale, J.A., Smith, A.E., Westphal, H. & Hennighausen, L. (1987) Production of human tissue plasminogen activator in transgenic mouse milk. *Bio/Technology* **5,** 1183–1187.

Goto, K., Kajihara, Y., Kosaka, S., Koba, M., Nakanishi, Y. & Ogawa, K. (1988) Pregnancies after co-culture of cumulus cells with bovine embryos derived from in-vitro fertilization of in-vitro matured follicular oocytes. *J. Reprod. Fert.* **83,** 753–758.

Hagen, D.R., Prather, R.S., Sims, M.M. & First, N.L. (1989) Development of pig embryos to blastocysts (BL) in sheep and pig oviducts and co-culture in vitro. *J. Anim. Sci.* **68** (Suppl. 1), 209–210, abstr.

Hammer, R.E., Pursel, V.G., Rexroad, C.E., Jr, Wall, R.J., Bolt, D.J., Ebert, K.M., Palmiter, R.D. & Brinster, R.L. (1985) Production of transgenic rabbits, sheep and pigs by microinjection. *Nature, Lond.* **315,** 680–683.

Herr, C., Matthaei, K.I. & Reed, K.C. (1989a) Accuracy of a rapid Y-chromosome-detecting bovine embryo sexing assay. *Proc. 8th Ann. Mtg, Aust. N.Z. Soc. Cell Biol., Melbourne,* p. 50, abstr.

Herr, C., Matthaei, K., Holt, N. & Reed, K. (1989b) Field implementation of a rapid Y-chromosome-detecting bovine embryo sexing assay. *Proc. 8th Ann. Mtg, Aust. N.Z. Soc. Cell Biol., Melbourne,* p. 51, abstr.

Hooper, M.L., Hardy, K., Handyside, A., Hunter, S. & Monk, M. (1987) HPRT-deficient (Lesch-Nyhan) mouse embryos derived from germline colonization by cultured cells. *Nature, Lond.* **326,** 292–295.

Jaenisch, R. (1988) Transgenic animals. *Science, NY* **240,** 1468–1475.

Kim, H.S. & Smithies, O. (1988) Recombinant fragment assay for gene targeting based on the polymerase chain reaction. *Nucleic Acids Res.* **16,** 8887–8903.

Krisher, R.L., Petters, R.M., Johnson, B.H., Bavister, B.D. & Archibong, A.E. (1989) Development of porcine embryos from the one-cell stage to blastocyst in mouse oviducts maintained in organ culture. *J. exp. Zool.* **249,** 235–239.

Kuehn, M.R., Bradley, A., Robertson, E.J. & Evans, M.J. (1987) A potential animal model for Lesch-Nyhan syndrome through introduction of HPRT mutations into mice. *Nature, Lond.* **326,** 295–298.

Lavitrano, M., Camaioni, A., Fazio, V.M., Dolch, S., Farace, M.G. & Spadafora, C. (1989) Sperm cells as vectors for introducing foreign DNA into eggs: genetic transformation of mice. *Cell* **57,** 717–723.

Leibfried-Rutledge, M.L., Critser, E.S., Eyestone, W.H., Northey, D.L. & First, N.L. (1987) Development potential of bovine oocytes matured in vitro or in vivo. *Biol. Reprod.* **36,** 376–383.

Leibfried-Rutledge, M.L., Critser, E.S., Parrish, J.J. & First, N.L. (1989) In vitro maturation and fertilization of bovine oocytes. *Theriogenology* **31,** 61–74.

Leibo, S.P. (1988) Bisection of mammalian embryos by micromanipulation. In *Hands-on IVF, Cryopreservation and Micromanipulation; American Fertility Society Regional Postgraduate Course,* Madison, WI.

Leonard, M., Kirszenbaum, M., Cotinot, C., Chesne, P., Heyman, Y., Stinnakre, M.G., Bishop, C., Delouis, C., Vaiman, M. & Fellous, M. (1987) Sexing bovine embryos using Y chromosome specific DNA probe. *Theriogenology* **27,** 248, abstr.

Loesch-Fries, L.S., Merlo, D., Zinnen, T., Burhop, L., Hill, K., Krahn, K., Jarvis, N., Nelson, S. & Halk, E. (1987) Expression of alfalfa mosaic virus RNA 4 in transgenic plants confers virus resistance. *EMBO J.* **7,** 1845–1851.

Lu, K.H., Gordon, I., Gallagher, M. & McGovern, H. (1987) Pregnancy established in cattle by transfer of embryos derived from in vitro fertilisation of oocytes matured in vitro. *Vet. Rec.* **121,** 259–260.

Maguire, J.E., Ehrlich, R., Frels, W. & Singer, D.S. (1990) Regulation of expression of a Class I major histocompatibility complex transgene. *J. Reprod. Fert., Suppl.* **41,** 59–62.

McGrane, M.M., de Vente, J., Yun, J., Bloom, J., Park, E., Wynshaw-Boris, A., Wagner, T., Rottman, F.M. & Hanson, R.W. (1988) Tissue specific expression and dietary regulation of a chimeric phosphoenolpyruvate carboxykinase/bovine growth hormone gene in transgenic mice. *J. biol. Chem.* **263,** 11443–11451.

Menino, A.R., Jr & Wright, R.W., Jr (1982) Development of one cell porcine embryos in two culture systems. *J. Anim. Sci.* **54,** 583–588.

Moor, R.M. & Trounson, A.O. (1977) Hormonal and follicular factors affecting maturation of sheep oocytes *in vitro* and their subsequent developmental capacity. *J. Reprod. Fert.* **49,** 101–109.

Murray, J.D., Nancarrow, C.D. & Ward, K.A. (1988) Techniques for the transfer of foreign genes into animals. *Proc. 11th Int. Congr. Anim. Reprod. & AI, Dublin,* vol. V, pp. 19–27.

Nagai, T., Takahashi, T., Masuda, H., Shioya, Y., Kuwayama, M., Fukushima, M., Iwasaki, S. & Harrada, A. (1988) In-vitro fertilization of pig oocytes by frozen boar spermatozoa. *J. Reprod. Fert.* **84,** 585–591.

Nakamura, Y., Leppert, M., O'Connell, P., Wolff, R., Holm, T., Culver, M., Martin, C., Fujimoto, E., Hoff, M., Kumlin, E. & White, R. (1987) Variable number of tandem repeat (VNTR) markers for human gene mapping. *Science, NY* **235,** 1616–1622.

Notarianni, E., Laurie, S., Moor, R.M. & Evans, M.J. (1990) Maintenance and differentiation in culture of pluripotential embryonic cell lines from pig blastocysts. *J. Reprod. Fert., Suppl.* **41,** 51–56.

O'Brien, S.J. (Ed.) 1987) *Genetic Maps*, Vol. 4. Cold Spring Harbor Laboratory, NY.
Osborn, L., Rosenberg, M.P., Keller, S.A. & Meisler, M.H. (1987) Tissue specific and insulin-dependent expression of a pancreatic amylase gene in transgenic mice. *Molec. cell. Biol.* **7,** 326–334.
Ozil, J.P., Heyman, Y. & Renard, J.P. (1982) Production of monozygotic twins by micromanipulation and cervical transfer in the cow. *Vet. Rec.* **110,** 126–127.
Palmiter, R.D., Brinster, R.L., Hammer, R.E., Trumbauer, M.E., Rosenfeld, M.G., Birnberg, N.C. & Evans, R.M. (1982) Dramatic growth of mice that develop from eggs microinjected with metallothionein-growth hormone fusion genes. *Nature, Lond.* **300,** 611–615.
Palmiter, R.D., Norstedt, G., Gelinas, R.E., Hammer, R.E. & Brinster, R.L. (1983) Metallothionein-human GH fusion genes stimulate growth of mice. *Science, NY* **222,** 809–814.
Parrish, J.J., Susko-Parrish, J.L., Handrow, R.R., Sims, M.M. & First, N.L. (1989) Capacitation of bovine spermatozoa by oviduct fluid. *Biol. Reprod.* **40,** 1020–1025.
Petrovskis, E.A., Meyer, A.L. & Post, L.E. (1988) Reduced yield of infectious pseudorabies virus and herpes simplex virus from cell lines producing viral glycoprotein GP50. *J. Virol.* **62,** 2196–2199.
Pinkert, C.A., Widera, G., Cowing, E., Heber-Katz, E., Palmiter, R.D., Flavell, R.A. & Brinster, R.L. (1985) Tissue specific, inducible and functional expression of the E alpha dMHC class II gene in transgenic mice. *EMBO J.* **4,** 2225–2230.
Popescu, C.P., Cotinot, C., Boscher, J. & Kirszenbaum, M. (1988) Chromosomal localization of a bovine male specific probe. *Annls Genet.* **31,** 39–42.
Prather, R.S. & First, N.L. (1990) Cloning embryos by nuclear transfer. *J. Reprod. Fert., Suppl.* **41,** 125–134.
Prather, R.S., Barnes, F.L., Sims, M.M. Robl, J.M., Eyestone, W.H. & First, N.L. (1987) Nuclear transplantation in the bovine: assessment of donor nuclei and recipient oocyte. *Biol. Reprod.* **37,** 859–866.
Prather, R.S., Sims, M.M. & First, N.L. (1989) Nuclear transfer in early pig embryos. *Biol. Reprod.* **41,** 414–418.
Pursel, V.G., Hammer, R.E., Bolt, D.J., Palmiter, R.D. & Brinster, R.L. (1990) Integration, expression and germ-line transmission of growth-related genes in pigs. *J. Reprod. Fert., Suppl.* **41,** 77–87.
Reed, K.C., Mathews, M.E. & Jones, M.A. (1988) Sex determination in ruminants using Y-chromosome specific polynucleotides. Published Patent Application. Patent Cooperative Treaty No. WO88/01300.
Rexroad, C.E. & Pursel, V.G. (1988) Status of gene transfer in domestic animals. *Proc. 11th Int. Congr. Anim. Reprod. & A.I. Dublin*, vol. V, pp. 28–35.
Robl, J.M. & First, N.L. (1985) Manipulation of gametes and embryos in the pig. *J. Reprod. Fert.* **33,** 101–114.
Rorie, R.W., Voelkel, S.A., McFarland, C.W., Southern, L.L. & Godke, R.A. (1985) Micromanipulation of day-6 porcine embryos to produce split-embryo piglets. *Theriogenology* **23,** 225, abstr.
Salter, D.W., Smith, E.J., Hughes, S.H., Wright, S.E. & Crittenden, L.B. (1987) Transgenic chickens: insertion of retroviral genes into the chicken germ line. *Virology* **157,** 236–240.
Shani, M. (1985) Tissue specific expression of the rat myosin light chain L gene in transgenic mice. *Nature, Lond.* **314,** 283–286.
Shani, M. (1986) Tissue specific and developmentally regulated expression of a chimeric actin/globin gene in transgenic mice. *Molec. cell. Biol.* **6,** 2624–2631.
Shelden, R.F., Skoskiewicz, M.J., Howie, K.B., Russell, P.S. & Goodman, H.M. (1986) Regulation of human insulin gene expression in transgenic mice. *Nature, Lond.* **321,** 525–528.
Simons, J.P., McClenaghan, M. & Clark, A.J. (1987) Alteration of the quality of milk by expression of sheep beta-lactoglobulin in transgenic mice. *Science, NY* **328,** 530–532.
Sirard, M.A., Lambert, R.D., Menard, D.P. & Bedoya, M. (1985) Pregnancies after in-vitro fertilization of cow follicular oocytes, their incubation in rabbit oviducts and their transfer to the cow uterus. *J. Reprod. Fert.* **75,** 551–556.
Sirard, M.A., Parrish, J.J., Ware, C.B., Leibfried-Rutledge, M.L. & First, N.L. (1988) The culture of bovine oocytes to obtain developmentally competent embryos. *Biol. Reprod.* **39,** 546–552.
Smith, L.C. & Wilmut, I. (1989) Influence of nuclear and cytoplasmic activity on the development in vivo of sheep embryos after nuclear transplantation. *Biol. Reprod.* **40,** 1027–1035.
Smithies, O., Gregg, R.G., Boggs, S.S., Koralewski, M.A. & Kucherlapati, R.S. (1985) Insertion of DNA sequences into the human chromosomal β globin locus via homologous recombination. *Nature, Lond.* **317,** 230–234.
Soriano, P., Cane, R.D., Mulligan, R.C. & Jaenisch, R. (1986) Tissue specific and ectopic expression of genes introduced into transgenic mice by retroviruses. *Science, NY* **234,** 1409–1413.
Staeheli, P., Haller, O., Boll, W., Lindenmann, J. & Weissmann, C. (1986) Mx protein: constitutive expression in 3T3 cells transformed with cloned Mx cDNA. *Cell* **44,** 147–158.
Staigmiller, R.B. & Moor, R.M. (1984) Effect of follicle cells on the maturation and developmental competence of ovine oocytes matured outside the follicle. *Gamete Res.* **9,** 221–229.
Stice, S.L. & Robl, J.M. (1988) Nuclear reprogramming in nuclear transplant rabbit embryos. *Biol. Reprod.* **39,** 657–664.
Tarkowski, A.K. & Wroblewska, J. (1967) Development of blastomeres of mouse eggs isolated at the 4- and 8-cell stage. *J. Embryol. exp. Morph.* **36,** 155–180.
Thomas, K.R. & Capecchi, M. (1987) Site-directed mutagenesis by gene targeting in mouse embryo-derived stem cells. *Cell* **51,** 503–512.
Thompson, S., Clarke, A.R., Pow, A.M., Hooper, M.L. & Melton, D.W. (1989) Germ line transmission and expression of a corrected HPRT gene produced by gene targeting in embryonic stem cells. *Cell* **56,** 313–321.
Wachtel, S.S., Nakamura, D., Wachtel, G., Fenton, W., Kent, M.G. & Jaswaney, V. (1988) Sex selection with monoclonal H-Y antibody. *Fert. Steril.* **50,** 355–360.
Ward, K.A., Murray, J.D., Nancarrow, C.D., Sutton, R. & Boland, M.P. (1984) The role of embryo gene transfer in sheep breeding programs. In *Reproduction in Sheep,*

pp. 279–285. Eds D. R. Lindsay & D. T. Pearce. Australian Academy of Science, Canberra.

Watkins, P.C. (1988) Restriction fragment length polymorphism (RFLP): applications in human chromosome mapping and genetic disease research. *Biotechniques* **6,** 310–314.

White, K.L. (1988) Identification of embryonic sex by immunological methods. In *Hands-on IVF, Cryopreservation, and Micromanipulation.* American Fertility Society of Regional Workshop, Madison, WI.

White, K.L., Anderson, G.B. & Bon Durant, R.H. (1987) Expression of a male-specific factor on various stages of preimplantation bovine embryos. *Biol. Reprod.* **37,** 867–873.

Willadsen, S.M. (1982) Micromanipulation of embryos of the large domestic species. In *Mammalian Egg Transfer*, pp. 185–210. Ed. C. E. Adams. CRC Press, Boca Raton.

Willadsen, S.M. (1986) Nuclear transplantation in sheep embryos. *Nature, Lond.* **320,** 63–65.

Wilmut, I., Archibald, A.L., Harris, S., McClenaghan, M., Simons, J.P., Whitelaw, C.B.A. & Clark, A.J. (1990) Modification of milk composition. *J. Reprod. Fert., Suppl.* **41,** 135–146.

Womack, J.E. & Moll, Y.D. (1986) Gene map of the cow: conservation of linkage with mouse and man. *J. Hered.* **77,** 2–7.

Wright, R.W. & Bondioli, K.R. (1981) Aspects of in vitro fertilization and embryo culture in domestic animals. *J. Anim. Sci.* **53,** 702–729.

Xu, K.P., Greve, T., Callesen, H. & Hyttel, P. (1987) Pregnancy resulting from cattle oocytes matured and fertilized *in vitro*. *J. Reprod. Fert.* **81,** 501–504.

DNA: GENERAL

Chairman
R. H. Foote

J. Reprod. Fert., Suppl. **41** (1990), 17–23
Printed in Great Britain

Developmental regulation and tissue-specific expression of a chimaeric phosphoenolpyruvate carboxykinase/bovine growth hormone gene in transgenic animals

Mary M. McGrane, June S. Yun*, W. J. Roesler, E. A. Park, T. E. Wagner* and R. W. Hanson

*Pew Center for Molecular Nutrition and Department of Biochemistry, Case Western Reserve University School of Medicine, Cleveland, Ohio 44106, USA; and *Edison Animal Biotechnology Center, Ohio University, Athens, Ohio 45701, USA*

Summary. Expression of the bovine growth hormone (bGH) gene, directed by the phosphoenolpyruvate carboxykinase (PEPCK) gene promoter, in transgenic animals was investigated. Different lengths of the 5′ PEPCK promoter–regulatory domain were utilized to control bGH expression; these included −2000/+73, −460/+73, −355/+73, and −174/+73. The chimaeric PEPCK/bGH gene containing −460/+73 of PEPCK 5′ flanking sequence (PEPCK/bGH(460)) is regulated by cAMP, insulin, and dexamethasone in the same manner as the endogenous PEPCK gene. This PEPCK promoter–regulatory domain also controls the tissue-specific expression of the bGH gene to liver, kidney, adipose tissue, jejunum and mammary gland. Furthermore, the correct developmental pattern of expression is observed in the mouse lines which contain PEPCK/bGH(460). The transgene mRNA is not detected during fetal development until Day 19. At Day 1 after birth, due to alterations in the insulin:glucagon ratio, the amounts of transgene mRNA are greatly increased, similar to the endogenous PEPCK gene.

Keywords: phosphoenolpyruvate carboxykinase; growth hormone; transgenic mice; tissue specificity; development; liver

Introduction

Transgenic animals have been used to: (1) characterize elements in the promoter of a gene which are responsible for hormonal regulation and tissue specificity, (2) examine the effects of the expression of a given gene in an intact animal, and (3) alter the phenotype of an animal by the production of a specific gene product, such as growth hormone. This technology has thus provided a powerful tool for using intact animals to examine numerous questions on the molecular level. We have examined the effect of the promoter for a regulatory enzyme in hepatic and renal gluconeogenesis, phosphoenolpyruvate carboxykinase (GTP) (EC 4.1.1.32) (PEPCK), on the tissue-specific expression and transcriptional regulation of a linked bovine growth hormone (bGH) structural gene.

PEPCK catalyses the first committed and the rate-limiting step in the hepatic gluconeogenic pathway (Rognstad, 1979). Alterations in the activity of the enzyme are due to changes in enzyme synthesis which are directly related to the rate of transcription of the PEPCK gene. The transcription of this gene is regulated by numerous hormones. The rapidity of the response of the gene to hormonal stimuli, as well as its high level of expression in liver and kidney, make it an ideal promoter in a variety of biological systems. It is a valuable tool for studying the regulated expression of linked structural genes in the intact animal. Transcription from the PEPCK promoter is regulated by glucagon, glucocorticoids, insulin, phorbol esters, and thyroid hormone in the specific tissues in which it is expressed (Lamers *et al.*, 1982; Granner *et al.*, 1983; Wynshaw-Bovis *et al.*, 1984; Loose *et al.*, 1985; Chu & Granner, 1986; Short *et al.*, 1986; Petersen *et al.*, 1988). PEPCK mRNA is present in liver, kidney cortex, adipose tissue, jejunum, and mammary gland (Tilghman *et al.*, 1976; Garcia-Ruiz *et al.*, 1983). The synthesis of the enzyme is controlled in a differential manner in each of these tissues. The levels of the hepatic enzyme are induced by cAMP and glucocorticoids and are decreased by insulin and phorbol esters. In contrast, renal PEPCK is induced by acidosis and glucocorticoids and decreased by alkalosis (Iynedjian *et al.*, 1975). The gene for PEPCK in kidney is not responsive to the normal changes in glucagon and insulin which control the hepatic enzyme (Iynedjian *et al.*, 1975). A third pattern of PEPCK expression is evident in adipose tissue, in which the rate of transcription is decreased (rather than increased as in liver and kidney) by glucocorticoids (Nechushtan *et al.*, 1988).

The diverse and complex regulatory features of the PEPCK promoter described above suggest that a number of *cis*-acting elements must be present to confer these various forms of transcriptional regulation. Some of these elements have been characterized, including those conferring responsiveness to cAMP and glucocorticoids. Additionally, regions of the promoter required for insulin responsiveness and tissue specificity have been localized to sequences spanning $-460/+73$. In support of these multiple regulatory properties of the PEPCK promoter, DNase I footprint analysis has identified 8 protein binding sites, excluding the TATA box, within the region $-460/+73$ (Fig. 1). Four of these sites, all of which contain some degree of sequence homology to cAMP-responsive elements, appear to bind the same protein and may provide the basis for the robust cAMP responsiveness of the intact PEPCK promoter. Footprinting with nuclear extracts prepared from various rat tissues further indicated that several of the binding sites show tissue specificity. In particular, site P3 is occupied by a protein limited to liver nuclear extracts, while sites P5 and P6 bind proteins relatively enriched in liver and kidney. Site P2, which contains the consensus binding sequence for Hepatic Nuclear Factor-1, binds a protein present in liver, kidney, and brain nuclei. However, the DNase I digestion pattern produced over this site by nuclear protein extracts from the liver was unique compared to the pattern noted for nuclear extracts from the kidney and brain, suggesting that various forms of a protein may exist in different tissues. Several *cis*-acting elements are therefore probably involved in determining the tissue-specific expression pattern of the PEPCK gene. In summary, the multiple protein binding sites in the PEPCK promoter reflect the complex transcriptional regulation of this gene.

Hormonal regulation of expression of a chimaeric PEPCK/bGH gene introduced into transgenic mice

Chimaeric genes containing the PEPCK promoter driving the expression of a selectable marker gene have been utilized extensively in cell culture to study the effect of hormones upon gene transcription. We have utilized the $-460/+73$ bp PEPCK promoter sequence ligated to the bGH structural gene (PEPCK/bGH(460)) (Fig. 2); this was introduced into the germ line of mice by microinjection (for details see McGrane *et al.*, 1988). The administration of Bt_2cAMP to these transgenic mice caused a 2-fold increase in serum bGH concentrations in 90 min (Fig. 3). Moreover, alterations in the carbohydrate content of the diet drastically changed serum bGH values in these same mice. When transgenic animals were fasted for 24 h and then placed on a high carbohydrate (81·5%) diet, the concentrations of serum bGH dropped to 5% of that noted in untreated transgenic mice after 1 week on the diet (Fig. 3). When the high carbohydrate diet was replaced with a high-protein (61%), carbohydrate-free diet, the serum bGH values were increased 30-fold over the same period. The decrease in serum bGH caused by the administration of a high carbohydrate diet is most probably mediated by increased insulin secretion and the consequent inhibitory effect of this hormone upon the bGH-linked PEPCK promoter. Dexamethasone treatment of mice containing PEPCK/bGH(460) decreased PEPCK/bGH mRNA values in liver by 75%, due to

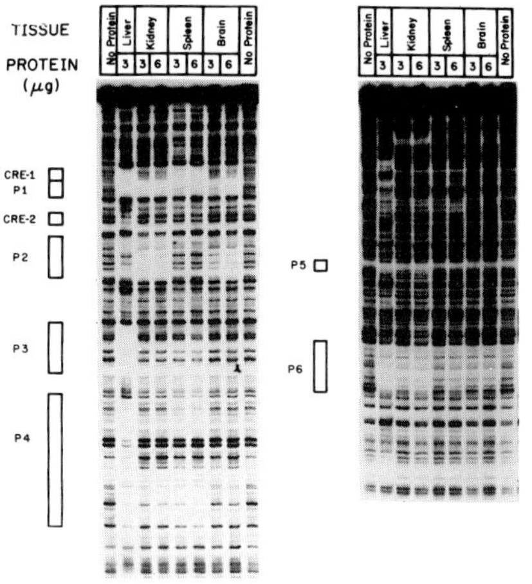

Fig. 1. DNase I footprint of the PEPCK promoter. The PEPCK BamHI/Bgl II probe, end-labelled with T4 polynucleotide kinase, was incubated with nuclear extracts prepared from the indicated rat tissues and subjected to DNase I digestion. The boxes to the left of the footprints are the DNase I-protected regions obtained with rat liver nuclear extracts. (From Roesler *et al.*, 1989.)

the increase in insulin elicited by glucocorticoid treatment (M. Hatzoglou & R. Hampson, unpublished observations). Taken together, our experiments with these mice indicate that the *cis*-responsive elements for the major hormonal influences upon endogenous PEPCK expression can be regulated in the predicted physiological manner in transgenic mice.

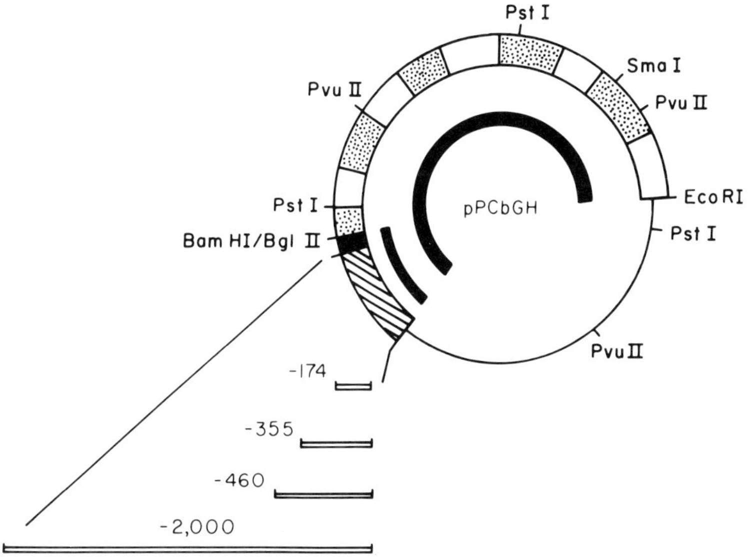

Fig. 2. The recombinant plasmid containing the PEPCK/bGH gene. A pBR322 based plasmid, pbGH, containing the entire structural gene for bGH was engineered such that bGH 5′-flanking sequence was replaced by the −2000, −460, −355 or −174 bp PEPCK promoter region (hatched segment) plus the 73 bp of the first exon of PEPCK (solid segment). The stippled areas of the plasmid represent the exons of bGH and the open segments, the introns of bGH. The solid line is pBR322 sequence. The probes utilized for Northern analysis are those indicated by the thick lines at the centre of the plasmid diagram. For microinjection, the chimaeric PEPCK/bGH gene was excised from the surrounding pBR322 plasmid.

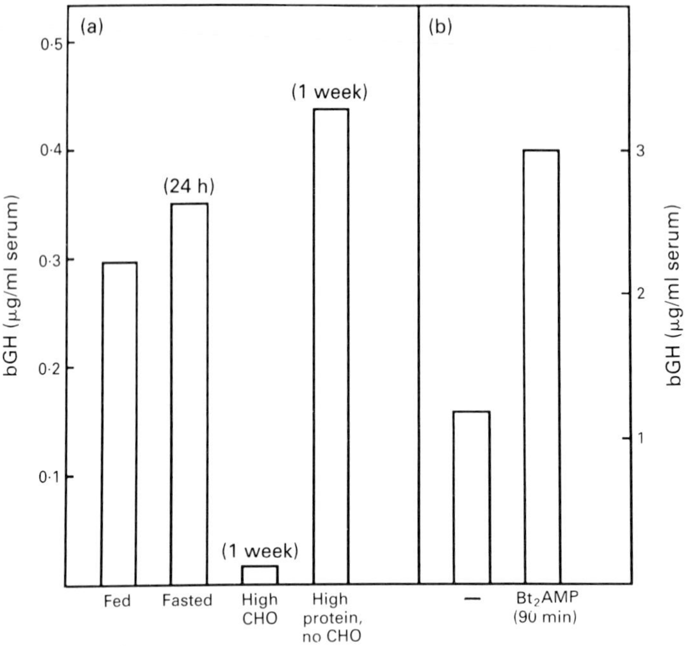

Fig. 3. Effect of high carbohydrate and high protein diets and Bt_2cAMP administration on the concentrations of serum bGH in transgenic mice. (a) A representative transgenic mouse, expressing high levels of serum bGH, was fasted for 24 h and then fed a high carbohydrate diet for 1 week. The animal was then fed a high protein diet devoid of carbohydrate for 1 week. Serum bGH concentrations were determined by an ELISA. (b) A representative transgenic mouse was treated with Bt_2cAMP and theophylline at three consecutive 30-min intervals. After 90-min the mouse was bled and the concentration of serum bGH determined (after McGrane *et al.*, 1988).

Tissue specificity of PEPCK/bGH gene expression

The tissue distribution of PEPCK/bGH mRNA has been examined in transgenic mice (McGrane *et al.*, 1988) and transgenic pigs (Wieghart *et al.*, 1990). As noted with numerous transgenic mouse lines, using different promoters and structural genes, factors regulating the tissue-specific expression of a transgene are a complex issue. Endogenous PEPCK is expressed primarily in liver, kidney cortex and adipose tissue, but lower levels are also detected in jejunum and mammary gland. The tissue distribution of PEPCK/bGH mRNA has been examined in mice with deletions in the PEPCK promoter as follows; −2000(PEPCK/bGH(2000)), −460(PEPCK/bGH(460)), −355(PEPCK/bGH(355)), and −174(PEPCK/bGH(174)). PEPCK/bGH mRNA was detected in the appropriate tissues in mice bearing the PEPCK/bGH(2000), PEPCK/bGH(460) and PEPCK/bGH(355) chimaeric genes (Table 1). All of the mice containing PEPCK/bGH(2000) also gave evidence of bGH expression in tissues in which PEPCK mRNA is extremely low; mRNA was detected in lung and skeletal muscle, to various degrees depending upon the line of animal. Numerous founder mice containing the PEPCK/bGH(174) gene were examined for expression of the transgene, but no PEPCK/bGH mRNA was detected in any tissue, nor was there detectable serum bGH. This promoter deletion (−174) appears to remove sequences necessary for basal gene transcription in the animal. In every transgenic mouse line examined, renal PEPCK/bGH mRNA was low relative to the level of endogenous PEPCK mRNA. This may be due to a decrease in stability of the transgene mRNA in the kidney, as compared to the liver. We have consistently found variability in the tissue distribution and the level of expression of PEPCK/bGH mRNA in different mouse lines.

Although liver is always the predominant site of PEPCK/bGH transgene expression, the level of expression in other tissues varies widely in these transgenic lines. Two mouse lines containing the PEPCK/bGH(460) gene provide an illustration of this. One line exhibits high levels of PEPCK/bGH mRNA in liver, kidney, jejunum and adipose tissue, whereas another exhibits high levels of mRNA in liver, kidney and mammary gland, with no detectable mRNA in jejunum or adipose tissue (data not shown). However, in all the lines examined which contain the PEPCK/bGH(460) transgene, no PEPCK/bGH mRNA was detected in tissues in which endogenous PEPCK is not expressed. Therefore, the tissue distribution which is observed in animals containing the PEPCK/bGH(460) chimaeric gene appears to be the result of: (1) promoter-determined tissue specificity and (2) integration site effects, which alter quantitatively the level of transgene expression (sometimes to below detectable limits) from line to line.

Table 1. Tissue distribution of PEPCK/bGHmRNA in transgenic mice with gene deletions

PEPCK deletion	Liver	Kidney	Adipose	Intestine	Mammary	Lung	Heart	Spleen	Skeletal muscle
−2000	+++	+	++	+	++	+	—	—	+
−460	++++	++	+++	+	++	—	—	—	—
−355	++	+	—	+	—	—	—	—	—
−174	—	—	—	—	—	—	—	—	—

Total RNA was isolated from the specified tissues as described by McGrane *et al.* (1988). Northern analysis was conducted utilizing the PEPCK/bGH BamHI-EcoRI probe (long probe in Fig. 2) and relative amounts of mRNA were determined by densitometric scan.

The tissue distribution of PEPCK/bGH mRNA in pigs containing the PEPCK/bGH(460) transgene is similar to that observed in mice (Wieghart *et al.*, 1990). The highest level of PEPCK/bGH expression is in liver, while the levels of kidney PEPCK/bGH are low. In one line, PEPCK/bGH mRNA has been detected in jejunum, but no mRNA has been detected in either adipose or mammary tissue. Therefore, *cis*-acting elements contained within −460 to +73 of the promoter of the PEPCK gene of the rat retain the ability to direct tissue-specific expression in the pig. This fact should prove useful when assessing the utility of a given promoter (such as that for PEPCK) for future use in agricultural animals.

Developmental regulation of PEPCK/bGH gene expression

The gene for PEPCK is not expressed until late in development, when low levels of PEPCK mRNA can be detected (Garcia-Ruiz *et al.*, 1978). In response to post-natal hypoglycaemia, plasma glucagon concentrations are increased and plasma insulin values are decreased (Girard *et al.*, 1973; DiMarco *et al.*, 1978). This alteration in the insulin:glucagon ratio at birth triggers a rapid increase in hepatic cAMP (Girard *et al.*, 1973) and an induction of PEPCK gene expression. The concomitant increase in PEPCK activity coincides with the initiation of hepatic gluconeogenesis (Ballard & Hanson, 1967). Therefore, this induction in transcriptional activity of the PEPCK gene is critical for glucose homeostasis in the newborn. The regulatory DNA sequences responsible for this transcriptional switch have yet to be identified, so we sought to determine those sequences sufficient for the correct development regulation of the PEPCK/bGH transgene. We have determined that the PEPCK/bGH(460) transgene exhibits the same pattern of developmental expression as noted with the endogenous PEPCK gene. We have examined RNA from livers of mouse fetuses at Days 15, 17, 19 and at Day 1 after birth. As indicated in Fig. 4, bGH mRNA is not detected until Day 19 of fetal development and the levels are induced approximately 200-fold after birth. The low levels of mRNA for the transgene noted at Day 19 are identical to the values observed for the endogenous PEPCK gene at Day 19. For comparison, we examined the expression of mouse metallothionein 1 (Mt)/bGH mRNA in the livers of 15-day fetal mice. High levels of bGH mRNA were observed in the livers of mice containing the transgene. It is evident that the Mt-directed bGH gene is transcriptionally active early in fetal development, while transcription from the PEPCK promoter is initiated during the perinatal period. This difference in the developmental patterning of expression of the bGH structural gene in these two different groups of transgenic mice may lead to later differences in the way the adult mice respond to high levels of ectopic bGH expression.

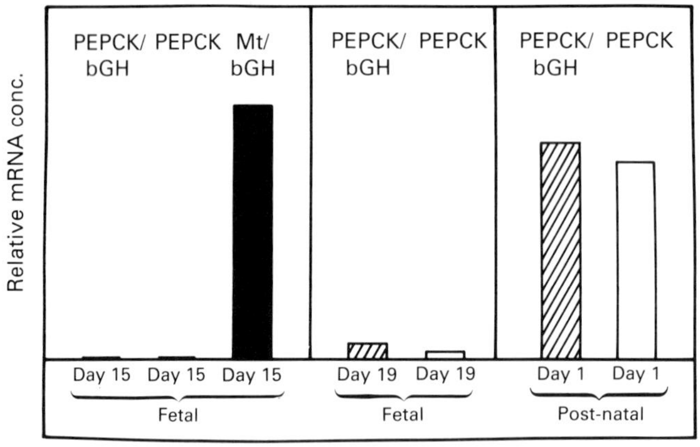

Fig. 4. Developmental profile of PEPCK/bGH transgene expression in mice. Total RNA was isolated from fetal and postnatal livers of transgenic mice as described by McGrane *et al.* (1988). Northern analysis was conducted utilizing either a bGH cDNA probe (to compare PEPCK/bGH and Mt/bGH mRNA levels) or a PEPCK Bam HI-Bgl II probe, short probe in Fig. 2 (to compare PEPCK/bGH and endogenous PEPCK mRNA levels). The relative amounts of mRNA were determined by densitometric scan.

Discussion

Our studies have demonstrated that the PEPCK promoter–regulatory domain (−460/+73) is potentially useful in transgenic animals since it is expressed in a tissue-specific manner and it can be readily regulated by diet. The promoter also has strong basal activity, directing bGH transcription at a level equivalent to that of the intact endogenous PEPCK promoter in liver. This active promoter can be controlled by altering the carbohydrate content of the diet, to the extent that it is virtually silent. The deleterious effects of high levels of GH expression have been reported in mice (Bartke *et al.*, 1988; Doi *et al.*, 1988; Quaife *et al.*, 1989) and agricultural animals (Pursel *et al.*, 1989). An effective switch to reduce the concentrations of GH, such as carbohydrate feeding, has the potential for eliminating these health problems related to excessive expression of GH. Another benefit is the accurate targeting of expression of linked structural genes to specific tissues such as liver or kidney (an advantage in gene therapy) and the targeted expression to mammary tissue (an advantage in molecular farming, to produce gene products in the milk of agricultural animals). Finally, the fact that the transgene remains dormant during fetal life, when linked to the PEPCK promoter, eliminates the risk of transgene expression interfering with normal fetal development. Although these potential advantages of the PEPCK promoter remain to be tested in the appropriate systems, we feel that this promoter will be a useful tool both in the expression of GH in transgenic animals and in the expression of other linked structural genes.

This work was supported by funds from grants DK-25541-12 and DK-21859-11 (R.W.H.) and DH-09042 (T.E.W.) from the National Institutes of Health and from funds from the Edison Program of the State of Ohio and the PEW Charitable trust. W.J.R. is a Postdoctoral fellow of the Medical Research Council of Canada.

References

Ballard, F.J. & Hanson, R.W. (1967) Changes in lipid synthesis in rat liver during development. *Biochem. J.* **102,** 952–958.

Bartke, A., Steger, R.W., Hodges, S.L., Parkening, T.A., Collins, T.J., Yun, J.S. & Wagner, T.E. (1988) Infertility in transgenic female mice with human growth hormone expression: evidence for luteal failure. *J. exp. Zool.* **248,** 121–124.

Chu, D.T.W. & Granner, D.K. (1986) The effect of phorbol esters and diacylglyercol on expression of the phosphoenolpyruvate carboxykinase (GTP) gene in rat hepatoma H4IIE cells. *J. biol. Chem.* **261,** 16848–16853.

DiMarco, P.N., Ghisalberti, A.V., Martin, C.E. & Oliver, I.T. (1978) Perinatal changes in liver corticosterone, serum insulin and plasma glucagon and corticosterone in the rat. *Eur. J. Biochem.* **87,** 243–247.

Doi, T., Striker, L.J., Quaife, C., Conti, F.G., Palmiter, R.D., Behringer, R., Brinster, R.L. & Striker, G.E. (1988) Progressive glomerulosclerosis develops in transgenic mice chronically expressing growth hormone and growth hormone releasing factor but not in those expressing insulinlike growth factor-1. *Am. J. Path.* **131,** 398–403.

Garcia-Ruiz, J.P., Ingram, R. & Hanson, R.W. (1978) Changes in hepatic messenger RNA for phosphoenolpyruvate carboxykinase (GTP) during development. *Proc. natn. Acad. Sci. USA* **75,** 4189–4193.

Garcia-Ruiz, J.P., Lobato, M.F., Ros, M. & Moreno, F.J. (1983) Presence of cytosolic phosphoenolpyruvate carboxykinase activity in rat mammary gland. *Enzyme* **30,** 265–268.

Girard, J.R., Cuendet, G.S., Marliss, E.B., Kervran, A., Rieutort, M. & Assan, R. (1973) Fuels, hormones, and liver metabolism at term and during the early post-natal period in the rat. *J. clin. Invest.* **53,** 3190–3200.

Granner, D., Andreone, T., Sasaki, K. & Beale, E. (1983) Inhibition of transcription of the phosphoenolpyruvate carboxykinase gene by insulin. *Nature, Lond.* **305,** 549–551.

Iynedjian, P.B., Ballard, F.J. & Hanson, R.W. (1975) The regulation of phosphoenolpyruvate carboxykinase (GTP) synthesis in rat kidney cortex. *J. biol. Chem.* **250,** 5596–5603.

Lamers, W.H., Hanson, R.W. & Meisner, H.M. (1982) cAMP stimulates transcription of the gene for cytosolic phosphoenolpyruvate carboxykinase in the rat liver nuclei. *Proc. natn. Acad. Sci. USA* **79,** 5137–5141.

Loose, D.S., Cameron, D.K., Short, H.P. & Hanson, R.W. (1985) Thyroid hormone regulates transcription of the gene for cytosolic phosphoenolpyruvate carboxykinase (GTP) in rat liver. *Biochemistry, NY* **24,** 4509–4512.

McGrane, M.M., deVente, J., Yun, J.S., Bloom, J., Park, E., Wynshaw-Boris, A., Wagner, T.E., Rottman, F.W. & Hanson, R.W. (1988) Tissue specific expression and dietary regulation of a chimaeric phosphoenolpyruvate carboxykinase/bovine growth hormone gene in transgenic mice. *J. biol. Chem.* **263,** 11443–11451.

Nechushtan, H., Benvenisty, N., Brandeis, R. & Reshef, L. (1988) Glucocorticoids control phosphoenolpyruvate carboxykinase activity in rat mammary gland. *Nucleic Acids Res.* **15,** 6405–6411.

Petersen, D.D., Magnuson, M.A. & Granner, D.K. (1988) Location and characterization of two widely separated glucocorticoid response elements in the phosphoenolpyruvate carboxykinase gene. *Molec. cell. Biol.* **8,** 96–104.

Pursel, V.G., Pinkert, C.A., Miller, K.F., Bolt, D.J., Campbell, R.G., Palmiter, R.D., Brinster, R.L. & Hammer, R.E. (1989) Genetic engineering of livestock. *Science, NY* **244,** 1281–1288.

Quaife, C.J., Mathews, L.S., Pinkert, C.A., Hammer, R.E., Brinster, R.L. & Palmiter, R.D. (1989) Histopathology associated with elevated levels of growth hormone and insulin-like growth factor-1 in transgenic mice. *Endocrinology* **124,** 40–48.

Roesler, W.J., Vandenbark, G.R. & Hanson, R.W. (1989) Identification of multiple protein binding domains in the promoter-regulatory region of the phosphoenolpyruvate carboxykinase (GTP) gene. *J. biol. Chem.* **264,** 9657–9664.

Rognstad, R. (1979) Rate limiting steps in metabolic pathways. *J. biol. Chem.* **254,** 1875–1878.

Short, J.M., Wynshaw-Boris, A., Lugo, T.G. Short, H.P. & Hanson, R.W. (1986) Characterization of the phosphoenolpyruvate carboxykinase (GTP) promoter-regulatory region. *J. biol. Chem.* **261,** 9721–9726.

Tilghman, S.M., Ballard, F.J. & Hanson, R.W. (1976) Hormonal regulation of phosphoenolpyruvate carboxykinase (GTP) in mammalian tissue. In *Gluconeogenesis: Its Regulation in Mammalian Species*, pp. 47–87. Eds R. W. Hanson & M. A. Mehlman. John Wiley & Sons, New York.

Wieghart, M., Hoover, J.L., McGrane, M.M., Hanson, R.W., Rottman, F.W., Holtzman, S.H., Wagner, T.E. & Pinkert, C.A. (1990) Production of transgenic pigs containing a rat phosphoenolpyruvate carboxykinase–bovine growth hormone fusion gene. *J. Reprod. Fert., Suppl.* **41,** 89–96.

Wynshaw-Boris, A., Lugo, T.G., Short, J.M., Fournier, R.E.K. & Hanson, R.W. (1984) Identification of a cAMP regulatory region in the gene for rat cytosolic phosphoenolpyruvate carboxykinase (GTP). *J. biol. Chem.* **259,** 12161–12170.

J. Reprod. Fert., Suppl. **41** (1990), 25–35

In-vitro mutagenesis of the bovine growth hormone gene

J. J. Kopchick, S. J. McAndrew, A. Shafer, W. T. Blue, J. S. Yun, T. E. Wagner and W. Y. Chen

Department of Zoology, Molecular and Cellular Biology Program, and Edison Animal Biotechnology Center, Ohio University, Athens, Ohio 45701, USA

Summary. The biological activities of bovine growth hormone (bGH) were studied in a transgenic mouse model system. The following experimental design was used: (1) in-vitro mutagenesis of the bGH gene; (2) expression of the mutated gene in cultured mouse cells under transcriptional regulation of the mouse metallothionein I promoter; (3) binding studies of the mutated and wild-type protein to mouse liver membrane preparations; (4) generation of transgenic mice which express the mutant hormone; and (5) growth rate analysis of transgenic mice. Removal of the alanine + 1 codon from the bGH gene or a substitution of serine for cysteine 189 does not affect the ability of the mutant protein to influence transgenic mouse growth. Also, mutations which increase the hydrophobicity within the bGH alpha helix 3 region (amino acid residues 109–126) do not alter the enhanced growth rate in transgenic mice which express these mutated bGH proteins.

Keywords: GH gene; in-vitro mutagenesis; cattle

Introduction

Bovine growth hormone (bGH), a protein produced in the anterior pituitary gland, contains 191 amino acids and has a molecular mass of approximately 22 000 (Andrews, 1966; Dellacha *et al.*, 1966; Ellis *et al.*, 1966; Miller *et al.*, 1980). Growth hormones found in a variety of species have been reported to be involved in growth promotion (Martin, 1978; Palmiter *et al.*, 1982; Raben, 1958) and are responsible for a variety of metabolic processes including lipid, nitrogen, mineral and carbohydrate metabolism (Milman & Russell, 1950; Raben & Hollenberg, 1959; Swislocki & Szego, 1965; Hjalmarson & Ahren, 1967; Swislocki, 1968; Swislocki *et al.*, 1970; Pandian *et al.*, 1971; Kostyo & Nutting, 1973, 1974; Goodman, 1978; Goodman *et al.*, 1986). Additionally, bGH has been shown to stimulate milk production in the cow (Folley, 1956; Peel *et al.*, 1981, 1983).

The mechanism by which the growth hormone molecule exerts its biological effects is not well understood. It has been reported that the growth hormone molecule may possess multiple bio-active domains (Kostyo, 1986), may bind to different receptors on different tissue (Smal *et al.*, 1986; Sigel *et al.*, 1981) or may contain distinct amino acid sequences within the molecule which are responsible for the various biological activities (Wade *et al.*, 1977, 1982; Ng & Bornstein, 1979; Paladini *et al.*, 1979, 1984; Ng *et al.*, 1974; Armstrong *et al.*, 1983; Ng & Harcourt, 1986; Salem, 1988).

Various growth hormone genes and complementary DNAs (cDNAs) have been cloned, including those derived from human (Goeddel *et al.*, 1979), rat (Seeburg *et al.*, 1977), cow (Woychick *et al.*, 1982; Seeburg *et al.*, 1983), pig (Seeburg *et al.*, 1983), and chicken (Lamb *et al.*, 1988). Most of the genes have been expressed in *Escherichia coli* (*E. coli*). The human and cow genes have been expressed in cultured mammalian cells (Robins *et al.*, 1982; Kopchick *et al.*, 1985). Also, growth hormone genes have been used to produce transgenic mice (Palmiter *et al.*, 1982; McGrane *et al.*, 1988) as well as other transgenic animals (Hammer *et al.*, 1985). When growth hormone genes are expressed and their products secreted into the serum of transgenic mice, the animals grow as much as two times larger than control littermates.

In this report, mutations targeted to three different regions of the bGH gene were generated. These mutations alter the bGH amino terminus, cysteine residues which are involved in disulphide bond formation, and residues within an alpha helix. Using the growth hormone transgenic mouse as a model system for growth, these mutated bGH genes were evaluated for their ability to enhance growth in transgenic mice.

Materials and Methods

Plasmid construction

Plasmids, pBGH-10Δ and pBGH-10Δ6, were derived from pBGH-10 which is a pBR322 based vector (Fig. 1). pBGH-10Δ contains deletions of 1922 base pairs from the NdeI to BamHI sites in pBR322. pBGH-10Δ6 was derived from pBGH-10Δ and contains deletions of introns B, C and D.

Mutagenesis

The plasmids pBGH-10Δ6-AΔ^{+1} and pBGH-10Δ6-K^{112}L,K^{114}W were both derived from pBGH-10Δ6 while pBGH-10Δ-C^{189}S was derived from pBGH-10Δ (Fig. 1). The former two mutations were constructed using oligonucleotide-directed mutagenesis of heteroduplex plasmid DNA (Vlasuk & Inouye, 1983; Morinaga *et al.*, 1984) while the latter was generated by segment-directed mutagenesis using complementary oligonucleotides to replace the DNA between the Tth111I site (found near the 3′ end of Exon IV) and the XmaI site (located near the 5′ end of Exon V).

In order to prepare heteroduplex plasmid DNAs, linear and gapped linear plasmid DNAs were generated. The plasmid pBGH-10Δ6 was digested with AatII and BalI plus SmaI separately, or AatII and SmaI plus HindIII separately to produce linear and gapped linear DNAs used in construction of the AlaΔ^{+1} or C^{189}S plasmids, respectively. The antisense oligonucleotides used to direct the deletion of the alanine codon at position +1 and the substitution of the cysteine to serine codon at position +189 were (5′GGCTGGGAAGCCCACCAC3′) and (5′CTAGAAGGCGGAG CTGGCTCCCCGAA3′), respectively. Approximately 300 ng of each linear and 1 pmol of oligonucleotide primer were mixed, heat denatured and allowed to cool in intervals (100°C, 6 min; 23°C, 20 min; 4°C, 20 min; 0°C, 10 min). Sequences containing gaps flanking the annealed oligonucleotide were made flush by use of either the Klenow fragment of *E. coli* DNA polymerase I (15°C, overnight) or the modified T7 DNA polymerase (23°C, 5 min: Sequenase, United States Biochemical Corp., Cleveland, OH, USA), in the presence of dNTPs and T4 DNA ligase (Boehringer-Mannheim, Indianapolis, IN, USA). Then 1–5 μl were used directly to transform competent MC1061 *E. coli*. Subsequently, colonies were transferred to nitrocellulose and hybridized with the oligonucleotide primer labelled with α-^{32}P-ATP using 1 μl T4 polynucleotide kinase (New England Biolabs, Beverley, MA, USA; 10 000 U/ml: sp. act. ~1–5 × 10^7 μCi/mmol). After differential washing of the filters and autoradiography, hybridization-positive colonies were selected, amplified, and plasmid DNA prepared for retransformation of competent MC1061 *E. coli*. A second screen by oligonucleotide-specific hybridization was carried out to ensure a homogeneous population of plasmid molecules containing the desired mutation.

The mutated plasmid pBGH-10Δ6-K^{112}L,K^{114}W used the complementary oligonucleotides (5′GTGTCTATGAG*CTC*CTG*TG*GGACCTGGAGGAAGG*G*ATCCTGGCCCTGATGCGG GAGCTGGAAGATGGCACCCC3′; 73-mer) and (5′CCGGGGGGTGCCATCTTCCAGCTCC CGCATCAGGGCCAGGAT*C*CCTTCCTCCAGGTCC*CA*CAG*GAG*CTCATAGACA3′; 76-mer) to replace the wild type bGH sequences between Tth111I and XmaI. The oligonucleotides encode DNA changes which result in the substitutions of leucine for lysine at position 112 and

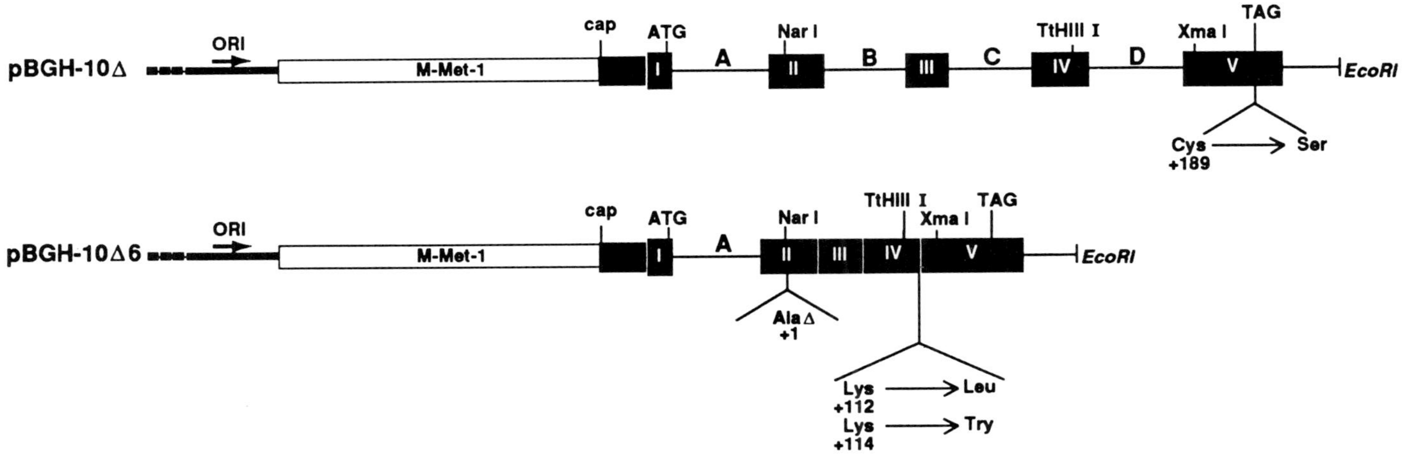

Fig. 1. Recombinant bGH plasmids. pBGH-10Δ contains the mouse metallothionein I transcriptional regulatory sequences fused to the bGH gene which contains five exons (black boxes I–V) and four introns (A–D). The translation initiation (ATG) and termination (TAG) signals are indicated. This fusion gene was incorporated into pBR322 at the EcoRI site. The pBR322 origin of replication is indicated (ORI). pBGH-10Δ6 is identical to pBGH-10Δ with the exception that introns B, C, and D have been removed.

tryptophan for lysine at position 114. In addition, the oligonucleotide duplex encodes a silent base-pair change designed to create a unique BamHI restriction site which simplified screening procedures. The oligonucleotides were annealed and subcloned between the Tth111I and XmaI sites. Mutated plasmid DNAs were identified by digestion with BamHI.

Sequence analysis

The nucleotide sequences of the mutated bovine growth hormone target regions were determined by using the dideoxy chain-termination method with modified T7 DNA polymerase (Sequenase, United States Biochemical; Sanger *et al.*, 1977). Oligonucleotide primers for manual DNA sequencing were synthesized using the DuPont Coder ♯300 DNA synthesizer and purified by denaturing polyacrylamide gel electrophoresis, passive elution and concentration by ethanol precipitation. The oligonucleotide primers used for the direct sequencing analyses of pBGH-10Δ6-AlaΔ^{+1}, pBGH-10Δ-C^{189}S and pBGH-10Δ6-K^{112}L,K^{114}W were a 22-mer (5′GCAGCCAGCTG ATGCAGGAGCT3′), 26-mer (5′TCAGGTACGTCTCCGTCTTATGCAGG3′), and an 18-mer (5′AAATTTGTCATAGGTCTG3′), respectively. Briefly, 1–3 μg double-stranded plasmid DNA were denatured in the presence of 0·2 N-NaOH, and 10–20 pmol oligonucleotide primer were allowed to anneal (65°C, 2 min, followed by 30 min slow cooling) to the denatured template. A two-step polymerization was performed by using the modified T7 DNA polymerase which extends the oligonucleotide-primed chain in the presence of dNTPs and deoxyadenosine 5′-[α-^{35}S]triotriphosphate (sp. act. >1000 Ci/mmol; Amersham) followed by transfer of equal aliquants into each of four specific dideoxynucleotide mixes which randomly terminate chain elongation. Following addition of a formamide termination buffer to each reaction, the samples were incubated at 80°C for 2 min and the DNA sequence was determined after size fractionation of the four sets of fragments by 8% polyacrylamide/8 M-urea electrophoresis and autoradiography.

Cell culture and transient expression system

Mouse L cells were maintained in DMEM (Gibco, Grand Island, NY, USA) plus 10% calf serum and 50 μg gentamicin/ml (Gibco). In this study, a modification of a previously described transfection procedure was used (Lopata *et al.*, 1984). Briefly, 2 μg plasmid DNA were added to 1·0 ml DMEM containing 0·2 mg DEAE–Dextran. This solution was added to approximately 10^6 cells in a 35-mm tissue culture plate which had been washed previously with 2·0 ml DMEM. After incubation of the cells for 1 h at 37°C, the DNA–DEAE–Dextran solution was removed and the cells ‘shocked’ for 90 sec with 2·0 ml 10% DMSO in Hepes-buffered saline, at room temperature. Subsequently, the ‘shock’ solution was removed and cells washed with 2·0 ml DMEM. Medium containing 10% Nu-Serum (Collaborative Research, Bedford, MA, USA) plus 50 μg gentamicin/ml was changed daily. Culture fluids were stored at −20°C. For bGH binding assays, transfected cells were incubated in DMEM minus serum for 16 h, after which the culture fluids were removed and frozen at −20°C.

Radioimmunoassay (RIA)

Culture fluids were assayed for bGH by a standard double-antibody RIA as described by Leung *et al.* (1984) and Kopchick *et al.* (1985).

Polyacrylamide-gel electrophoresis (PAGE)

Sodium dodecyl sulphate (SDS) PAGE analyses of secreted bGH have been described (Kopchick *et al.*, 1985; Kelder *et al.*, 1989). Briefly, transfected cells were metabolically labelled for 16 h with

20 μCi L-[^{35}S]methionine. Culture fluids were collected and incubated with polyclonal rabbit anti-bGH serum which had been previously absorbed with normal mouse cell extract. After overnight incubation at 4°C, *Staphylococcus aureus*, Protein A (Pansorb: CalBiochem, La Jolla, CA, USA) was added, immunoprecipitates collected, and proteins resolved by 15% SDS-PAGE (Kopchick *et al.*, 1979).

Binding studies

Membrane binding studies were performed as previously described (Smith & Talamants, 1987). Liver membrane preparations from C57BL/6J × SJL hybrid mice of either sex (60–120 days old) were homogenized with a Brinkman Polytron in 4 volumes (w/v) of 0·3 M-sucrose, 10 mM-EDTA, 50 mM-Hepes, 0·1 mM-TPCK and 1 mM-PMSF at pH 8·0. The above step and all the following protocols were carried out at 4°C. The homogenate was centrifuged at 20 000 ***g*** for 30 min and the supernatant was centrifuged at 100 000 ***g*** for 1 h. The pellets were washed once with 10 mM-Hepes, pH 8·0, and recentrifuged. These pellets were resuspended in 10 mM-Hepes, pH 8·0, to a protein concentration of approximately 50 mg/ml. Samples of the membranes were frozen on solid CO_2, and stored at −20°C. Membrane protein concentrations were determined by the method of Lowry *et al.* (1951).

Competitive binding assays were performed using the following protocol. Microsomal membranes corresponding to 3 mg protein were incubated with 30 000 c.p.m. ^{125}I-labelled bGH/tube (Cambridge Medical Diagnostics) and unlabelled bGH ranging from 0·07 to 70 nM (0·1 to 500 ng/tube) in a total volume of 0·3 ml assay buffer (20 mM-Hepes, 10 mM-$CaCl_2$, 0·1% BSA, and 0·05% NaN_3 pH 8·0). All assays were performed in triplicate. After overnight incubation at room temperature, membrane-bound hormone was separated from free hormone by the addition of 1 ml ice-cold assay buffer followed by centrifugation at 1000 ***g*** for 20 min. Membrane pellets were then assayed for radioactivity. Specifically bound radioactivity was determined by subtraction from the value produced by incubation of membranes with 5 μg unlabelled bGH (Smith & Talamants, 1987).

Transgenic mouse production

The procedure for production of transgenic mice by direct microinjection of DNA into the male pronucleus of fertilized mouse eggs was as described previously (McGrane *et al.*, 1988). DNA extraction from mouse tails, dot blots, and serum determinations was as described (McGrane *et al.*, 1988).

Results

In-vitro mutagenesis

Using the in-vitro mutagenesis protocols described above, 3 mutated bGH genes were generated. The mutations, pBGH-10Δ6-AlaΔ^{+1}, pBGH-10Δ-C^{189}S, and pBGH-10Δ6-K^{112}L,K^{114}W, encode the following respective amino acid changes: a deletion of Ala1, a substitution of serine for cysteine 189, and substitutions of leucine and tryptophan for lysine 112 and lysine 114, respectively. The mutations were confirmed by Sanger dideoxy sequence analyses shown in Fig. 2.

In-vitro expression

The plasmids encoding the mutations as well as wild-type bGH DNA (pBGH-10Δ) were transiently introduced into cultured mouse L cells, pulse labelled with [^{35}S]methionine and culture fluids immunoprecipitated using a rabbit polyclonal bGH antiserum. After SDS-PAGE and fluorography, radioactive protein bands of approximate molecular weight of 22 000 were observed

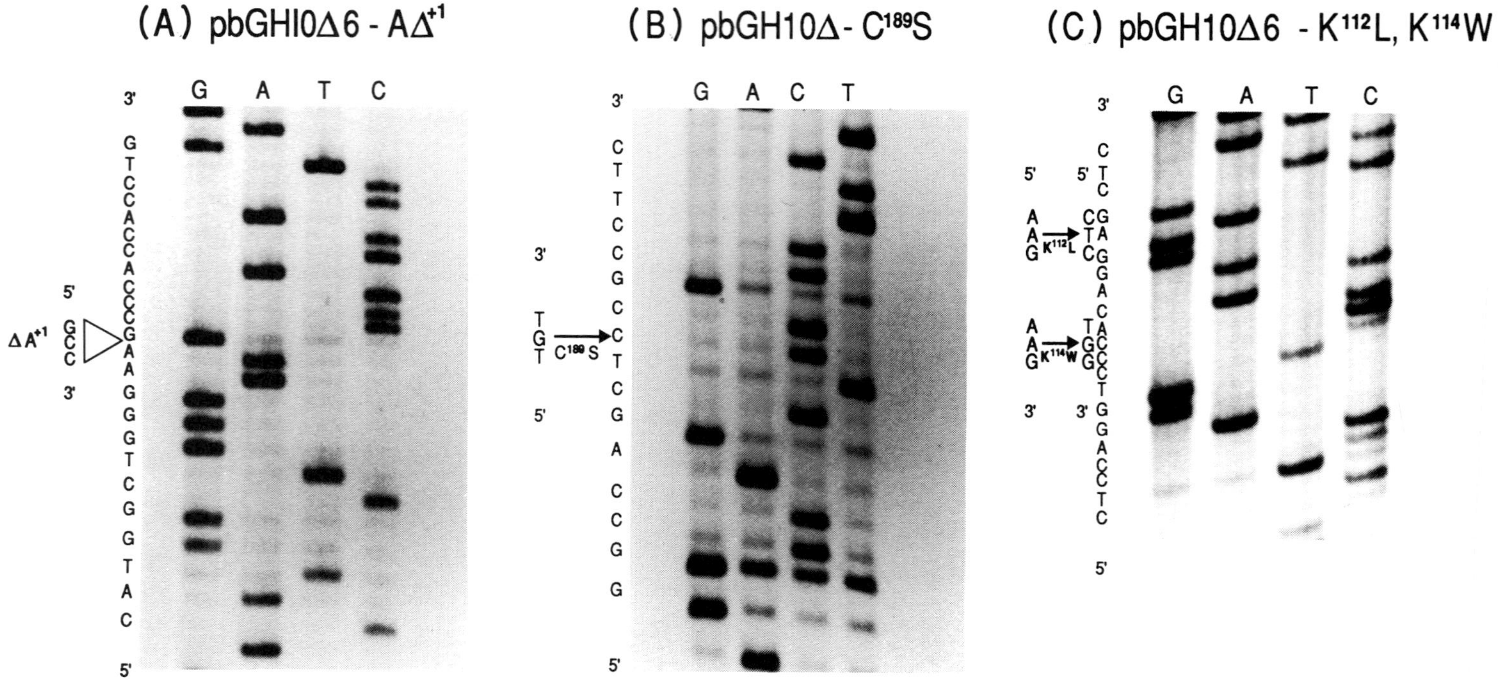

Fig. 2. DNA sequence analysis of the target regions of the bGH gene altered by mutagenesis. Mutated plasmid DNAs were sequenced by the dideoxy chain termination method (see 'Materials and Methods') and the reaction products analysed by 8% polyacrylamide/8 M-urea gel electrophoresis and autoradiography. The mutated target regions of pBGH-10Δ6-AΔ^{+1}, pBGH-10Δ-C^{189}S and pBGH-10Δ6-K^{112}L,K^{114}W are shown in Panels A, B and C, respectively. The lanes corresponding to each nucleotide reaction set, G, A, T, or C are shown across the top and the determined sequence (3′ to 5′) shown on the left. The wild-type and mutant sequences are depicted for the sense strand only.

(Fig. 3). Wild-type bGH (Lane C, Fig. 3) derived from pBGH-10Δ and bGH mutant proteins (Lanes E, F, G, Fig. 3) derived from pBGH-10Δ6-AlaΔ^{+1}, pBGH-10Δ6-K^{112}L,K^{114}W, and pBGH-10Δ-C^{189}S revealed discrete protein bands. The concentrations of bGH in these culture fluids are given in Table 1. Although the concentration of bGH in culture fluids derived from the mutant genes appears to be significantly less than wild-type bGH, these results have not been confirmed. However, bGH of the proper apparent molecular mass is secreted by the transfected cells.

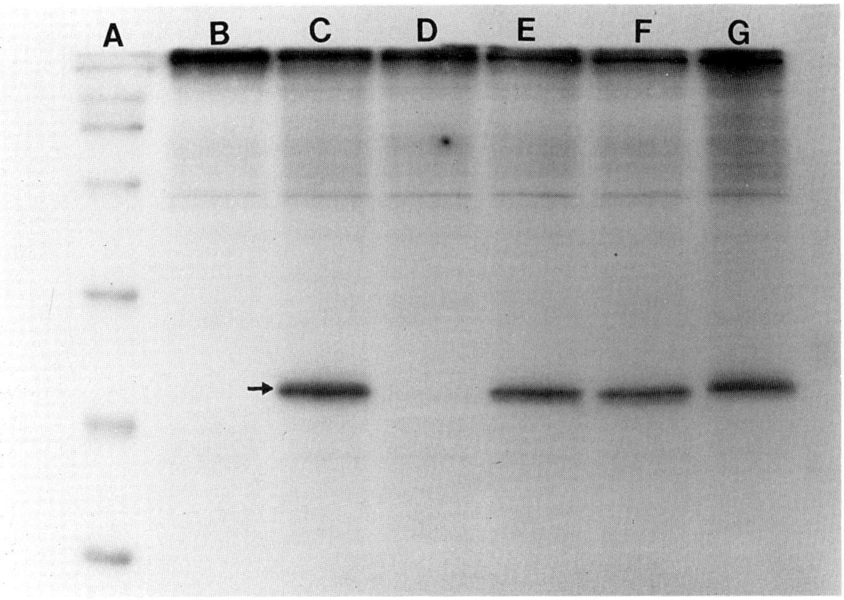

Fig. 3. Immunoprecipitations of [^{35}S]methionine-labelled culture fluids derived from transfected mouse L cells. Day 5 cell culture fluids were exposed to rabbit anti-bGH antibody, and analysed by SDA 15% PAGE, followed by fluorography. Lanes: A, molecular weight markers containing approximately 2 μg/protein standard; B, culture fluids from non-transfected cells; C, culture fluids from pBGH-10Δ6 transfected cells; D, culture fluids from pBGH-6-AΔ^{+1} transfected cells; E, culture fluids from pBGH-10Δ6-AΔ^{+1} transfected cells; F, culture fluid from pBGH-10Δ6-K^{112}L,K^{114}W transfected cells; and G, culture fluids from pBGH-10Δ-C^{189}S transfected cells. The apparent molecular mass of wild-type bGH is approximately 22 000. The arrow indicates 'wild-type' bGH.

Table 1. Bovine growth hormone expression in mouse L cells

Plasmid DNA	bGH (ng/ml)
No DNA	4·96
pBGH-10Δ6	603·09
pBGH-10Δ6-AΔ^{+1}	288·14
pBGH-10Δ6-K^{112}L,K^{114}W	243·67
pBGH-10Δ6-C^{189}S	216·18

Mouse L cells were transfected with 2·0 μg plasmid DNA (see Materials and Methods). Secretion of bGH into culture fluid was assayed 120 h after transfection by a standard double-antibody radioimmunoassay (Kopchick *et al.*, 1985). The values represent the mean of two experiments.

Binding studies

Culture fluids lacking serum were collected from cells transfected by pBGH-10Δ6 (wild-type bGH) and mutated bGH DNAs (pBGH-10Δ6-AΔ^{+1}, pBGH-10Δ6-K^{112}L,K^{114}W, pBGH-10Δ-C^{189}S). Following lyophilization of the culture media and bGH concentration determinations, competitive membrane binding studies were carried out. Effective doses which resulted in 50% displacement (ED50) of ^{125}I-labelled bGH from the membrane preparations were determined. The bGH mutations encoded by pBGH-10Δ6-K^{112}L,K^{114}W (lysine 112 to leucine and lysine 114 to tryptophan) and pBGH-10Δ-C^{189}S (cysteine 189 to serine) revealed ED_{50} values similar to wild-type bGH (Table 2). The bGH mutations encoded by pBGH-10Δ6-AΔ^{+1} revealed an ED_{50} value approximately 2 times that of wild-type bGH (Table 2).

Table 2. Binding of wild-type (wt) and mutant (mut) bGH to mouse liver membrane preparations

bGH	ED_{50} (mM)	ED_{50} (mut) / ED_{50} (wt)
Wild type	0·11	—
K^{112}L,K^{114}W	0·09	0·82
AΔ^{+1}	0·28	0·22
C^{189}S	0·14	1·27

Growth hormone was collected from transfected mouse cells (performed in triplicate) in serum-free medium and concentrated by lyophilization. The values of effective dose of 50% displacement of ^{125}I-labelled bGH (ED_{50}) were calculated from displacement curves of competitive binding of ^{125}I-labelled bGH to the membrane preparations.

Production of transgenic mice

A series of transgenic mouse lines which contain wild-type and mutated bGH genes were produced by standard microinjection techniques (McGrane *et al.*, 1988). The genes contain the transcriptional regulatory sequences of the mouse metallothionein I promoter which has been shown to be active in liver tissue as well as other tissues of the transgenic mouse (Palmiter *et al.*, 1982). Offspring generated by the microinjection procedure were assayed for bGH DNA by slot blot hybridization analysis (data not shown). Mouse lines were generated which contain approximately one copy of the bGH transgene sequences derived from pBGH-10Δ6, pBGH-10Δ6-AlaΔ^{+1}, and pBGH-10Δ6-K^{112}L,K^{114}W, and approximately 1–5 copies from pBGH-10Δ-C^{189}S. Serum from transgenic animals was assayed for bGH concentrations. Growth rates are shown in Table 3. All mice which expressed the wild-type or mutated bGH transgene in serum also possessed a corresponding enhanced growth rate. The growth ratio between the transgenic animals and control littermates was approximately 1·5 for all animals (Table 3). This is a typical value for transgenic animals which express wild-type hGH, rGH, or bGH (Palmiter *et al.*, 1982; McGrane *et al.*, 1988).

Discussion

In attempting to define regions of the bGH molecule which are involved in growth, we have used an experimental strategy which addresses gene structure/biological function relationships. In particular, the following general protocol was used: (1) generation of mutations within the bGH gene; (2) expression of the gene in cultured mouse cells under the transcriptional direction of the mouse metallothionein promoter; (3) binding assessment of the corresponding protein to mouse liver

Table 3. Transgenic mouse growth rates

Weeks after birth	bGH DNA (g)			
	Wild type	$A^{+1}\Delta$	$L^{112}S,L^{114}W$	$C^{189}S$
4	26·5 (22·0)	26·9 (19·6)	23·4 (19·5)	23·4 (20·4)
5		30·8 (22·4)	29·8 (23·4)	24·4 (21·1)
7	36·5 (26·0)	27·2 (26·21)	38·6 (27·0)	27·3 (22·9)
10	46·2 (30·0)	47·5 (27·3)	43·2 (29·2)	33·2 (28·3)
14		55·4 (29·0)	52·5 (36·6)	36·4 (28·4)
17		58·9 (34·0)	53·8 (38·2)	45·6 (31·1)
Transgenic control (growth ratio; week 17)	1·37	1·73	1·41	1·47

*Values in parentheses represent weights of non-transgenic litter mates.

membrane preparations; and (4) determination of the growth properties of transgenic mice which express the mutant gene.

In this study, we have found that three distinct alterations of the bGH gene do not alter the ability of the respective bGH protein to bind to mouse liver membrane preparations or the ability to generate large transgenic mice. The first mutated bGH gene, pBGH-10Δ6-AlaΔ^{+1}, encodes a bGH molecule in which the first amino acid (alanine) of the post-translationally processed protein is deleted. The amino terminal structure of bGH has been shown to be heterogeneous, i.e. approximately an equal mixture of molecules beginning with ala^{+1} or phe^{+2} (Leung *et al.*, 1986). Removal of ala^{+1} generates a bGH protein containing only phenylalanine at position +1 as deduced from amino terminal sequencing analysis (data not shown). This bGH molecule binds to mouse liver membrane preparations with less affinity than wild-type bGH, but it enhances transgenic mouse growth to the same extent as wild-type bGH. Apparently, alanine +1 in bGH is dispensible for its biological activity in the enhanced growth phenotype of transgenic mice. Further experiments are required to substantiate the membrane binding studies.

Cysteine residues within proteins are important in the overall structure of proteins (Thornton, 1981; Creighton, 1988). Four cysteine residues are found within bGH (amino acid residues 53, 164, 181 and 189). In bGH, cysteine residues form disulphide bridges between residues 53 and 164, and 181 and 189. In this study, cysteine 189 was replaced with serine, thereby generating a bGH molecule with 3 cysteine residues. The bGH molecule encoded by this gene retains its ability to bind to liver membrane preparations and to enhance transgenic mouse growth. Thus, the disulphide bridge between cysteines 181 and 189 apparently is not necessary for either of these activities. We presume that the disulphide bridge between cysteine residues 53 and 164 remains intact in this mutated protein, but no data have been generated to support this belief.

The growth hormone molecule contains 4 alpha-helical regions (Abdel-Meguid *et al.*, 1987). The third alpha-helical region of bGH exists in the form of an amphipathic helix. Amphipathic helical regions of other peptide hormones, e.g. calcitonin (Kaiser & Kezdy, 1984) and growth hormone-releasing factor (Tou *et al.*, 1986), have been shown to be important in the biological activities of the hormones. In this study, we have altered two charged amino acid residues, lysines 112 and 114, in order to define their importance in the biological activities of bGH. Substitution of leucine for lysine 112 and tryptophan for lysine 114 does not alter the ability of these molecules to bind to mouse liver membrane preparations or to enhance transgenic mouse growth. However, transgenic mice which contain a bGH mutation in which glycine^{119} has been changed to arginine do not exhibit enhanced growth rate (data not shown). Other alterations affecting this amphipathic helical region of bGH are in progress.

The above results indicate that a wide variety of changes can be generated in the bGH molecule without affecting its ability to stimulate growth in a transgenic mouse model. The effect of structural gene mutations and their relationship to growth hormone receptor binding and enhancement of animal growth should provide further clues in defining the biologically important regions in the growth hormone molecule.

We thank Diana Guthrie for typing and Dr Joe Cioffi for reviewing this manuscript. J.J.K. is supported in part by the State of Ohio's Eminent Scholar program which includes a grant from Milton and Lawrence Goll.

References

Abdel-Meguid, S.S., Shieh, H.S., Smith, W.W., Dayringer, H.E., Violand, B.N. & Bental, L.A. (1987) Three dimensional structure of a genetically engineered variant of porcine growth hormone. *Proc. natn. Acad. Sci. USA* **84,** 6434–6437.

Andrews, P. (1966) Molecular weights of prolactin and pituitary growth hormones estimated by gel filtration. *Nature, Lond.* **209,** 155–157.

Armstrong, J.McD., Bornstein, J., Bromley, J.O., Macaulay, S.L. & Ng, F.M. (1983) Parallel insulin-like actions of human growth hormone and its part sequence hGH 7-13. *Acta endocr., Copenh.* **102,** 492–498.

Creighton, T.E. (1988) Disulfide bonds and protein stability. *BioEssays* **8,** 57–62.

Dellacha, J.M., Santome, J.A. & Faiferman, L. (1966) Molecular weight of bovine growth hormone. *Experientia* **22,** 16–17.

Ellis, G.J., Marler, E., Chen, H.C. & Wilhelm, A.E. (1966) Molecular weight of bovine, porcine and human growth hormone by sedimentation equilibrium. *Fedn Proc. Fedn Am. Socs exp. Biol.* **25,** 348, abstr.

Folley, S.J. (1956) *The Physiology and Biochemistry of Lactation*, pp. 45–72. C. C. Thomas, Springfield.

Goeddel, D.V., Heyneker, H.L., Hozumi, T., Arentzen, R., Itakura, K., Yansura, D.G., Ross, J.J., Miozari, G., Crea, R. & Seeburg, P.H. (1979) Direct expression in *Escherichia coli* of a DNA sequence coding for human growth hormone. *Nature, Lond.* **281,** 544–548.

Goodman, H.M. (1978) Effects of growth hormone on the utilization of L-leucine in adipose tissue. *Endocrinology* **102,** 210–217.

Goodman, H.M., Grichting, G. & Coiro, V. (1986) Growth hormone action on adipocytes. In *Human Growth Hormone*, pp. 499–512. Eds S. Raiti & R. H. Tolman. Plenum Press, New York.

Hammer, R.E., Pursel, V.G., Rexroad, C.E., Wall, R.J., Bolt, D.J., Ebert, K.M., Palmiter, R.D. & Brinster, R.L. (1985) Production of transgenic rabbits, sheep, and pigs by microinjection. *Nature, Lond.* **315,** 680–683.

Hjalmarson, A. & Ahren, K. (1967) Sensitivity of the rat diaphragm to growth hormone. Early and late effects of growth hormone on amino acid and pentose uptake. *Acta endocr., Copenh.* **56,** 347–358.

Kaiser, E.T. & Kezdy, F.J. (1984) Amphophilic secondary structure: design of peptide hormones. *Science, NY* **223,** 249–255.

Kelder, B., Chen, H. & Kopchick, J.J. (1989) Activation of the mouse metallothionein-I promoter in transiently transfected avian cells. *Gene* **76,** 75–89.

Kopchick, J.J., Karshin, W.L. & Arlinghaus, R.B. (1979) Tryptic peptide analysis of 'gag' and 'gag-pol' gene products of Rauscher murine leukemia virus. *J. Virol.* **30,** 610–623.

Kopchick, J.J., Malavarca, R.H., Livelli, T.J. & Leung, F.C. (1985) Use of avian retroviral bovine growth hormone DNA recombinants to direct expression of biologically active growth hormone by cultured fibroblasts. *DNA* **4,** 23–31.

Kostyo, J.L. (1986) The multipotent nature of growth hormone. In *Human Growth Hormone*, pp. 449–461. Eds S. Raiti & R. H. Tolman. Plenum Press, New York.

Kostyo, J.L. & Nutting, D.F. (1973) Acute *in vivo* actions of growth hormone on protein synthesis in various tissues of the hypophysectomized rats and their relationship to the levels of thymidine factor and insulin in the plasma. *Horm. Metab. Res.* **5,** 167–172.

Kostyo, J.L. & Nutting, D.F. (1974) Growth hormone and protein metabolism. In *Handbook of Physiology*, Sec. 7, Vol. 4, pp. 187–210. Eds E. Krobel & W. H. Sawyer. American Physiological Society, Washington, D.C.

Lamb, J.C., Galehouse, D.M. & Foster, D.N. (1988) Chicken growth hormone cDNA sequence. *Nucleic Acids Research* **16,** 9339.

Leung, F.C., Taylor, J.G., Steelman, S.L., Bennett, C.D., Rodkey, J.A., Long, R.A., Serio, R., Weppelman, R.M. & Olson, G. (1984) Purification and properties of chicken growth hormone and the development of a homologous radioimmunoassay. *Gen. comp. Endocrinol.* **56,** 389–400.

Leung, F.C., Jones, B., Steelman, S.L., Rosemblum, C.I. & Kopchick, J.J. (1986) Purification and physiochemical properties of a recombinant bGH produced by cultured murine fibroblasts. *Endocrinology* **119,** 1489–1496.

Lopata, M.A., Cleveland, D.W. & Soelner-Webb, B. (1984) High levels of transient expression of a chloramphenicol acetyl transferase gene by a DEAE-dextran mediated DNA transfection coupled with a dimethyl sulfoxide or glycerol shock treatment. *Nucl. Acids Res.* **12,** 5707–5717.

Lowry, O.H., Rosebrough, J.H., Farr, A.L. & Randall, R.J. (1951) Protein measurement with the Folin phenol reagent. *J. biol. Chem.* **193,** 265–275.

Martin, J.B. (1978) Neural regulation of growth hormone secretion. *New Engl. J. Med.* **288,** 1384–1393.

McGrane, M.M., deVente, J., Yun, J., Bloom, J., Park, E., Wynshaw-Boris, A., Wagner, T., Rottman, F.M. & Hanson, R.W. (1988) Tissue specific expression and dietary regulation of a chimeric phosphoenolpyruvate carboxykinase/bovine growth hormone gene in transgenic mice. *J. biol. Chem.* **263,** 11443–11451.

Miller, W.L., Martial, J.A. & Baxter, J.D. (1980) Molecular cloning of DNA complementary to bovine growth hormone mRNA. *J. biol. Chem.* **255,** 7521–7524.

Milman, A.E. & Russell, J.A. (1950) Some aspects of purified pituitary growth hormone on carbohydrate metabolism. *Endocrinology* **47,** 114–128.

Morinaga, Y., Fraceschini, T., Inouye, S. & Inouye, M. (1984) Improvement of oligonucleotide-directed site-specific mutagenesis using double-stranded plasmid DNA. *Bio/Technology* July, 636–639.

Ng, F.M. & Bornstein, J. (1979) Insulin-potentiating action of a synthetic amino-terminal fragment of human growth hormone (hGH 1-15) in streptozotocin-diabetic rats. *Diabetes* **28,** 1126–1130.

Ng, F.M. & Harcourt, J.A. (1986) Stimulation of 2-deoxyglucose uptake in rat adipocytes by a human growth hormone fragment (hGH4-15). *Diabetologia* **29,** 882–887.

Ng, F.M., Bornstein, J., Welder, C., Zimmet, P.Z. & Taft, P. (1974) Insulin potentiating action of synthetic peptides relating to the amino terminal sequence of human growth hormone. *Diabetes* **23,** 943–949.

Paladini, A.C., Pena, C. & Retegui, L.A. (1979) The intriguing nature of the multiple actions of growth hormone. *Trends Biochem. Sci.* **4,** 256–260.

Paladini, A.C., Pena, C. & Poskus, E. (1984) Molecular biology of growth hormone. *CRC Crit. Rev. Biochem.* **15,** 25–56.

Palmiter, R.D., Brinster, R.L., Hammer, R.E., Trumbauer, M.E., Rosenfeld, M.G., Birnberg, N.C. & Evans, R.M. (1982) Dramatic growth of mice that develop from eggs microinjected with metallothionein-growth hormone fusion genes. *Nature, Lond.* **300,** 611–615.

Pandian, M.R., Gupta, S.L. & Talwar, G.P. (1971) Studies on the early interactions of growth hormone: effect *in vitro* on lipogenesis in adipose tissue. *Endocrinology* **88,** 928–935.

Peel, C.J., Bauman, D.E., Gorewit, R.C. & Snifton, C.J. (1981) Effect of exogenous growth hormone on lactational performance in high yielding dairy cows. *J. Nutr.* **111,** 1662–1671.

Peel, C.J., Frank, T.J., Bauman, D.E. & Gorewit, R.C. (1983) Effect of exogenous growth hormone in early and late lactation on lactation performance of dairy cows. *J. Dairy Sci.* **66,** 776–782.

Raben, M.S. (1958) Treatment of a pituitary dwarf with human growth hormone. *J. clin. Endocr.* **18,** 901–904.

Raben, M.S. & Hollenberg, C.H. (1959) Effect of growth hormone on plasma fatty acids. *J. clin. Invest.* **38,** 484–487.

Robins, D.M., Paek, I., Seeburg, P.H. & Axel, R. (1982) Regulated expression of human growth hormone genes in mouse cells. *Cell* **29,** 623–631.

Salem, M.A.M. (1988) Effects of the amino-terminal portion of human growth hormone on glucose clearance and metabolism in normal, diabetic, hypophysectomized, and diabetic-hypophysectomized rats. *Endocrinology* **123,** 1565–1576.

Sanger, F., Miklen, S. & Coulson, A.R. (1977) DNA sequencing with chain-terminating inhibitors. *Proc. natn. Acad. Sci. USA* **74,** 5463–5467.

Seeburg, P.H., Shine, J., Martial, J.A., Baxter, J.D. & Goodman, H.M. (1977) Nucleotide sequence and amplification in bacteria of the structural gene for rat growth hormone. *Nature, Lond.* **270,** 486–494.

Seeburg, P.H., Sias, S., Adelman, J., DeBoer, H.A., Hayflick, J., Jhurani, P., Goeddel, D.V. & Heyneker, H.L. (1983) Efficient bacterial expression of bovine and porcine growth hormones. *DNA* **2,** 37–45.

Sigel, M.B., Thorpe, N.A., Kobrin, M.S., Lewis, U.J. & VanderLaan, W.P. (1981) Binding characteristics of a biologically active variant of human growth hormone (20K) to growth hormone and lactogen receptors. *Endocrinology* **108,** 1600–1603.

Smal, J., Closset, J., Hennen, G. & DeMeyts, P. (1986) The receptor binding properties of the 20K variant of human growth hormone explain its discrepant insulin-like and growth promoting activities. *Biochem. Biophys. Res. Commun.* **134,** 159–165.

Smith, W.C. & Talamants, F. (1987) Identification and characterization of a heterogeneous population of growth hormone receptors in mouse hepatic membranes. *J. biol. Chem.* **262,** 2213–2219.

Swislocki, N.I. (1968) Effects of nutritional status and the pituitary on the acute plasma free fatty acid and glucose responses of rats to growth hormone administration. *Metabolism* **17,** 174–180.

Swislocki, N.I. & Szego, C.M. (1965) Acute reduction of plasma nonesterified fatty acid by growth hormone in hypophysectomized and Houssay rats. *Endocrinology* **76,** 665–672.

Swislocki, N.I., Sonnenberg, M. & Yamasaki, N. (1970) *In vitro* metabolic effects of bovine growth hormone fragment in adipose tissue. *Endocrinology* **87,** 900–904.

Thornton, J.M. (1981) Disulfide bridges in globular proteins. *J. molec. Biol.* **151,** 261–287.

Tou, J.S., Kaempfe, L.A., Vinegard, B.D., Buonomo, F.C., Della-Fera, M.A. & Baile, C.A. 1986) Amphophilic growth hormone releasing factor (GKF) analogs: peptide design and biological activity *in vivo*. *Biochem. Biophys. Res. Commun.* **139,** 763–770.

Vlasuk, G.P. & Inouye, S. (1983) Site-specific mutagenesis using synthetic oligodeoxyribonucleotides as mutagens. In *Experimental Manipulation of Gene Expression*, pp. 291–303. Ed. M. Inouye. Academic Press, New York.

Wade, J.D., Pullin, C.O., Ng, F.M. & Bornstein, J. (1977) The synthesis and hyperglycemic activity of the amino acid sequence 172-191 of human growth hormone. *Biochem. Biophys. Res. Commun.* **78,** 827–832.

Wade, J.D., Ng, F.M., Bornstein, J., Pullin, C.O. & Pearce, J.S. (1982) Effect of C-terminal chain shortening on the insulin-antagonistic activity of human growth hormone 177-191. *Acta endocr., Copenh.* **101,** 10–14.

Woychick, R.P., Camper, S.A., Lyons, R.H., Horowitz, S., Goodwin, E.G. & Rottman, F.M. (1982) Cloning and nucleotide sequencing of the bovine growth hormone gene. *Nucl. Acids Res.* **10,** 7197–7210.

GENE DELIVERY

Chairman
D. D. McGregor

J. Reprod. Fert., Suppl. **41** (1990), 39–49
Printed in Great Britain

Vectors and genes for improvement of animal strains

S. H. Hughes, C. J. Petropoulos, M. J. Federspiel, P. Sutrave, Suzanne Forry-Schaudies and J. A. Bradac

BRI-Basic Research Program, NCI-Frederick Cancer Research Facility, P.O. Box B, Frederick, Maryland 21701, USA

Summary. Strain improvement of agriculturally important animals will require efficient techniques for gene delivery, the ability to regulate the expression of the newly introduced genes and, most important, the identification of genes whose appropriate expression could cause improvement of the animal. We have developed a series of avian retroviral vectors that can be used to introduce new genetic information into the germ line of chickens, for which transgenics cannot be created by direct microinjection of DNA into fertilized eggs.

We have identified a 220-bp segment of the chicken skeletal muscle α-actin gene that can cause other genes to be expressed specifically in striated muscle. This chicken promoter shows correct tissue specificity in transgenic mice and presumably could be used in other mammalian species. The skeletal muscle α-actin promoter has been inserted into the avian retroviral vectors and the promotor is functional in cultured cells infected by these retroviral vectors. The tissue specificity of the expression of the skeletal muscle α-actin promoter carried by the retroviral vectors will soon be tested *in vivo*.

We are studying two types of genes that might be useful in strain improvement; genes that could produce dominant resistance to infection by pathogenic viruses, and genes that could play critical roles in muscle development. Expression of the envelope glycoprotein of retroviruses can specifically block the cellular receptor that viruses use to infect a susceptible cell. Expression of the avian leukosis virus subgroup A envelope in transgenic chickens prevents infection by pathogenic viruses of the same subgroup. We are attempting to block reticuloendotheliosis virus infection by expressing the reticuloendotheliosis envelope glycoprotein. We have shown that we can block infection in cultured cells, and we are now creating retroviral vectors for experiments *in vivo*.

We have also begun to study the cellular homologue of the *ski* oncogene, which has been shown to stimulate the differentiation of quail myoblasts *in vitro*. Biologically active cDNAs have been isolated; we have now begun to analyse the effects of expressing the c-*ski* proteins in the whole animal.

Keywords: avian leukosis viruses; enhancers; c-*ski*; actin; tropomyosin; reticuloendotheliosis virus

Introduction

There are three basic requirements that must be met before biotechnology can be applied to the improvement of agriculturally important animals. First, an efficient and reliable means of gene delivery must be devised. Second, it is important in most cases to be able to control the expression of the gene in the animal so that the gene is expressed in the appropriate tissues and at the appropriate stage(s) of development. Third, and most important, it is essential to identify and isolate genes whose expression in the animal would cause an improvement in the strain. This report gives a brief description of our progress in each of these three areas of research.

Gene delivery

In agriculturally important animals, it will be desirable to introduce the gene into germ cells or germ cell precursors. This not only makes it possible to introduce new genetic information into individual animals, but also to endow the animals with the ability to pass on the newly acquired genetic information to their offspring. In mammals, new genes can be introduced into the germ line by injecting DNA into fertilized eggs *in vitro*. Transgenic animals created by this technique can and do pass the newly acquired gene to their offspring (Gordon *et al.*, 1980; Palmiter *et al.*, 1982, 1983, 1984; Hammer *et al.*, 1984; Pursel *et al.*, 1989). Because the reproductive physiology of birds is sufficiently different from mammals, the direct application of microinjection techniques to the creation of transgenic birds has not been possible. However, retroviral vectors have been used to produce transgenic mammals and transgenic birds (Jaenisch, 1976; Harbers *et al.*, 1981; Jaenisch *et al.*, 1983; Salter *et al.*, 1986, 1987; Bosselman *et al.*, 1989; Crittenden *et al.*, 1989; Salter & Crittenden, 1989).

Given the successes that have been obtained in creating transgenic mammals by direct microinjection, the general usefulness of retroviral vectors for this task is questionable. In the avian system there is as yet no reasonable alternative to retroviral vectors. Fortunately, it has been shown that vectors that derive from reticuloendotheliosis viruses and from avian leukosis viruses can both be used to create transgenic birds, with reasonable efficiency (Salter *et al.*, 1986; Bosselman *et al.*, 1989; Crittenden *et al.*, 1989; Salter & Crittenden, 1989).

We have a long-standing interest in the creation of efficient avian retroviral vectors. Initially, we concentrated on the construction and development of replication-competent vectors (Sorge & Hughes, 1982a, b, c; Sorge *et al.*, 1983; Hughes & Kosik, 1984; Hughes *et al.*, 1987; Greenhouse *et al.*, 1988). This phase of the work is now essentially complete. Our current goal is the development of replication-defective vectors that can be used *in vivo*; in particular we want to build vectors that can be used to create transgenic chickens.

Vectors

We have developed a series of replication-competent retroviral vectors that can be used with cells in culture and in the intact animal. A series of adaptor plasmids have been created that can be used to convert virtually any DNA segment of appropriate size into a segment that can be easily inserted into these retroviral vectors (Hughes *et al.*, 1987). Since some segments lack appropriate signals for expression in the viral vector system, some of the adaptor plasmids supply functional elements including eukaryotic translational initiation codons and leader sequences (see Table 2).

The vector constructions are based on two naturally occurring avian retroviruses that have acquired and express the oncogene, *src*, Rous sarcoma virus (RSV) and Bryan high titre virus. RSV, and the vectors that derive from it, are replication-competent and grow to high titre (see Table 1). The vectors that derive from RSV come in two forms, RCAS which contains a splice acceptor that permits expression of an inserted gene from the LTR promoter via a spliced subgenomic mRNA, and RCAN which lacks the splice acceptor. To express the sequences inserted into RCAN, an internal promoter is included in the construction. The basic vectors, RCAN and RCAS, are available in envelope subgroups A, B and D (Hughes *et al.*, 1987). The availability of vectors with different subgroups makes it possible to introduce two vectors into the same cell, and extends the host range of the vectors. The subgroup D derivatives can infect mammalian cells, although with an efficiency that is significantly lower than the efficiency in avian cells.

The Bryan high titre virus is deleted for *env* and is, therefore, replication defective (Weiss *et al.*, 1985). Recombination between Bryan and the replication-competent helper virus does not occur at detectable frequency. We are now constructing replication-defective vectors that are conceptually derived from Bryan high titre virus. These vectors actually derive from RCAN and RCAS and have

Table 1. Retroviral vectors

Vector	Subgroups available	Viral genes in vector	LTR	Purpose	Titre	Comments
RCAS	A, B, D	*gag, pol, env*	ALV (enh$^+$)	High level expression from LTR via a spliced mRNA	$>10^6$	Replication competent
RCAN	A, B, D	*gag, pol, env*	ALV (enh$^+$)	Expression of genes from an internal promoter	$>10^6$	Replication competent
RCOS	A, B, D	*gag, pol, env*	RAV-O (enh$^-$)	Low level expression from LTR via spliced mRNA	$\sim 10^5$	Replication-competent, low oncogenicity: suitable for transgenics, long-term infections *in vivo*
RCON	A, B, D	*gag, pol, env*	RAV-O (enh$^-$)	Expression of genes from an internal promoter	$\sim 10^5$	Replication-competent, low oncogenicity: suitable for transgenics, long-term infections *in vivo*
BOS	(A)*	*gag, pol*	ALV (enh$^+$)	High level expression of genes from LTR via a spliced mRNA	?	Replication-defective
BAN	(A)*	*gag, pol*	ALV (enh$^+$)	Expression from internal promoter	?	Replication-defective
BAS	(A)*	*gag, pol*	RAV-O (enh$^-$)	Low level expression from LTR via spliced mRNA	?	Now being constructed
BON	(A)*	*gag, pol*	RAV-O (enh$^-$)	Low level expression from an internal promoter	?	Now being constructed
neo/LTR	none	none	ALV (enh$^+$)	Direct expression from LTR	Does not apply	Allows sequences in adaptors to be expressed in mammalian cell lines following transfection and *neo* selection

*Subgroup-A *env* complementing line from H.-J. Kung.

the *env* gene removed. Although the vectors are still in the developmental stages, we are constructing two versions; BAS, which will have a splice acceptor and express an insert from the LTR, and BAN, which will lack the splice acceptor. Workers in the laboratory of Dr H.-J. Kung have constructed permanent lines of quail cells that express high levels of the subgroup A envelope. Preliminary data suggest that the helper cell system works. We are currently attempting to refine the vectors to obtain both high titres of virus and good levels of expression of inserted genes while avoiding problems caused by recombination between the vector and the helper.

To facilitate the introduction of the vector into the helper cell line, the plasmid carrying the BAN and BAS vectors contains the selectable marker *neo*; however, this selectable marker resides outside the viral segment. In addition, the plasmids carrying BAN and BAS also carry sites that make it simple to convert the plasmid to a defined linear DNA molecule which improves the efficiency of isolating cell lines that express the vector.

The RCAN, RCAS, BAN and BAS vectors all have ALV LTRs. Viruses that carry the ALV LTR can activate the endogenous oncogenes c-*myc* and c-*erbB* and, as a consequence, are moderately oncogenic (Hayward *et al.*, 1981; Neel *et al.*, 1981; Payne *et al.*, 1981, 1982; Fung *et al.*, 1983). When a reporter gene under the control of an internal promoter is inserted into the RCAN vector, the promoter is functional and the reporter gene is expressed. Although we still need to test these vectors *in vivo*, in-vitro experiments have shown that the transcriptional enhancer in the ALV LTR substantially increases the level of transcription from the internal promoter.

A second series of vectors has been developed that are more suitable for applications *in vivo*. These vectors have been used to make transgenic birds, and they are suitable for a variety of long-term in-vivo applications. The LTRs of the endogenous avian virus RAV-O, which lack the powerful enhancer found in the LTRs of the RSV-ALV viruses (Cullen *et al.*, 1983; Luciw *et al.*, 1983; Cullen *et al.*, 1985; Weber & Schaffner, 1985), have been substituted for the ALV LTRs of the RCAN and RCAS vectors. The RAV-O LTR vectors that correspond to RCAN and RCAS are called RCON and RCOS. They replicate much less efficiently (~10%) than the corresponding ALV LTR vectors, and express a test gene (CAT) from the LTR at approximately 2–4% of the level of the corresponding ALV LTR vectors (Greenhouse *et al.*, 1988).

In parallel with the experiments in which an internal promoter and a reporter gene were inserted into the ALV LTR vector, RCAN, we have also inserted the same promoter–reporter gene cassette into the RAV-O LTR vector, RCON. Although this relieves some of the problems seen with the RCAN vector, we still need to do some experiments *in vivo* to determine whether a tissue-specific promoter (such as the skeletal muscle α-actin promoter described in a later section of this report) will retain its tissue specificity in the context of a retroviral vector.

The replication-competent RAV-O LTR vectors have been used to make transgenic chickens by injection of viruses into fertilized eggs (Salter *et al.*, 1986, 1987; Crittenden *et al.*, 1989; Salter & Crittenden, 1989). The process is relatively efficient with replication-competent vectors (>10%), and we are constructing and developing replication-defective RAV-O LTR vectors (BON and BOS) based on the Bryan vectors, BAN and BAS (see Table 2).

Table 2. Adaptor plasmids

Adaptor	Polylinker	Purpose	Comments
*Cla*12	=pUC12	To convert a variety of DNA pieces into *Cla*I fragments	Does not interfere with transcription or translation in either orientation
*Cla*BB	*Bam*HI, *Xho*I, *Mlu*I, *Apa*I, *Bgl*II	To convert a variety of DNA pieces into *Cla*I fragments	Does not interfere with transcription or translation in either orientation
*Cla*12 *Nco*	=pUC12N	To provide insert with a higher eukaryotic leader and initiator ATG	Prototype *Nco*I/ATG adaptor for higher eukaryotic expression vectors
SA *Cla*12 *Nco*	=pUC12N	To provide insert with a higher eukaryotic leader and initiator ATG and a splice acceptor	

Adaptors

The retroviral vectors are relatively large and contain sites for most of the commonly used restriction enzymes so that the choices of restriction sites for insertion of new sequences into the vectors are quite limited. All of our retroviral vectors have a unique *Cla*I site for insertion of foreign DNA. To simplify the insertion of DNA into the vectors, we have constructed a series of plasmids called adaptors that permit the conversion of essentially any DNA segment into a fragment with *Cla*I ends (Hughes *et al.*, 1987; see also Table 2). In their simplest form, adaptors contain a polylinker flanked by *Cla*I sites, with the sequences of the polylinker chosen not to interfere with transcription or translation. Two adaptors of this type are described in Table 2 (*Cla*12 and *Cla*BB). More complicated versions of the adaptors exist and these supply, in addition to a polylinker, sequences of functional significance for higher eukaryotes, including an initiator ATG, an untranslated 5′ leader and, if desired, a splice acceptor. When an initiator ATG is present in an adaptor, it is contained within an *Nco*I recognition site (CCATGG). Using short synthetic DNA segments, it is possible to insert virtually any open reading frame into the *Nco*I adaptor and express that open reading frame wihout resorting to truncations or to gene fusions.

Regulation of gene expression

Meat is composed of striated muscle; the ability to express proteins specifically in striated muscle could be used to affect either the quantity or the quality of meat. We are studying two genes, actin and tropomyosin, that are expressed in muscle. We are now working on projects designed to understand how the expression of the actin and tropomyosin genes are regulated at the level of transcription. We have defined a small segment of the chicken skeletal muscle α-actin gene that is sufficient to direct muscle-specific expression in transgenic mice (Petropoulos *et al.*, 1989).

Since this segment is derived from the chicken genome and functions appropriately in mice, it is reasonable to expect that this same segment can be used to direct expression in the muscle tissue of birds and mammals. We also plan to use this segment to develop and test tissue-specific expression retroviral vectors for the avian system. Parallel experiments are being done with a tropomyosin gene that has two promoters that have different patterns of tissue-specific expression. In addition, we are using the tropomyosin genes to explore a second type of tissue-specific regulation: tissue-specific RNA splicing (Bradac *et al.*, 1989; Forry-Schaudies *et al.*, in press).

α-actin

Our studies using transgenic mice that carry hybrid α-skeletal actin promoters linked directly to the reporter gene chloramphenicol acetyl transferase (α_{sk}/CAT) indicate that the 5′ flanking sequences upstream of the transcription start site and the first 27 bp of the 5′ untranslated region of the chicken α_{sk} actin gene leads to the specific expression of a heterologous gene in striated muscle (Petropoulos *et al.*, 1989). The expression of CAT transgenes linked to the 2·2 kb or 191 bp of 5′ flanking sequence is restricted to striated muscle. In a majority of the transgenic mice tested, the pattern of transgene expression is very similar to that of the endogenous murine α_{sk} actin gene. The principal site of transgene expression is skeletal muscle; expression in cardiac muscle is much lower (see Fig. 1). We find that the expression of α_{sk} actin/CAT transgenes that contain 191 bp of 5′ flanking sequence does not differ significantly from the expression of transgenes that contain 2·2 kb of 5′ flanking sequence. These findings define the location of tissue-specific, *cis*-acting transcriptional regulatory elements within a 218 bp region that spans from position −191 to +27 immediately surrounding the transcription start site of the gene. This same region may also regulate lower levels of expression of α_{sk} actin in cardiac muscle. The data suggest that it may be possible to direct/restrict the expression of any protein to the skeletal muscle of transgenic animals by simply fusing its coding sequence to the α_{sk} actin promoter.

RNase protection experiments have been used to characterize the 5′ ends of the α_{sk} actin/CAT transcripts generated in several transgenic lines. These results indicate that the site of transcription initiation of the transgene in striated muscle corresponds to the transcription initiation site of the endogenous chicken α_{sk} actin gene. This observation implies that *trans*-acting factors in the muscle cells of the mouse can interact appropriately with the *cis*-acting regulatory sequences of the chicken α_{sk} actin gene to correctly initiate transcription, implying that the chicken α_{sk} gene can be used to direct expression of genes in the muscle of both mammals and birds.

Although expression of CAT was, in almost every case, appropriately directed to striated muscle by the α_{sk} actin promoter, the levels of expression (measured both at the RNA and protein levels) varied significantly amongst the transgenic lines. We have been able to show that higher

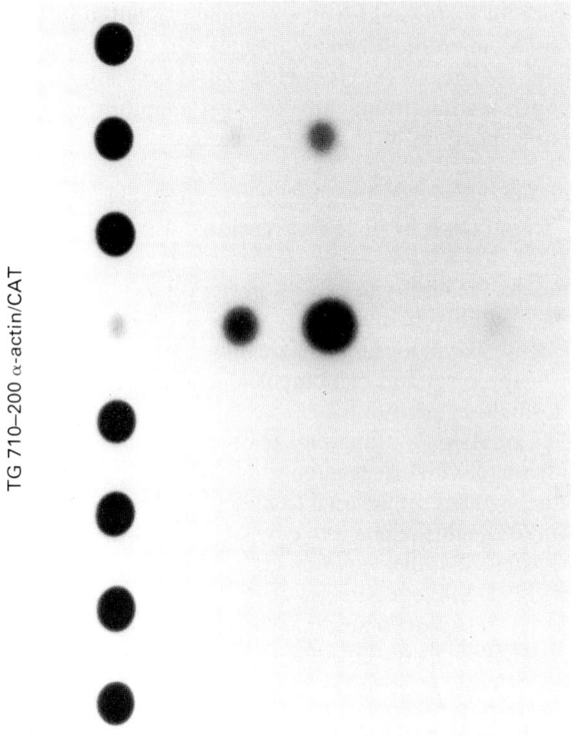

Fig. 1. The expression of CAT in a transgenic mouse containing a CAT gene linked to a chicken skeletal α-actin promoter. A mouse carrying a 220-base fragment of the chicken skeletal muscle α-actin promoter linked to CAT was identified. The animal was killed and its tissues homogenized. Equal amounts of protein from each tissue were assayed for CAT activity. Dilution experiments showed that the specific activity of CAT was approximately 16-fold higher in skeletal muscle than in the heart.

levels of α_{sk} actin/CAT transgene expression are not the direct result of increased copy number. However, we cannot rule out the possibility that the levels of expression do not reflect the number of transcriptionally active transgene copies in the genome. Another possible explanation for this quantitative variation is that sequences at, or near, the site of transgene integration may influence the levels of transgene expression, or that a *cis*-acting transcriptional regulatory element is absent from our constructs.

Tropomyosin

We have isolated and sequenced full-length cDNAs for all three major transcripts of the chicken tropomyosin I gene (Bradac *et al.*, 1989), and have completed the sequencing of the genomic sequence (13 kb) that gives rise to these three transcripts (Forry-Schaudies *et al.*, in press). There are two specific projects we are pursuing that depend on the knowledge we have obtained from the structural analyses. The tropomyosin gene contains two promoters, one of which is specifically expressed both in smooth and skeletal muscle, the other is active in several non-muscle tissues. The pattern of expression of these promoters is significantly different from the corresponding tissue-specific actin promoters and we are attempting to define what segments of the chicken tropomyosin I gene are required for appropriate regulation of these promoters.

In addition to regulation at the transcriptional level, the three major mRNAs that derive from the chicken tropomyosin I gene are differentially spliced in a tissue-specific manner. We are attempting to use the transgenic mouse system to investigate the regulation of the tissue-specific alternate splicing in this system.

Identification of useful genes

The most difficult problem in animal biotechnology is the identification and isolation of genes that will have some direct beneficial effect when introduced into the germ line and appropriately expressed. We have focussed on two types of genes that could be used to improve strains of agriculturally important animals; genes that should confer resistance to pathogenic agents, and genes that may play a key role in controlling muscle development. In the experiments designed to confer genetic resistance to viral pathogens, we are attempting to extend the initial observations, made by Dr L. B. Crittenden and his colleagues, that it is possible to induce experimentally dominant resistance to infection by the subgroup A strain of avian leukosis virus (ALV) *in vivo*. Salter & Crittenden (1989) introduced a retroviral vector into the chickens that expressed high levels of a viral protein that blocks the cellular receptor required for entry of the subgroup A ALV. Birds that express the protein are completely resistant to challenge with a pathogenic subgroup-A ALV; sibling birds that lack the gene are susceptible. We are currently testing whether these techniques can be extended to other pathogenic avian retroviruses.

In experiments designed to explore the regulation of muscle development and differentiation, we have initially focussed on the *ski* oncogene, which may play a fundamental role in regulating muscle differentiation. In these experiments, our initial goals are to study the gene and to determine what role it plays in development. The long-term goal is to try to use the gene to increase muscle mass in transgenic animals.

Genetic resistance to viral infection

It is known that certain pathogenic viruses depend on a specific interaction between a virally encoded protein and a host receptor to penetrate susceptible cells efficiently. If the host receptor is physically blocked, the virus is prevented from entering the cell. Retroviral genomes encode an envelope glycoprotein, the product of the *env* gene, that binds to the host receptor. Cells that express high levels of the envelope glycoprotein are resistant to infection by retroviruses of the same envelope subgroup (Weiss *et al.*, 1985). This blockade can be used to induce genetic resistance to pathogenic retroviruses *in vivo*. Salter & Crittenden (1989) have used retroviral vectors to introduce the subgroup-A *env* gene into the germ line of chickens. Chickens that have acquired the subgroup-A *env* gene and express it at a high level are genetically resistant to challenge from pathogenic subgroup-A viruses (Salter & Crittenden, 1989). We are now testing whether we can extend their original observations to another type of pathogenic avian retrovirus, reticuloendotheliosis virus (REV). We first wished to demonstrate that expression of high levels of the REV envelope glycoprotein is sufficient to produce high levels of resistance to REV in cultured cells *in vitro*.

The envelope genes of two extensively studied REV isolates, REV-A and spleen necrosis virus (SNV), were compared for their ability to confer resistance to challenge by REV-A and SNV viruses. We found that, by expressing REV envelope, the relative resistance to REV infection was increased up to 25 000-fold in some cell lines (Federspiel *et al.*, 1989). This level of interference approaches the resistance of REV-infected D17 cells to reinfection (see D17/REV-A, Table 3). Delwart & Panganiban (1989) have also shown that expression of the REV envelope glycoprotein produces viral resistance *in vitro* although they report substantially lower levels of resistance.

We were initially surprised that the level of expression of the envelope glycoproteins, as measured by immunoprecipitation, did not always correlate with the level of resistance to infection. However, since there is no transfer of resistance if cells that produce the glycoprotein are

Table 3. Virological assays to determine the resistance of the REV envelope cell lines (adapted from Federspiel *et al.*, 1989)

Cell line	Hygromycin colony no. after infection with JD215HYRAM*	Relative resistance†
R1	10	10 000
R2	3×10^3	33
R3	1×10^3	100
R4	4×10^3	25
R5	90	1 100
R6	4	25 000
R7	30	3 300
R8	2×10^3	50
R9	30	3 300
R10	9×10^4	0
S1	7×10^5	0
S2	1×10^4	10
S3	3×10^3	33
S4	4	25 000
S5	7×10^3	14
S6	3×10^3	33
S7	1×10^4	10
S8	1×10^5	0
S9	1×10^5	0
S10	1×10^4	10
D17	1×10^5	—
D17/REVA‡	1	100 000§

*The cell lines were infected at a multiplicity of infection of 1·0.

†The hygromycin resistance data were used to determine the relative resistance of the REV envelope cell lines compared to the parental D17 cell line.

‡D17 cells were infected with REV-A at a multiplicity of infection of 1·0 six days prior to JD215HYRAM challenge.

§The relative resistance of D17/REV-A was estimated from only one colony but was similar to other reported levels (Delwart & Panganiban, 1989).

co-cultivated with cells that do not, we considered the possibility that some of the cell lines were heterogeneous with respect to envelope expression. Immunofluorescence studies showed that resistance to viral infection correlates with the uniformity of expression of the envelope glycoprotein on the surface of the cells (Federspiel *et al.*, 1989). Lines that make relatively large amounts of envelope glycoprotein but show minimal resistance contain significant numbers of cells that express little or no envelope glycoprotein. This heterogeneity could arise in at least two ways. It is possible that either the cell lines are not truly clonal or that the lines are clonal and that the expression is unstable. Instability of this type is not uncommon for genes introduced into cultured cells by transfection.

We do not anticipate similar problems of heterogeneity of expression if we introduce the REV envelope glycoprotein into cells (or the intact animal) using a retroviral vector. Although transfection allowed us to derive envelope-expressing cell lines whose resistance to infection approached levels seen in infected cells (25 000-fold when compared with the parental D17 line), there is variability in the resistance displayed by the subclones that lack envelope DNA. We cannot yet conclude that the 25 000-fold resistance we have seen is exclusively due to envelope glycoprotein expression and not partly due to variation in the cells themselves. However, the data do show that envelope glycoprotein expression can account for at least a 1000-fold increase in resistance, and we suspect, although cannot yet prove, that the envelope glycoprotein is responsible for most, if not all, of the resistance induced by viral infection.

Our next goal will be to determine whether expressing REV envelope in chickens using retroviral vectors will increase resistance to REV infection. The expression of either the REV-A or the SNV *env* protein will block infection by either REV-A or SNV virus *in vitro*, so it is possible that the expression of one of the REV *env* genes could block infection from all of the REV isolates. However, since the expression of the *env* gene in one cell does not confer resistance to neighbouring cells, all the potential REV target cells must express the REV envelope glycoproteins. Once REV envelope gene is introduced into a retroviral vector, it should be possible to use the vectors to introduce the REV *env* gene into the germ line of chickens.

Skeletal muscle development

The major product produced by domesticated livestock is meat; an important goal of biotechnology is to gain an understanding of, and eventually to be able to alter muscle development. Several genes have been described that appear to play fundamental roles in regulating muscle differentiation. Two of the best characterized, myoD (Davis *et al.*, 1987; Tapscott *et al.*, 1988) and myogenin (Edmondson & Olson, 1989; Wright *et al.*, 1989), have been shown to promote skeletal muscle differentiation *in vitro*.

We have begun the analysis of another gene, *ski*, that may also play a direct role in muscle differentiation. *Ski* was first discovered as the oncogene in an acutely transforming retrovirus (Stavnezer *et al.*, 1981). Under proper conditions the *ski*-containing retrovirus can induce myogenic differentiation in cultured quail embryo cells.

The transforming product produced by these viruses was identified as a *gag–ski* fusion protein that localizes in the nucleus (Barkas *et al.*, 1986; Stavnezer *et al.*, 1986). We have isolated and characterized three distinct c-*ski* cDNAs from a cDNA library made from the body wall of a 10-day chick embryo (Sutrave & Hughes, 1989). Nucleic acid sequence analysis of these cDNAs suggested that the c-*ski* gene is alternately spliced. S1 nuclease protection analysis of total cellular mRNAs has confirmed the existence of RNAs corresponding to 2 of the 3 cDNAs. The cDNA sequence analysis and comparison with the available chicken genomic c-*ski* sequences indicated that the coding regions of the c-*ski* are contained within 7 exons (Sutrave & Hughes, 1989; Stavnezer *et al.*, 1989). We have introduced the 3 cDNAs into replication competent avian retroviral vectors and introduced these vectors into chicken embryonic fibroblasts. Viruses containing two of the three distinct types of cDNAs transform chick embryo fibroblasts in culture. We have developed antisera that specifically recognize that c-*ski* proteins made by these two types of viruses. Like the viral *gag–ski* fusion protein, the c-*ski* proteins are located primarily in the nucleus (Barkas *et al.*, 1986; P. Sutrave & S. H. Hughes, unpublished).

We are now in the process of evaluating the biological effects of expressing the c-*ski* cDNAs *in vivo*, using both the retroviral vector system and transgenic mice. The well-characterized α-actin promoter should be useful as a means for directing the expression of *ski* (and other genes) to skeletal muscle.

This research was carried out during the tenure of a postdoctoral fellowship from the US Department of Agriculture, Grant No. 87-CRCR-1-2584 to J.A.B., and Grant No. 88-37266-4141 from the US Department of Agriculture to M.J.F. Research sponsored by the National Cancer Institute, DHHS, under contract No. NO1-CO-74101 with BRI. The contents of this publication do not necessarily reflect the views or policies of the Department of Health & Human Services, nor does mention of trade names, commercial products, or organizations imply endorsement by the US Government.

References

Barkas, A.E., Brodeur, D. & Stavnezer, E. (1986) Polyproteins containing a domain encoded by the v-*ski* oncogene are located in the nuclei of SKV transformed cells. *Virology* **151**, 131–138.

Bosselman, R.A., Hsu, R.-Y., Boggs, T., Hu, S., Bruszewski, J., Ou, S., Kozar, L., Martin, F., Green, C., Jacobsen, F., Nicolson, M., Schultz, J.A., Semon, K.M., Rishell, W. & Stewart, G. (1989) Germline transmission of exogenous genes in the chicken. *Science, NY* **243**, 533–535.

Bradac, J.A., Gruber, C.E., Forry-Schaudies, S. & Hughes, S.H. (1989) Isolation and characterization of related cDNA clones encoding skeletal muscle β-tropomyosin and a low-molecular-weight nonmuscle tropomyosin. *Molec. cell. Biol.* **9**, 185–192.

Crittenden, L.B., Salter, D.W. & Federspiel, M.J. (1989) Segregation, viral phenotype, and proviral structure of 23 avian leukosis virus inserts in the germ line of chickens. *Theoret. appl. Genet.* **77**, 505–515.

Cullen, B.R., Skalka, A.M. & Ju, G. (1983) Endogenous avian retroviruses contain deficient promoter and leader sequences. *Proc. natn. Acad. Sci. USA* **80**, 2946–2950.

Cullen, B.R., Raymond, K. & Ju, G. (1985) Transcriptional activity of avian retroviral long terminal repeats directly correlates with enhancer activity. *J. Virol.* **53**, 515–521.

Davis, R.L., Weintraub, H. & Lassar, A.B. (1987) Expression of a single transfected cDNA converts fibroblasts to myoblasts. *Cell* **51**, 987–1000.

Delwart, E.L. & Panganiban, A.T. (1989) Role of reticuloendotheliosis virus envelope glycoprotein in superinfection interference. *J. Virol.* **63**, 273–280.

Edmondson, D.G. & Olson, E.N. (1989) A gene with homology to the *myc* similarity region of MyoD1 is expressed during myogenesis and is sufficient to activate the muscle differentiation program. *Genes and Development* **3**, 628–640.

Federspiel, M.J., Crittenden, L.B. & Hughes, S.H. (1989) Expression of avian reticuloendotheliosis virus envelope confers host resistance. *Virology* (in press).

Forry-Schaudies, S., Maihle, N.J. & Hughes, S.H. (1990) Generation of skeletal, smooth, and low molecular weight nonmuscle tropomyosin isoforms from the chicken tropomyosin 1 gene. *J. molec. Biol.*, (in press).

Fung, Y.-K.T., Lewis, W.G., Crittenden, L.B. & Kung, H.-J. (1983) Activation of the cellular oncogene *c-erbB* by LTR insertion: molecular basis for induction of erythroblastosis by avian leukosis virus. *Cell* **33**, 357–368.

Gordon, J.W., Scangos, G.A., Plotkin, D.J., Barbosa, J.A. & Ruddle, F.H. (1980) Genetic transformation of mouse embryos by microinjection of purified DNA. *Proc. natn. Acad. Sci. USA* **77**, 7380–7384.

Greenhouse, J.J., Petropoulos, C.J., Crittenden, L.B. & Hughes, S.H. (1988) Helper-independent retrovirus vectors with Rous-associated virus type O long terminal repeats. *J. Virol.* **62**, 4809–4812.

Hammer, R.E., Palmiter, R.D. & Brinster, R. L. (1984) Partial correction of murine hereditary growth disorder by germ line incorporation of a new gene. *Nature, Lond.* **311**, 65–67.

Harbers, K., Jahner, D. & Jaenisch, R. (1981) Microinjection of cloned retroviral genomes into mouse zygotes: integration and expression in the animal. *Nature, Lond.* **293**, 540–542.

Hayward, W.S., Neel, B.G. & Astrin, S.M. (1981) ALV-induced lymphoid leukosis: activation of a cellular *onc* gene by promoter insertion. *Nature, Lond.* **290**, 475–480.

Hughes, S. & Kosik, E. (1984) Mutagenesis of the region between *env* and *src* of the SR-A strain of Rous sarcoma virus for the purpose of constructing helper-independent vectors. *Virology* **136**, 89–99.

Hughes, S.H., Greenhouse, J.J., Petropoulos, C.J. & Sutrave. P. (1987) Adaptor plasmids simplify the insertion of foreign DNA into helper-independent retroviral vectors. *J. Virol.* **61**, 3004–3012.

Jaenisch, R. (1976) Germline integration and Mendelian transmission of the exogenous Moloney leukemia virus. *Proc. natn. Acad. Sci. USA* **73**, 1260–1264.

Jaenisch, R., Harbers, K., Schnieke, A., Lohler, J., Shumakov, I., Jahner, D., Grotkopp, D. & Hoffman, E. (1983) Germline integration of Moloney murine leukemia virus at the MOV13 locus heads to recessive lethal mutation and early embryonic death. *Cell* **32**, 209–216.

Luciw, P.A., Bishop, J.M., Varmus, H.E. & Capecchi, M.R. (1983) Location and function of retroviral and SV-40 sequences that enhance biochemical transformation after microinjection of DNA. *Cell* **33**, 705–716.

Neel, B.G., Hayward, W.S., Robinson, H.L., Fang, J. & Astrin, S.M. (1981) Avian leukosis virus-induced tumors have common proviral integration sites and synthesize discrete new RNAs: oncogenesis by promoter insertion. *Cell* **23**, 323–334.

Palmiter, R.D., Brinster, R.L., Hammer, R.E., Trumbauer, M.E., Rosenfeld, M.G., Birnberg, N.C. & Evans, R.M. (1982) Dramatic growth of mice that develop from eggs microinjected with metallothionein-growth hormone fusion genes. *Nature, Lond.* **300**, 611–615.

Palmiter, R.D., Norstedt, G., Gelinas, R.E., Hammer, R.E. & Brinster, R.L. (1983) Metallothionein-human GH fusion genes stimulate growth of mice. *Science, NY* **222**, 809–814.

Palmiter, R.D., Wilkie, T.M., Chen, H.Y. & Brinster, R.L. (1984) Transmission distortion and mosaicism in an unusual transgenic mouse pedigree. *Cell* **36**, 869–877.

Payne, G.S., Courtneidge, S.A., Crittenden, L.B., Fadly, A.M., Bishop, J.M. & Varmus, H.E. (1981) Analysis of avian leukosis virus DNA and RNA in bursal tumors: viral gene expression is not required for maintenance of the tumor state. *Cell* **23**, 311–322.

Payne, G.S., Bishop, J.M. & Varmus, H.E. (1982) Multiple arrangements of viral DNA and an activated host oncogene (c-*myc*) in bursal lymphomas. *Nature, Lond.* **295**, 209–214.

Petropoulos, C.J., Rosenberg, M.J., Jenkins, N.A., Copeland, N.G. & Hughes, S.H. (1989) The chicken skeletal muscle α-actin promoter is tissue-specific in transgenic mice. *Molec. cell. Biol.* (in press).

Pursel, V.G., Pinkert, C.A., Miller, K.F., Bolt, D.J., Campbell, R.G., Palmiter, R.D., Brinster, R.L. &

Hammer, R.E. (1989) Genetic engineering of livestock. *Science, NY* **244,** 1281–1287.

Salter, D.W. & Crittenden, L.B. (1989) Artificial insertion of a dominant gene for resistance to avian leukosis virus into the germ line of the chicken. *Theoret. appl. Genet.* **77,** 457–461.

Salter, D., Smith, E., Hughes, S.H., Wright, S. & Crittenden, L.B. (1986) Gene insertion by retroviruses. *Poult. Sci.* **65,** 1445–1458.

Salter, D.W., Smith, E.J., Hughes, S.H., Wright, S.E. & Crittenden, L.B. (1987) Transgenic chickens: insertion of retroviral genes into the chicken germline. *Virology* **157,** 236–240.

Sorge, J. & Hughes, S.H. (1982a) The polypurine tract adjacent to the U_3 region of the Rous sarcoma virus genome provides a Cis-acting function. *J. Virol.* **43,** 482–488.

Sorge, J. & Hughes, S.H. (1982b) Retrovirus vectors independent of selectable markers. In *Eukaryotic Viral Vectors*, pp. 127–133. Ed. Y. Gluzman. Cold Spring Harbor Laboratory, Cold Spring Harbor.

Sorge, J. & Hughes, S.H. (1982c) Splicing of intervening sequences introduced into an infectious retrovirus vector. *J. molec. appl. Genet.* **1,** 547–559.

Sorge, J., Ricci, W. & Hughes, S.H. (1983) *cis*-acting RNA packaging locus in the 115-nucleotide direct appeal of Rous sarcoma virus. *J. Virol.* **48,** 667–675.

Stavnezer, E., Gerhard, D.S., Binari, R.C. & Balzas, I. (1981) Generation of transforming viruses in culture of chicken fibroblasts infected with avian leukosis virus. *J. Virol.* **39,** 920–934.

Stavnezer, E., Barkas, A.E., Brennan, L.A., Brodeur, D. & Li, Y. (1986) Transforming Sloan-Kettering viruses generated from the cloned v-*ski* oncogene by *in vitro* and *in vivo* recombination. *J. Virol.* **57,** 1073–1083.

Stavnezer, E., Brodeur, D. & Brennan, L.A. (1989) The v-*ski* oncogene encodes a truncated set of c-*ski* coding exons with limited sequence and structural relatedness to v-*myc*. *Molec. cell. Biol.* (in press).

Sutrave, P. & Hughes, S.H. (1989) Isolation and characterization of three distinct cDNAs for the chicken c-*ski* gene. *Molec. cell. Biol.* (in press).

Tapscott, S.J., Davis, R.L., Thayer, M.J., Cheng, P.-F., Weintraub, H. & Lassar, A.B. (1988) A nuclear phosphoprotein requiring a *myc* homology region to convert fibroblasts to myoblasts. *Science, NY* **242,** 405–411.

Weber, F. & Schaffner, W. (1985) Enhancer activity correlates with the oncogenic potential of avian retroviruses. *EMBO J.* **4,** 949–956.

Weiss, R., Teich, N., Varmus, H. & Coffin, J. (Eds) (1985) *RNA Tumor Viruses*. Cold Spring Harbor Laboratory, Cold Spring Harbor.

Wright, W.E., Sassoon, D.A. & Lin, V.K. (1989) Myogenin, a factor regulating myogenesis, has a domain homologous to MyoD. *Cell* **56,** 607–617.

J. Reprod. Fert., Suppl. 41 (1990), 51–56

Maintenance and differentiation in culture of pluripotential embryonic cell lines from pig blastocysts

E. Notarianni*, S. Laurie†, R. M. Moor† and M. J. Evans*

*Department of Genetics, University of Cambridge, Cambridge, CB2 3EH, UK; and
†Institute of Animal Physiology and Genetics Research, Babraham Cambridge, CB2 4AT, UK

Summary. Cell lines were established from explanted blastocysts of domestic pigs; the cells could be maintained indefinitely when grown on mouse fibroblast feeder cell layers. They differentiate spontaneously at high densities, or when allowed to form aggregates when cultured on a non-adhesive substratum. Their appearance and differentiative behaviour resembles that of mouse embryonic stem cell lines. We are currently attempting to establish whether these cultures represent primary ectodermal lineages which would be of particular relevance to developmental and transgenic studies.

Keywords: embryonic stem cells; pig; ungulate; tissue culture

Introduction

The advent of mouse embryonic stem cell lines (Evans & Kaufman, 1981; Martin, 1981) and in particular the subsequent demonstration of their functional totipotency (Bradley *et al.*, 1984) expanded the range of techniques which are available for gene transfer and especially gene modification in mammals. In mice this provides an alternative route for transgenesis to that of zygote pronuclear microinjection which has the advantage that a large range of in-vitro genetic transformation and especially screening or selective techniques may be applied before reconstitution into an animal. Essentially this allows some of the techniques of somatic cell genetics to be used to introduce genetic manipulations into normal and fertile animals. This approach would offer unique practical opportunities for experimental genetic manipulation of domestic animals which would include targeted mutagenesis by selective inactivation or replacement of endogenous genes, the introduction of clonal alterations to the germ line and improved control of expression of transgenes through the sceening of phenotypes of cells in culture.

To date the only species other than the mouse for which the derivation of embryonic stem cells has been described is the hamster (Doetschman *et al.*, 1988). It would be useful to carry out similar work on domestic animals, particularly ungulates of commercial importance in livestock breeding such as cattle, sheep and pigs. There are, however, significant differences between early embryo development in rodents and in domestic animals such as ungulates, which means that it is not a trivial task to achieve similar results in these other species. We report here the isolation and some preliminary characterization of cell lines from pig embryos.

Methods

Hatched blastocysts were recovered by retrograde uterine flushing from Large British White gilts at 7–9 days after oestrus. The animals were naturally mated and not superovulated. Intact blastocysts or the inner cell masses manually dissected from them were explanted onto mitotically-inactivated feeder STO fibroblasts in a manner similar to that described by Evans & Kaufman (1981) for mouse blastocysts. The medium used was Dulbecco's modified Eagle's medium supplemented with 10% calf serum and 5% fetal calf serum. Neither conditioned medium nor exogenous growth factors were added.

To stimulate organized differentiation, cells were disaggregated by trypsinization and then seeded onto culture dishes which had been coated with a layer of 0·5% agarose as described by Magrane (1982) for stimulation of formation of embryoid bodies by human teratocarcinoma cell cultures. To enhance cell differentiation, mercaptoethanol and fetal calf serum were both omitted from the medium.

Results

Establishment of cultures

We have by these methods of culture been able routinely to establish cultures from explanted pig embryos. Although the success rate is variable it is often high with as many as 6 successful cultures being derived from 8 explanted blastocysts in one experiment.

When blastocysts or embryonic discs are brought into culture, they attach within 1 day. The primary outgrowths consist of colonies of large flat, highly-translucent epithelial cells (Fig. 1). These are clearly of an appearance and culture morphology very different from those of mouse embryonic stem cells, or EK cells. Portions of trophectoderm dissected from blastocysts between the ages of 7 and 10 days and cultured in the same way were unable to form colonies or outgrowths.

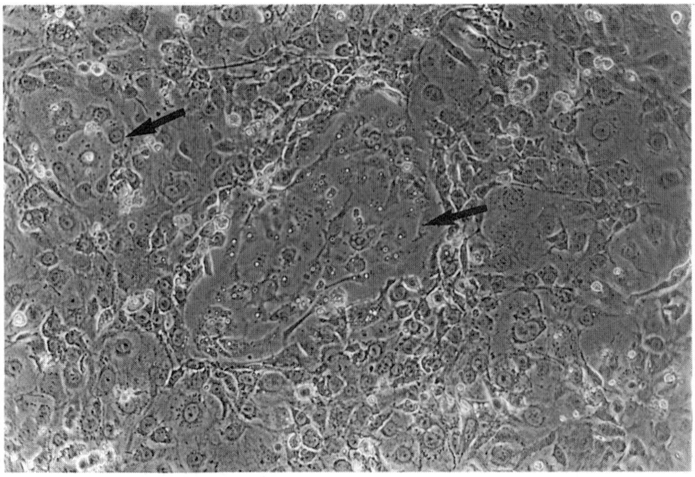

Fig. 1. Appearance of primary colony resulting from attachment of inner cell mass from an 8-day pig blastocyst. The flattened, translucent colonies are arrowed. × 100.

The primary outgrowths were disaggregated 7–14 days after explantation and passaged to fresh feeder layers. Progressively growing colonies were formed which grew as a monolayer with very distinct colony boundaries. The cells are epithelioid with large clear nuclei containing 2–4 prominent nucleoli, and relatively sparse cytoplasm. Some differences in the appearance of these cells have been noticed between different isolates, and these are mainly related to cell size.

Figure 2(a) shows small colonies of large, undifferentiated cells which continuously produce cells with morphological characteristics similar to those of trophoblast giant cells. Such cultures have been maintained for 4 months in continuous culture. Cultures of this type have only been observed to arise from 7-day embryos.

Figure 2(b) shows a cell type which is more stable and able to grow in larger colonies. These cells are the most common isolate. Several cell lines of the type shown in Fig. 2(b) have been derived from 7- and 9-day embryos. One cell line has been maintained in continuous culture for more than 1 year without change of cell phenotype. It therefore appears to be immortal. Differentiation of these cells occurs spontaneously when the cells are permitted to reach high density (Fig. 3b).

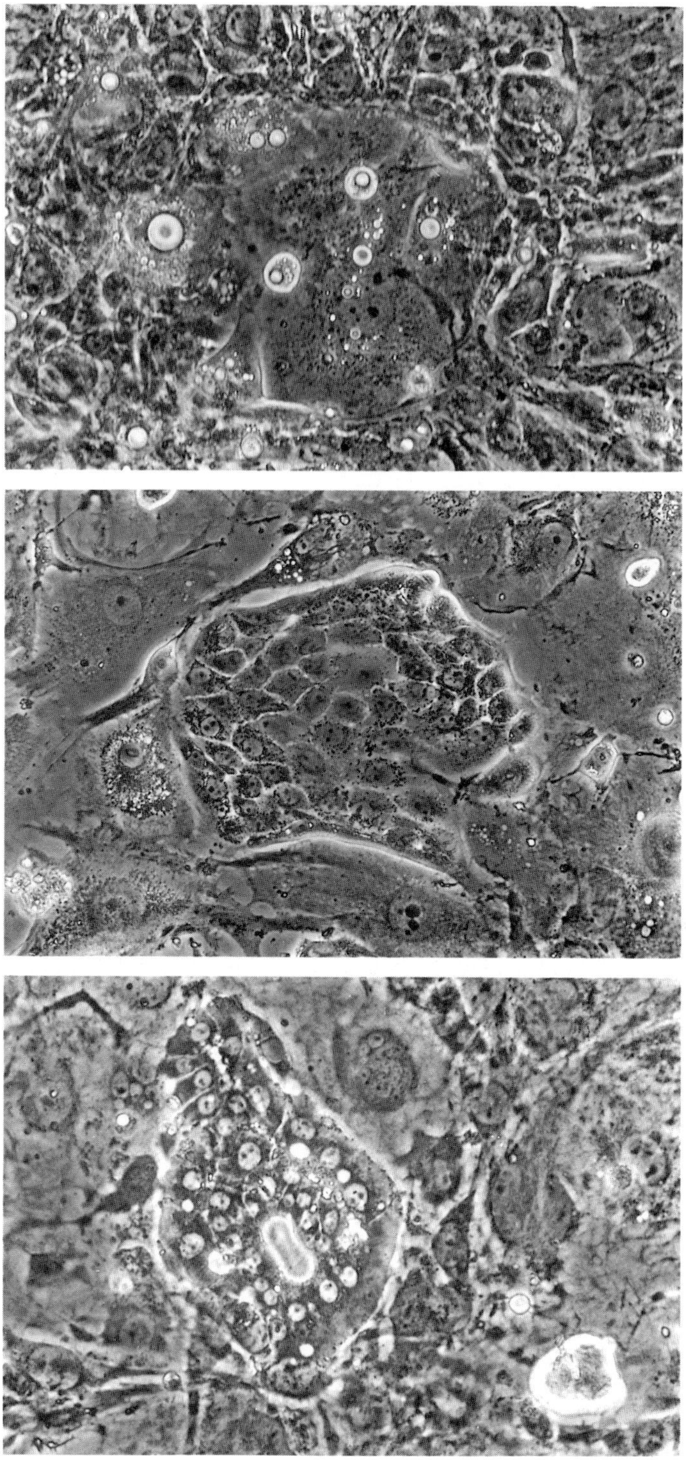

a b c

Fig. 2. Morphologies of colonies resulting from disaggregated primary outgrowths of inner cell masses. **(a)** Colony of cells producing large, trophoblast-like cells, which are visible at the perimeter. This culture was derived from a 7-day blastocyst. **(b, c)**, Colonies of stem-like cells, which are epithelial, adherent and have large nuclei and prominent nucleoli. The colony shown in (b) was derived from a 7-day blastocyst, and that in (c) from an 8-day blastocyst. ×250.

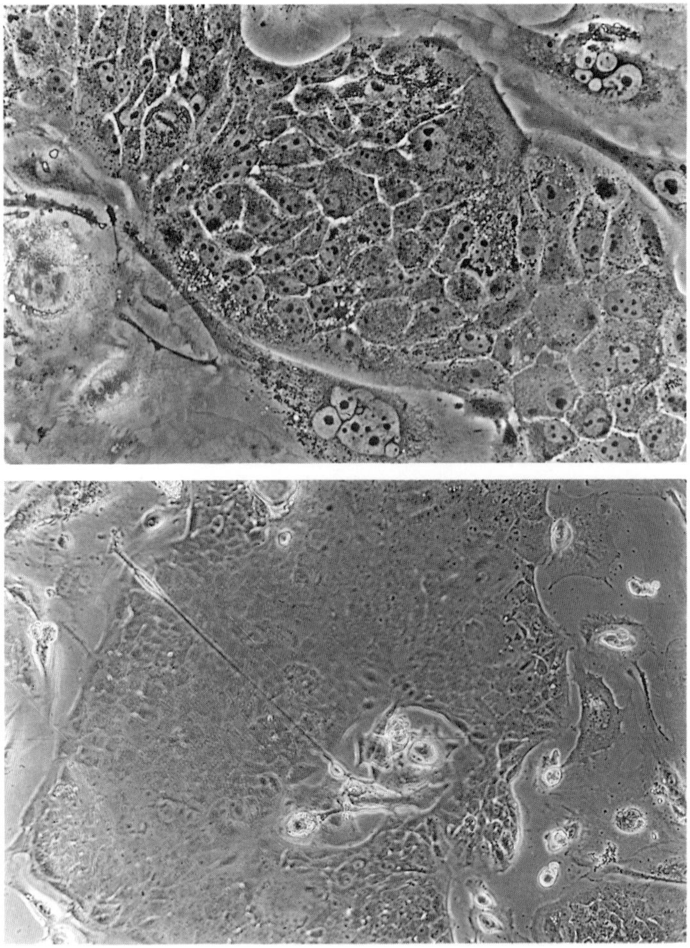

Fig. 3. Established pig cell line showing morphological differentiation. **(a)** Nest of undifferentiated cells. ×250. **(b)** Confluent monolayer showing morphological differentiation into neurone-like cells. ×100.

Overtly differentiated cells fail to reattach after passage, leading to regeneration of undifferentiated cultures.

Figure 2(c) shows a colony of smaller cells which are a more rarely isolated form.

All of these cell types grow more slowly than and differ in appearance from mouse embryonic stem cells. Pig cells show spontaneous differentiation in culture, mainly into trophoblast-like cells or endoderm-like cells, and other differentiated (mainly fibroblastoid) cell types are also formed.

Differentiation into embryoid bodies

As a test of differentiation potential (Martin & Evans, 1975) cells from the cell line which had been maintained for 12 months were induced to form aggregates by seeding onto a non-adhesive substratum. After several days, an outer smooth layer of cuboidal epithelial cells appeared at one end of the aggregates and at the other pole there were more loosely attached cells of a more ragged appearance. Extensive differentiation occurred when the embryoid bodies were permitted to attach to the substratum, by replating onto tissue-culture dishes (Fig. 4). Cells migrated and multiplied to form dense cultures, with several types visible, including epithelium, endoderm, muscle and neural

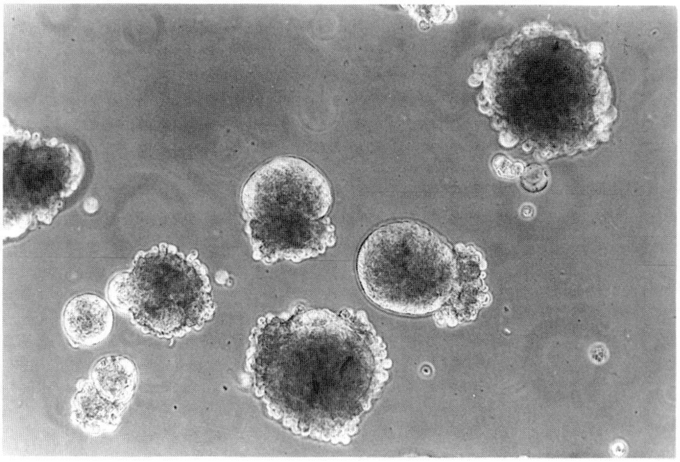

Fig. 4. Aggregates formed by a pig cell line after culture for 7 days on a non-adhesive substratum. × 100.

cells. These differentiated cells are representative derivatives of all three embryonic germ layers, and suggest that the stem-cell-like culture represents a primary ectodermal lineage of the pre-somite embryo.

Discussion

The cell lines isolated have an appearance considerably different from, although slightly reminiscent of, mouse embryonic stem cells. In appearance and form of growth the pig cell lines are more similar to some cell lines derived from human testicular teratocarcinomas. The form of development of their aggregates when maintained in suspension is very similar to that of a human teratocarcinoma cell line, Hutt KEB (as described by Magrane, 1984). We tentatively conclude that these differentiating structures are indeed homologous to the mouse embryoid bodies and this conclusion is strengthened by the observations reported here of a more extensive in-vitro differentiation following their re-explantation onto a tissue-culture surface. Evans (1983) has previously speculated that the asymmetric form of the Hutt KEB embryoid bodies reflects the development of the human embryo via an embryonic disc in contrast to the mouse egg cylinder. It is therefore interesting that, in the pig in which early development is via an embryonic disc with a clearly epithelioid embryonic epiblast, the isolated cells grow more as a monolayer than in the piling colonies typical of mouse embryonal carcinoma (EC) and EK cells, and differentiation of their embryoid bodies is clearly asymmetric. This distinctly different behaviour from that of mouse EK cells may be a general feature of non-rodent embryos in which embryonic development is via an embryonic disc.

We conclude that pluripotent embryonic lineages may be derived from the pig and can be maintained in culture. We are currently evaluating the potential of the pig embryonic cells to determine (a) their relationship to normal embryonic cells, and their distinctive features compared to the mouse EK lineages, and (b) whether there exist stem cell populations of restricted potency which may be transitory in nature, and (c) the origin and nature of those cells that are capable of expressing pluripotency under certain conditions. It is notable that the rate of proliferation of undifferentiated, stem-like cells in explants of pig embryos is slower than those of explants derived from mouse embryos. This may reflect another important difference in pre-implantational development in these species, i.e. the period of quiescence of the inner cell mass in ungulates up to the time of gastrulation.

As these new cell lines may be considered homologous to mouse EK cells, they have potential as a vector for genetic manipulation by their incorporation into a normal fertile pig via embryo chimaerism, thus contributing to the germ cell line.

References

Bradley, A., Evans, M.J., Kaufman, M.H. & Robertson, E. (1984) Formation of germ-line chimaeras from embryo-derived teratocarcinoma cells lines. *Nature, Lond.* **309,** 255–256.

Doetschman, T., Williams, P. & Maeda, N. (1988) Establishment of hamster blastocyst-derived embryonic stem (ES) cells. *Devl Biol.* **127,** 224–227.

Evans, M.J. (1983) Experimental teratomas. In *Clinics in Oncology*, vol. 2, pp. 77–91. W.B. Saunders, Philadelphia.

Evans, M.J. & Kaufman, M.H. (1981) Establishment in culture of pluripotential cells from mouse embryos. *Nature, Lond.* **292,** 154–156.

Magrane, G.G. (1982) *A comparative study of human and mouse teratocarcinomas*. Ph.D. thesis, University of London.

Martin, G.R. (1981) Isolation of a pluripotent cell line from early mouse embryos cultured in medium conditioned by teratocarcinoma stem cells. *Proc. natn. Acad. Sci. USA* **78,** 7634–7638.

Martin, G.R. & Evans, M.J. (1975) Differentiation of clonal lines of teratocarcinoma cells: formation of embryoid bodies in vitro. *Proc. natn. Acad. Sci. USA* **72,** 1441–1445.

DISEASE RESISTANCE

Chairman
D. F. Antczak

J. Reprod. Fert., Suppl. **41** (1990), 59–62
Printed in Great Britain

Regulation of expression of a Class I major histocompatibility complex transgene

Jean E. Maguire*, Rachel Ehrlich*, W. I. Frels† and Dinah S. Singer*

**Experimental Immunology Branch, Building 10, Room 4B-17, NIH, Bethesda, MD 20892, USA; and †Reproduction Laboratory, USDA-ARS, BARC-E, Beltsville, MD 20705, USA*

Keywords: class I; major histocompatibility complex; regulation; transgene; SLA

Introduction

The major histocompatibility gene complex (MHC) comprises two major multigene families, the Class I genes and the Class II genes. Both the Class I and Class II MHC families consist of a set of highly homologous genes that encode cell surface molecules playing a central role in cell-mediated immune responses. The Class I genes can be divided into two sub-families, the classical Class I genes, which are discussed in this paper, and the non-classical genes, which are not well understood functionally, but are characterized by the unique distribution of their products (Singer & Maguire, 1989). The products of classical Class I genes are polymorphic transmembrane glycoproteins of M_r 45 000 that associate on the cell surface with β_2-microglobulin. The function of these Class I MHC molecules is to act as a receptor for viral and other peptide antigens. In addition, host cytotoxic T lymphocyte recognition of classical Class I molecules on the surface of an allograft is a major factor in transplantation rejection.

Class I molecules are expressed on the surface of virtually all nucleated cells, although their levels of expression vary according to cell type and the presence or absence of certain inducing agents (Singer & Maguire, 1989). Analyses of the 5′ flanking region of several Class I genes have identified a number of DNA regulatory sequence elements within this region that can effect constitutive and/or inducible Class I gene transcription in cultured cells (Fig. 1) (Israel *et al.*, 1986; Kimura *et al.*, 1986; Miyazaki *et al.*, 1986; Collins *et al.*, 1986; Sugita *et al.*, 1987; Ehrlich *et al.*, 1988). The in-vivo function of these various elements has not yet been demonstrated. This paper discusses studies undertaken to identify Class I regulatory sequence elements functioning *in vivo*. Using a model system consisting of a series of transgenic mice containing various Class I gene 5′ deletion mutants, we demonstrate the ability of a silencer element, previously identified *in vitro* (Ehrlich *et al.*, 1988), to function *in vivo* in the transgenic animal.

Results

Positive and negative DNA regulatory elements are located in the 5′ flanking region of Class I genes (Kimura *et al.*, 1986; Miyazaki *et al.*, 1986; Ehrlich *et al.*, 1988) (Fig. 1). The initial in-vivo characterization of these regulatory elements was undertaken by studying transgenic mice containing the pig Class I transgene, PD1. The PD1 transgenic mouse was created by introducing into the genome of a B10 mouse a 9 kb fragment of pig DNA containing 1·1 kb of 5′ flanking sequences, 3·5 kb of the Class I PD1 gene, and 4·4 kb of 3′ flanking sequence (Frels *et al.*, 1985).

Initial analysis of the PD1 transgenic animal revealed a conservation of the patterns of tissue-specific expression of PD1 between the transgenic mouse and pig (Ehrlich *et al.*, 1989). This conservation of tissue-specific expression was exemplified by the fact that in the pig and transgenic

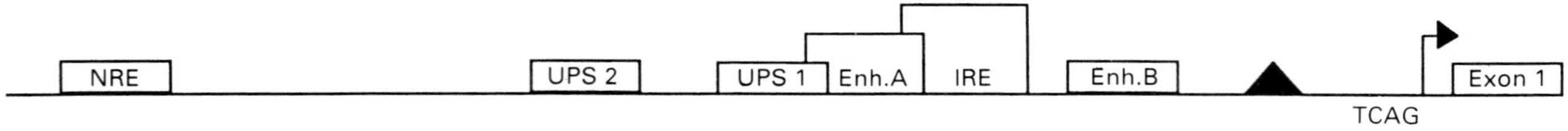

Fig. 1. DNA regulatory elements associated with Class I MHC genes. Several regulatory elements have been identified in the 5′ flanking region of Class I genes. These elements include: a basic promoter (▲) consisting of a TATA and CCAAT boxes, two enhancers (Enh. A and Enh. B), a negative regulatory element (NRE), and an interferon-responsive element (IRE). ↱ Indicates location of initiation of transcription occurring approximately 20 bp downstream from the sequence TCAG. Two conserved sequences (UPS1 and UPS2) have been identified in all pig Class I genes but their function is unknown.

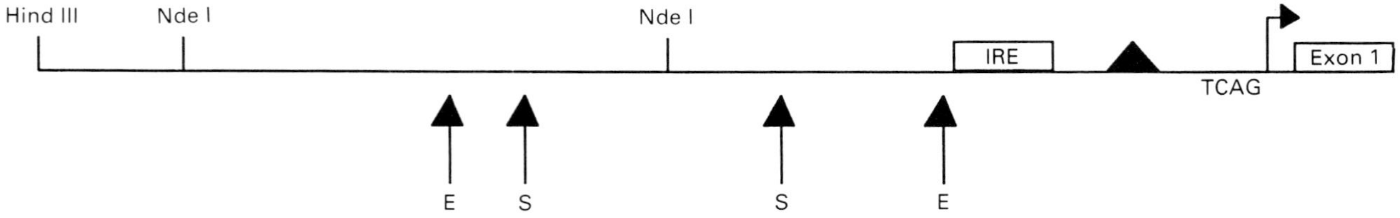

Fig. 2. Positive and negative regulatory elements identified in the 5′ flanking region of the PD1 gene. Identification of enhancer (E) and silencer (S) elements associated with PD1 was made *in vitro* by transient transfection of 5′ constructs into L cells. IRE = interferon responsive element; ▲ = basic promoter.

mouse, PD1 RNA is expressed at higher levels in lymphoid organs relative to non-lymphoid organs, and at higher levels in the testis than in the kidney. Furthermore, it was found that, both at the level of RNA and at the cell surface, PD1 was expressed at higher levels in B cells as opposed to T cells in the pig and the PD1 transgenic. These data indicate that the DNA regulatory elements necessary for normal PD1 expression are contained within the 9 kb pig DNA segment introduced into the transgenic mouse. In addition, it appears that there is sufficient conservation between pig and mouse species of *trans* acting factors that interact with these sequences to allow for their normal function *in vivo*.

One of the characteristics of Class I genes is that they respond to various immunomodulating agents. For example, treatment *in vitro* or *in vivo* with α/β interferon results in an increased Class I transcript and consequent cell surface expression of Class I antigens (Satz & Singer, 1984; Maguire *et al*., 1989). An interferon-responsive element has been identified in the 5′ flanking region of PD1 and other Class I genes (Kimura *et al*., 1986; Korber *et al*., 1987; Sugita *et al*., 1987; Singer *et al*., 1988). Consequently, it was of interest to determine whether the interferon-responsive elements associated with the PD1 gene were functional in the PD1 transgenic mouse. This was tested by treating transgenic mice with mouse α/β interferon, and determining the resulting changes in PD1-specific RNA expression (Ehrlich *et al*., 1989). The α/β interferon treatment resulted in an increase in PD1-specific RNA in the spleen, lymph node, thymus, lung, heart, testis and brain of the transgenic mouse. These results indicate that there is sufficient conservation of *trans* acting factors between the pig and mouse species to allow mouse *trans* acting factors, induced by mouse interferon, to interact with interferon-responsive DNA sequences of a pig Class I gene. In addition, the results demonstrated that the amount of PD1 RNA increase varies with tissue type. For example, treatment with α/β interferon resulted in an increase in PD1-specific RNA in the brain to levels over 6 times higher than the increase found in lymph nodes. These results indicate that fine regulation is occurring, and suggest that α/β interferon may actually affect multiple regulatory elements.

Analysis of the PD1 transgenic mouse demonstrated that, at a maximum, 1 kb of 5′ flanking sequences were sufficient to conserve tissue-specific expression of PD1. Consequently, in-vitro and in-vivo studies on the regulation of PD1 focussed on identifying DNA regulatory elements within the 1·1 kb 5′ flanking region.

To identify functionally the location of any regulatory elements within the PD1 gene, a series of deletion mutants was generated from the 5′ gene region. The mutants, all sharing a common 3′ boundary spanning the promoter, were ligated to a reporter gene encoding the enzyme chloramphenicol acetyl transferase, or CAT. The ability of each of these constructs to direct CAT transcription, and ultimately CAT enzyme synthesis, was first assessed *in vitro* by transient transfection into L cells. The results of these transfection studies demonstrated that the 5′ flanking region of PD1 contained an alternating array of both positive and negative regulatory elements (Fig. 2).

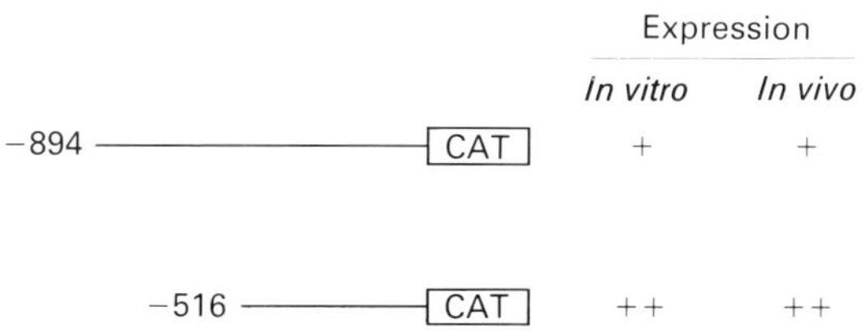

Fig. 3. Expression of −894 and −516 CAT constructs *in vitro* and *in vivo*. The expression of these constructs, consisting of 894 and 516 bp respectively of 5′ PD1 flanking sequences ligated to the gene encoding CAT, was examined *in vitro* after transient transfection into L cells and *in vivo* after introduction into a mouse genome.

Transgenic mice were made of certain of the deletion mutants so as to assess the ability of regulatory elements identified *in vitro* to function and determine tissue specific patterns of PD1 expression *in vivo*. The L-cell transfection studies identified the presence of a silencer occurring between −894 and −516 bp upstream from the initiation of translation. The ability of the silencer element to function *in vivo* was demonstrated by comparing the expression of CAT in the various tissues of transgenics containing either the −894 construct or the −516 construct (Fig. 3). This comparison demonstrated a consistently higher level of CAT activity in the tissues of the −516 transgenic mice relative to the −894 transgenics, thus demonstrating the in-vivo function of the silencer element.

Concluding remarks

We have found 5 regulatory elements associated with the pig Class I MHC gene PD1: 2 enhancers, 2 silencers, and an α/β interferon-responsive element. Using a transgenic mouse model we have been able to demonstrate, for the first time, the ability of one of the silencing elements to function *in vivo*, in a variety of tissues. Further studies are currently being conducted to characterize the tissue specificity of this silencing element, and to identify other regulatory elements functioning *in vivo*.

The ability to treat diseases through the introduction of exogenous genes is rapidly becoming a reality. To achieve maximal therapeutic benefit, with minimal side effects, it is critical to understand the mechanisms involved in tissue-specific expression of these genes. The studies we have presented serve as a model for identifying factors which regulate normal tissue-specific gene expression.

References

Collins, T., Lapierre, L., Fiers, W., Strominger, J. & Prober, J. (1986) Recombinant human tumor necrosis factor increases mRNA levels and surface expression of HLA-A,B antigens in vascular endothelial cells and dermal fibroblasts in vitro. *Proc. natn. Acad. Sci. USA* **83,** 446–450.

Ehrlich, R., Maguire, J.E. & Singer, D.S. (1988) Identification of negative and positive regulatory elements associated with a class I major histocompatibility complex gene. *Molec. cell. Biol.* **8,** 695–703.

Ehrlich, R., Sharrow, S.O., Maguire, J.E. & Singer, D.S. (1989) Expression of a class I MHC transgene: effects of in vivo α/β-interferon treatment. *Immunogenetics* **30,** 18–26.

Frels, W., Bluestone, J., Hodes, R., Capecchi, M. & Singer, D.S. (1985) Expression of a microinjected porcine class I major histocompatibility complex gene in transgenic mice. *Science, NY* **228,** 577–580.

Israel, A., Kimura, A., Fournier, A., Fellous, M. & Kourilsky, P. (1986) Interferon response sequence potentiates activity of an enhancer in the promoter region of a mouse H-2 gene. *Nature, Lond.* **322,** 743–746.

Kimura, A., Israel, A., LeBail, O. & Kourilsky, P. (1986) Detailed analysis of the mouse H-2Kb promoter: enhancer-like sequences and their role in the regulation class I gene expression. *Cell* **44,** 261–272.

Korber, B., Hood, L. & Stroynowski, I. (1987) Regulation of murine class I genes by interferons is controlled by regions located both 5′ and 3′ to the transcriptional initiation site. *Proc. natn. Acad. Sci. USA* **84,** 3380–3384.

Maguire, J.E., Gresser, I., Williams, A., Kielpinsky, G.L. & Colvin, R.B. (1989) Murine interferon α/β modulates the expression of MHC antigens in the kidneys of mice. *Transplantation* (In press).

Miyazaki, J., Appella, E. & Ozato, K. (1986) Negative regulation of the MHC class I gene in undifferentiated embryonal carcinoma cells. *Proc. natn. Acad. Sci. USA* **83,** 9537–9541.

Satz, M. & Singer, D.S. (1984) Effect of mouse interferon on the expression of a porcine MHC gene introduced into mouse L-cell. *J. Immunol.* **132,** 496–501.

Singer, D.S. & Maguire, J.E. (1989) Regulation of the expression of class I MHC genes. *CRC Crit. Reviews in Immunology* (In press).

Singer, D.S., Ehrlich, R., Golding, H., Satz, L., Parent, L. & Rudikoff, S. (1988) Structure and expression of class I MHC genes in the miniature swine. In *Molecular Biology of the Major Histocompatibility Complex of Domestic Animal Species*, pp. 53–62. Eds C. Warner, M. Rothschild & S. Lamont. Iowa State University Press, Ames.

Sugita, K., Miyazaki, J., Appella, E. & Ozato, K. (1987) Interferons increase transcription of a major histocompatibility class I gene via a 5′ interferon consensus sequence. *Molec. cell. Biol.* **7,** 2625–2630.

J. Reprod. Fert., Suppl. **41** (1990), 63–74

Printed in Great Britain

Interferons at the placental interface

R. M. Roberts, J. C. Cross, C. E. Farin, T. R. Hansen, S. W. Klemann and K. Imakawa

Departments of Animal Sciences and Biochemistry, University of Missouri-Columbia, Columbia, MO 65211, USA

Summary. The antiluteolytic factors secreted by sheep and cattle conceptuses are closely related structurally to alpha-interferons (IFN-αs). They are known as ovine and bovine trophoblast protein-1 (oTP-1 and bTP-1), respectively. The mRNAs for oTP-1 and bTP-1 are transcribed from multiple genes and are the major translatable messages of Day 13–17 sheep conceptuses and Day 15–20 cattle conceptuses. The proteins belong to the 172-amino acid IFN-α_{II} (or IFN-ω) subfamily and have the typical antiviral and antiproliferative properties of the 166-residue IFN-α_Is. These embryonic interferons also bind to the IFN-α receptor, which is present in uterine endometrium in high concentrations, and can influence the production of prostaglandin F-2α and the pattern of protein secretion in that tissue. Through use of in-situ hybridization procedures on tissue sections and Northern and dot blot analyses of extracted conceptus RNA, ovine oTP-1 mRNA has been shown to increase markedly around Day 13 and to decrease after about Day 15 of pregnancy. The mRNA is confined entirely to cells of the trophectoderm. Significant induction of mRNA that hybridizes to an oTP-1 cDNA occurs in response to exposure to polyI:polyC in Day 11 sheep blastocysts which normally have low levels of oTP-1 expression. However, the basis for induction in the normal progression of embryonic development remains unclear. The fact that preimplantation conceptuses of other species, e.g. pig, release substances with antiviral activity suggests that IFNs may have an important role in pregnancy that extends beyond the domestic ruminants.

Keywords: trophoblast protein-1; interferon; sheep; cattle; pregnancy; virus

Introduction

Ovine trophoblast protein-1 (oTP-1) and bovine trophoblast protein-1 (bTP-1) were initially recognized as the major polypeptide products secreted into the medium when early elongating, pre-attachment sheep and cattle conceptuses were placed into short-term in-vitro culture (Godkin *et al.*, 1982; Bartol *et al.*, 1985; Helmer *et al.*, 1987). Both products had a molecular weight of approximately 20 000 and consisted of multiple isoelectric variants. In addition, bTP-1 existed in at least two different size forms which were later shown to be the result of differential degrees of glycosylation of a smaller precursor form (Anthony *et al.*, 1988). Both proteins have been strongly implicated in initiating maternal responses to the presence of a conceptus in the uterus and in preventing a return to ovarian cyclicity (Bazer *et al.*, 1986). Introduction of oTP-1 into the uteri of non-pregnant ewes between Days 12 and 20 of the oestrous cycle, for example, extended the inter-oestrous interval and was associated with continued secretion of ovarian progesterone (Godkin *et al.*, 1984a; Vallet *et al.*, 1988). Essentially similar results were obtained by using either bTP-1 or crude conceptus culture medium in which bTP-1 was the major component (Knickerbocker *et al.*, 1986a; Plante *et al.*, 1988).

Whatever the mechanism whereby oTP-1 and bTP-1 affect the corpus luteum, the proteins themselves are not directly luteotrophic. Their action is probably indirect, with extension of luteal function achieved by reducing the pulsatile release of the uterine luteolysin prostaglandin F-2α from the endometrium (Fincher *et al.*, 1986; Knickerbocker *et al.*, 1986b). However, it has also become clear that neither oTP-1 nor bTP-1 are sufficient in themselves to prolong the oestrous cycle for more than a few days. It may be that other conceptus components, not necessarily proteins, play a part in maintaining the lifespan of the corpus luteum of sheep and cattle as pregnancy progresses.

In this paper recent progress on the molecular biology and biological activities of oTP-1 and bTP-1 is reviewed. The discovery that these compounds are interferons has raised numerous questions regarding the manner in which interferon-like proteins might interact with the maternal system and how the production of oTP-1 and bTP-1 might be controlled. It has also provided a new and totally unsuspected role for interferons, agents known primarily for their antiviral and growth inhibiting effects, in reproduction.

Molecular biology of oTP-1 and bTP-1

Identification of oTP-1 and bTP-1 as interferons

Several cDNAs to both oTP-1 and bTP-1 mRNAs have been cloned and sequenced (Imakawa *et al.*, 1987a, 1989; Stewart *et al.*, 1989a). All showed a number of common features (Fig. 1). They were approximately 1 kb in length and possessed a 585-base open reading frame that coded for a polypeptide of 195 amino acids. Analysis of the deduced amino acid sequences by the method of von Heijne (1986) predicted 23-residue signal sequences, with position-1 of the mature proteins being in each case a cysteine residue. Multiple mRNAs for oTP-1 (Imakawa *et al.*, 1987a; Anthony *et al.*, 1988; Stewart *et al.*, 1989a) and bTP-1 (Imakawa *et al.*, 1989) have been shown to exist during the period that synthesis of the two proteins was maximal. These mRNAs coded for polypeptides that differed as much as 10% in sequence within species. In contrast, the oTP-1 and bTP-1 polypeptides showed 75–80% sequence identity (Imakawa *et al.*, 1989).

Computer analysis of the predicted amino acid sequences for oTP-1 and bTP-1 showed that both proteins possessed between 45 and 55% sequence identity to interferons of the alpha family (IFN-α) (Imakawa *et al.*, 1989). This degree of similarity was considered to be significant as the IFN-αs show a similar degree of sequence diversity between mammalian species. Moreover, the four cysteines at positions 1, 29, 99 and 139 known to be involved in intramolecular disulphide bridges (Cys 1/Cys 99; Cys 29/Cys 139) (Wetzel, 1981) and a Cys-Ala-Trp-Glu-Ile-Val-Arg sequence (residues 139–145) that is conserved in most IFN-αs were found on both oTP-1 and bTP-1. There was only a low degree of similarity to the fibroblast IFNs (IFN-β) and no homology with immune IFN (IFN-γ). That oTP-1 and bTP-1 were indeed interferons has been confirmed in a number of functional tests (see below), which include their ability to protect cells from viral lysis, to inhibit cell proliferation and to bind to IFN-α/β receptors.

oTP-1 and bTP-1 most closely resemble a little known group of IFNs that are 172 amino acids in length (in contrast to the 166-residue polypeptides found typically in the IFN-α family). These unusual molecules have been named either IFN-ω (Hauptmann & Swetly, 1985; Himmler *et al.*, 1987) or IFN-α_{II}s, with the term IFN-α_I reserved for the more common 166-residue forms (Capon *et al.*, 1985). Unfortunately, IFN nomenclature is still in some disarray, and there has been considerable confusion in the reproductive literature regarding the classification of oTP-1 and bTP-1. In particular, several recent studies have utilized human IFN-α_I2 and considered it to represent an IFN-α_{II}. In fact, human IFN-α_I2 is unique even among the IFN-α_Is in that it is 165 amino acids in length (see DeMaeyer & DeMaeyer-Guignard, 1988). The other mature IFN-α_Is contain 166 amino acids. IFN-α_I2 is but one of a total of at least 14 non-allelic genes that express functional IFN-α_Is in the human.

```
oTP-1     CTGAGATGGCATCAGAGAACCTACCTGAAGGTTCCCCCTGACCCCATCTCAGCCAGCCCAGCAGCAGCCGCATCTTCCCC
bTP-1     GATCCCCGCAAAC TACAATTCACCTGAAGGTTCACCCAGACCCCATCTCAGCCAGCCCAGCAGCAGCCACATCTTCCCC
          *******  * * **    ***                 *  *                               *

s1                                                                               s23  1   2
met ala phe val leu ser leu leu met ala leu val leu val ser tyr gly pro gly gly ser leu gly cys tyr
ATG GCC TTC GTG CTC TCT CTA CTG ATG GCC CTG GTG CTG GTC AGC TAT GGC CCA GGA GGA TCT CTG GGT TGT TAC
ATG GCC TTC GTG CTC TCT CTA CTG ATG GCC CTG GTG CTG GTC AGC TAC GGC CAG GGA CGA TCT CTG GGT TGT TAC
                                                             *      gln     arg

3                                                                                               27
leu ser arg lys leu met leu asp ala arg glu asn leu arg leu leu asp arg met asn arg leu ser pro his
CTA TCT CGC AAA CTC ATG CTG GAT GCC AGG GAG AAC CTC ACG CTC CTG GAC CGA ATG AAC AGA CTG TCA CCT CAT
CTG TCT GAG GAC CAC ATG CTA GGT GCC AGG GAG AAC CTC ACG CTC CTG GCC CGA ATG AAC AGA CTC TCT CCT CAT
 *      glu asp his      * gly                              ala                      *   *

28                                                                                              52
ser cys leu gln asp arg lys asp phe gly leu pro gln glu met val glu gly asp gln leu gln lys asp gln
TCC TGT CTG CAG GAC AGA AAA GAC TTT GGT CTT CCC CAG GAG ATG GTG GAG GGC GAC CAG CTC CAG AAG GAC CAG
CCC TGT CTG CAG GAC AGA AAA GAC TTT GGT CTT CCT CAG GAG ATG GTG GAG GGC AAC CAG CTC CAG AAG GAT CAG
pro                                          *                      asn                      *

53                                                                                              77
ala phe ser val leu tyr glu met leu gln gln ser phe asn val phe his thr glu arg ser ser ala ala trp
GCC TTC TCT GTG CTC TAC GAG ATG CTC CAG CAG AGC TTC AAC GTC TTC CAC ACA GAG CGC TCC TCT GCT GCC TGG
GCT ATC TCT GTG CTC CAC GAG ATG CTC CAG CAG TGC CTC AAC CTC TTC TAC ACA GAG CAC TCG TCT GCT GCC TGG
 * ile              his                     cys leu     leu     tyr         his  *

78                                                                                             102
asn thr thr leu leu glu gln leu cys thr gly leu gln gln gln leu asp his leu asp thr cys arg gly pro
AAC ACC ACC CTC CTG GAG CAG CTC TGC ACT GGA CTC CAA CAG CAG CTG GAC CAC CTG GAC ACC TGC AGA GGT CCC
AAC ACC ACC CTC CTG GAG CAG CTC TGC ACT GGG CTC CAA CAG CAG CTG GAG GAC CTG GAC GCC TGC CTG GGC CCA
 #                                       *                      glu asp         ala     leu   *

103                                                                                            127
val met gly glu lys asp ser glu leu gly lys met asp pro ile val thr val lys lys tyr phe gln gly ile
GTG ATG GGA GAG AAA GAC TCT GAA CTG GGA AAG ATG GAC CCC ATT GTG ACC GTG AAG AAG TAC TTC CAG GGC ATC
GTG ATG GGA GAG AAA GAC TCT GAC ATG GGA AGC ATG GGC CCC ATT CTG ACT GTG AAG AAG TAC TTC CAG GGT ATC
                            asp met     arg     gly         leu  *                           *

128                                                                                            152
his asp tyr leu gln glu lys gly tyr ser asp cys ala trp glu ile val arg val glu met met arg ala leu
CAT GAC TAC CTG CAA GAG AAG GGA TAC AGC GAC TGC GCC TGG GAA ATC GTC AGA GTC GAG ATG ATG AGA GCC CTC
CAT GTC TAC CTG AAA GAA AAA GAA TAC AGT GAC TGC GCC TGG GAA ATC ATC AGA GTG GAG ATG ATG AGA GCC CTC
    val         lys  *   * glu       *                      ile      *

153                                                                 172
thr ser ser thr thr leu gln lys arg leu thr lys thr gly gly asp leu asn ser pro end
ACT TCA TCA ACC ACC TTG CAA AAA AGG TTA ACA AAG ACC GGT GGA GAT CTG AAC TCA CCT TGA   TGACTCTTGCCGAC
TCT TCA TCA ACC ACC TTG CAA AAA AGG TTA AGA AAG ATG GGT GGA GAT CTG AAC TCA CTT TGA GATGACTCTCGCTGAC
ser                                     arg     met                         leu end **        *  *

TAAGATGCCACATCACCCTCGTACACCCGCCTGTGTTCATTTCAGAAGACTCTGATTTCTGCTCCAGCCACCAAATTCATTGAATTACTTTAGCTGATAC
TAAGATGCCACATCACCTTCCGTACACTCACCTGTGTTCATTTCAGAAGACTCTGATTTCTGCTTCAGCCACCGAAATCATTGAATTACTTTAACTGATAC
                *     * *                                       *      *  *               *

TTTGTCAGTAGTAAAAAGCAAGTAGATATAAAAGTATTCAGCTGTAGGGGCATGAGTCCTGAAATGATGCCTTCCCTGATGTTATCTGTTGCTGATTTAT
TTTGTCAGCAGTAATAAGCAAGTAGATATAAAAGTACTCAGCTGTAGGGGCATGAGTCCTTAAGTGATGCCTGCCCTGATGTTATCTGTTGTTGATTTAT
        *    *                  *                    *  *      *                *

TTATACCTTCTTGCATTTAACATACTTAAAATATTAGGAAATTTGTTAAGTTAC
GTATTCCTTCTTGCATCTAACATACTTAAAATATTAGGAAATTTGTAAAGTTACATTTCATTTGTACATC TATTAAAATTTCTAAAACATGTTTACCA
*   *     *   *                              *

TTTTGTGTTATTAAATTTGTCCTTTGTTCTATTTATTAAATCAAAGAAAATGAAAAAAAAAAAAAAAAA
```

Fig. 1. A comparison of the nucleotide sequence for a previously undescribed oTP-1 cDNA with that of a full-length bTP-1 cDNA previously published by Imakawa *et al.* (1989). Nucleotide and amino acid differences with the published sequence of another oTP-1 cDNA (Imakawa *et al.*, 1987a) are indicated by *underlines*. An asterisk (*) indicates nucleotide changes that are silent, i.e. provide no amino acid sequence differences between the oTP-1 and bTP-1 isoforms. Amino acid sequence differences between the oTP-1 and bTP-1 are shown beneath the bTP-1 nucleotide sequence. A potential glycosylation site (Asn[78]) is marked (♯). Amino acid residue positions are shown above the oTP-1 sequence; S1-S23 is the signal sequence; 1-172 represents the mature protein. Data for the bTP-1 cDNA have been previously published (Imakawa *et al.*, 1989).

It is suspected that oTP-1 and bTP-1 may even form a special subclass within the IFN-α_{II} subfamily since their mRNAs show remarkable conservation of sequence at their 3′ untranslated ends (Imakawa *et al.*, 1989). Such conservation is not noted between the mRNAs for bTP-1 and bovine IFN-α_{II}s.

It has been estimated that there are numerous IFN-α_{II} genes in cattle (Capon *et al.*, 1985). There is no information in sheep, because, until oTP-1 mRNA was cloned no ovine IFN gene had ever been cloned or sequenced. Therefore, it has been of interest to determine whether there are several genes for oTP-1, as suggested from the multiplicity of mRNAs and oTP-1 isoforms, and whether these genes are intronless, as are all other known IFN-α_I and IFN-α_{II} genes.

Southern blotting experiments using a range of restriction endonucleases and probes derived from an oTP-1 cDNA have revealed that there are several genes for this interferon (unpublished results), although it remains unclear how many of these genes are actually transcribed. Preliminary experiments have indicated there may be fewer genes for bTP-1 than for oTP-1 and that the bTP-1 genes may be a separate subset within the total IFN-α_{II} gene family in cattle.

Developmental changes in oTP-1 and bTP-1 gene expression

Immunocytochemical localization of oTP-1 (Godkin *et al.*, 1984b) and bTP-1 (Lifsey *et al.*, 1989) by specific antisera has shown the two proteins to be localized in the trophectoderm of Day-16 sheep and Day-20 cattle conceptuses. In both cases mononuclear rather than the rarer binucleate cells appeared to be the dominant sites of production. These results have been confirmed and extended by use of in-situ hybridization with labelled cDNA probes to localize oTP-1 mRNAs on tissue sections (Farin *et al.*, 1989). The mRNA for oTP-1 in Day-15 conceptuses was found almost exclusively in trophectoderm cells (Figs 2a & b). The underlying extraembryonic endoderm, the yolk sac and the embryo proper did not give a significant signal. By contrast each of these tissues was strongly positive for actin mRNA (Fig. 2c).

Biosynthetic studies with cultured sheep conceptuses have shown that high rates of oTP-1 synthesis only occurred between Days 13 and 21 of pregnancy with peak values between Days 14 and 17 (Godkin *et al.*, 1982; Hansen *et al.*, 1985) when greater than 100 μg of the protein were released from some conceptuses in a 24-h period. However, low amounts of oTP-1 production by conceptuses as early as Days 8–10 has been detected by using a sensitive radioimmunoassay (Ashworth & Bazer, 1989). The values may be underestimates of the amounts synthesized *in vivo*, since production of oTP-1 and IFN activity begins to tail off markedly after 12 h in culture (Hansen *et al.*, 1989a).

Based on Northern and dot blot analyses of total cellular RNA from conceptuses, oTP-1 mRNA was shown to be present in peak amounts per unit of DNA at about Day 14 (Hansen *et al.*, 1988; Stewart *et al.*, 1989a). At this stage its mRNA was several times more abundant than actin mRNA. oTP-1 mRNA concentration on a per cell basis then declined so that by Day 20 it was only about 1% of its peak value (Hansen *et al.*, 1988). Since the total mass of conceptus tissue increased markedly during this period, the total quantity of mRNA per conceptus, and oTP-1 synthesis itself, dropped less dramatically.

In-situ hybridization (Farin *et al.*, 1989) has provided more detailed information about changes in mRNA concentrations. There were detectable amounts of oTP-1 mRNA at Day 11, but the signal increased markedly around Day 13 when the conceptuses were still spherical. This increase may be significant as Day 12–13 has generally been recognized as the time that the signal for maternal recognition of pregnancy must be perceived by the ewe if she is to remain pregnant (see Bazer *et al.*, 1986). After Day 15 amounts of oTP-1 mRNA had fallen, and by Day 23 the label was localized to a few indistinct foci in the chorionic membrane. Conceivably these sites represented developing cotyledons.

In-situ hybridization techniques have also been used to measure the relative amounts of oTP-1 mRNA present in Day-17 conceptuses at the beginning and end of a 24-h culture period (Hansen *et al.*, 1989a). During culture, the concentration of oTP-1 mRNA fell to about 20%, while actin mRNA was scarcely affected. By labelling RNA in conceptuses with uridine and isolating oTP-1 mRNA at subsequent time points, Hansen *et al.* (1989a) also concluded that oTP-1 mRNA turned over rapidly in cultured conceptuses. It will be of interest to know whether oTP-1 mRNA is similarly unstable in conceptuses developing *in utero*. If it is, rates of oTP-1 gene transcription must be remarkably high to account for the large amounts of oTP-1 mRNA present in the trophectoderm cells. Alternatively, the mRNA may be stabilized in conceptuses developing *in utero*.

Experiments designed to examine bTP-1 mRNA levels in bovine conceptuses have shown somewhat similar changes with time (unpublished results). Amounts of mRNA per unit DNA were highest in the period immediately following the Day 15–17 window for maternal recognition of pregnancy. However, some small spherical embryos as early as Day 12 of development also gave strong positive signals when examined by in-situ hybridization.

Induction of oTP-1 mRNA *in vitro*

One of the most urgent questions regarding oTP-1 and bTP-1 is how the expression of their respective genes is controlled, since an insight into this process may provide a means of manipulating fertility of sheep, cattle and possibly other species by pharmacological intervention. At present it is

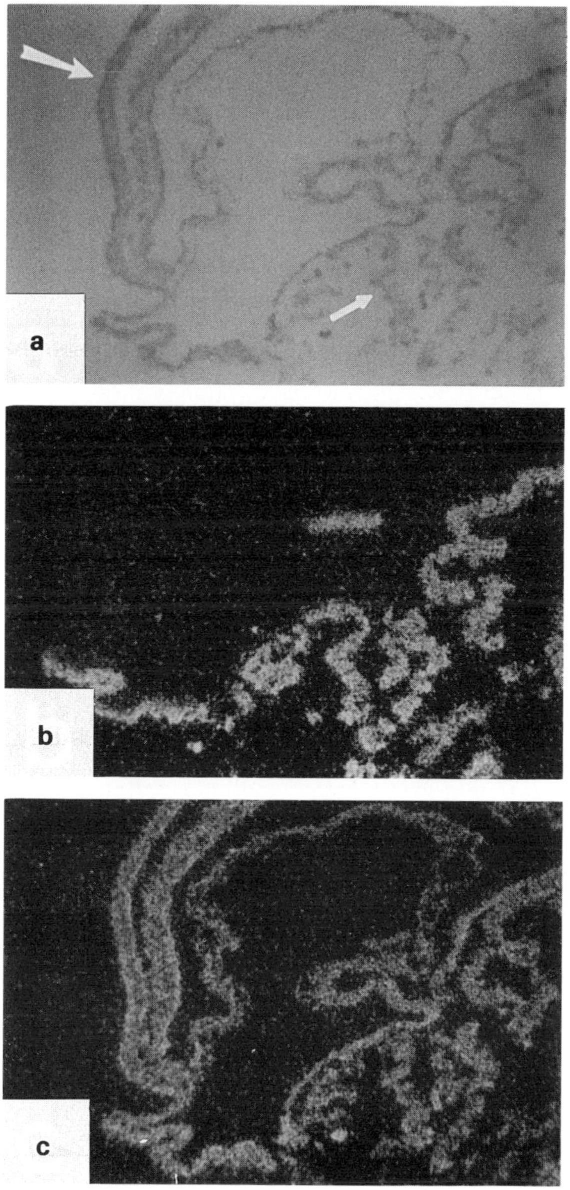

Fig. 2. In-situ localization of oTP-1 and actin mRNA in a Day 15 sheep embryo. (a) Bright-field micrograph showing portions of the embryonic disc (large arrow) and trophectoderm (small arrow) from an embryo fixed in 4% paraformaldehyde and embedded in paraffin wax. (b) Dark-field micrograph of an adjacent section hybridized with a [^{35}S]oTP cDNA probe specific for the unique 3′-untranslated region of the oTP-1 mRNA. Bright white areas indicate positive hybridization. Note that the oTP-1 mRNA is found only in the trophectoderm and is not associated with the embryonic disc. (c) Dark-field micrograph of an adjacent section hybridized with [^{35}S]γ-actin cDNA. Note that, in contrast to oTP-1, actin mRNA is localized in all embryonic tissues including both the embryonic disc and trophectoderm. × 100.

not known whether induction is the result of a naturally unfolding developmental pattern or whether gene transcription is activated by 'inducers' produced locally, for example, by the surrounding maternal endometrium.

We have recently examined whether Day 11 sheep conceptuses, which normally have only low amounts of oTP-1 mRNA, can be induced to express oTP-1 mRNA by using a synthetic double-stranded RNA, polyI:polyC, a compound known to induce IFNs in other cell types (Fig. 3). The study clearly demonstrated that treated conceptuses showed a significant increase in oTP-1 mRNA compared with controls. This increased signal was probably not the result of activation of IFN-α_Is, since highly stringent hybridization conditions were used and the 3′-end probe was highly specific for oTP-1 mRNA.

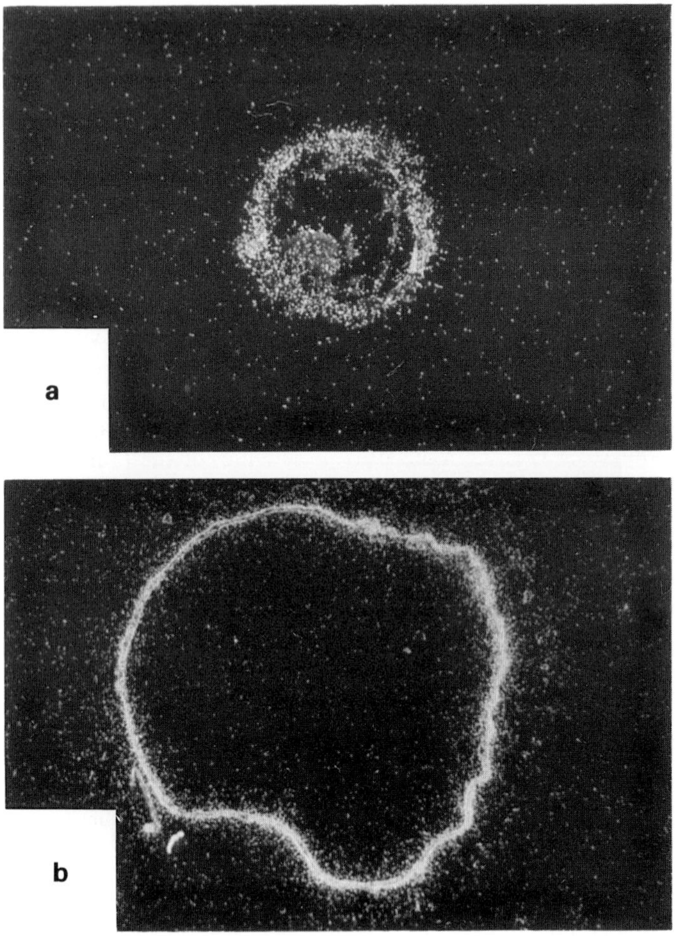

Fig. 3. Effect of culture in the presence of polyI:polyC (PIPC) on mRNA levels of oTP-1 in Day-11 sheep embryos. oTP-1 mRNA, indicated by the presence of silver grains (individual white dots), was localized by using in-situ hybridization with a ^{35}S-labelled cDNA probe. (**a**) Dark-field micrograph of a control embryo which was cultured for 8 h in Ham's F12 medium supplemented with 5% (v/v) heat-inactivated fetal calf serum plus sodium pyruvate, L-glutamine and antibiotics. (**b**) Dark-field micrograph of a PIPC-treated embryo which was cultured for 8 h in the same medium supplemented with 100 µg PIPC/ml. The greater density of silver grains over the trophectoderm of the PIPC-treated embryo compared to the control embryo indicates an increase in the amount of oTP-1 mRNA present. × 100.

Although it seems unlikely that oTP-1 gene transcription in trophectoderm *in utero* is the result of induction by double-stranded RNAs or viruses, these experiments do suggest that oTP-1 may be responsive to some of the same external signals known to influence expression of other IFN-α genes. Experiments are underway to determine whether oTP-1 can be induced by components secreted by the maternal uterus after Day 12 of pregnancy and to sequence the upstream region of the oTP-1 and bTP-1 genes to see whether these structures possess the same *cis*-response elements believed to be involved in the viral induction of other IFN-α/β genes (DeMaeyer & DeMaeyer-Guignard, 1988).

Biological activities of oTP-1 and bTP-1

Antiviral and antiproliferative activities

Both oTP-1 (Imakawa *et al.*, 1987b; Pontzer *et al.*, 1988; Roberts *et al.*, 1989) and bTP-1 (Godkin *et al.*, 1988) have potent antiviral activity and thus provide a potential barrier at the placental interface to viral infection of the embryo. By using a cytopathic inhibition assay in which kidney epithelial cells were challenged by vesicular stomatitis virus (VSV), Roberts *et al.* (1989) showed that oTP-1 had an activity greater than 10^7 relative units/mg protein. The activity demonstrated by Pontzer *et al.* (1988) was even higher ($2–3 \times 10^8$ units/mg). These values fell within the range shown by other known IFN-αs (Pestka *et al.*, 1987). The oTP-1 antiviral activity was not restricted to vesicular stomatitis virus but was also effective against the alphagatavirus, Semliki forest virus, a picornavirus, bovine enterovirus-1, and the alphaherpes viruses, pseudorabies and bovine herpes (Roberts *et al.*, 1989). The results of one such study are shown in Table 1. This experiment was positively controlled by comparing the antiviral activity of oTP-1 with that of an ovine IFN (obtained from the serum of a polyI:polyC injected sheep) and with human leucocyte and human lymphoblastoid IFN-αs. The ratios of activities of the four IFN preparations were similar for each virus.

Table 1. Comparison of the antiviral activities of ovine polyI:polyC-induced IFN, oTP-1, human leukocyte IFN-α, and human lymphoblastoid IFN-α against four different viruses on GBK-2 cells relative to vesicular stomatis virus (from Roberts *et al.*, 1989)

		IFNs (relative antiviral activity, %)			
Virus*	Plaque forming units/cll	Ovine IFN†	oTP-1	Human leucocyte IFN-α	Human lymphoblastoid IFN-α
VSV‡	1	100	100	100	100
SFV	3	7·2	26	11	11
BEV-1	0·2	6·3	34	11	26
PRV	1	8·6	6·5	4·9	3·8
BH-2	1	5·8	33	33	19·5

Assays were carried out in GBK-2 cells by the methods described by Roberts *et al.* (1989).

*VSV, vesicular stomatitis virus; SVF, Semliki forest virus; BEV-1, bovine enterovirus type 1; PRV, pseudorabies virus; BH-2, bovine herpes type 2.

†The source of this IFN was serum from a sheep that had been injected with poly(I): poly(C).

‡The antiviral activities against VSV were as follows: ovine poly(I):poly(C)-induced IFN, 360 U/ml; oTP-1, 150 000 U/ml; human leukocyte IFN-α, 17 000 U/ml; human lymphoblastoid IFN-α, 200 000 U/ml.

It would seem unlikely, however, that the function of oTP-1 is to provide antiviral protection to the conceptus since the protein seems to be produced in significant amounts for only a few days in early pregnancy. Even though IFNs have historically been associated with their ability to render cells resistant to viral lysis, they are also potent cytokines and have a range of activities on their target cells, including effects on cell proliferation, cell differentiation and protein synthesis (Rossi, 1985; Tamm *et al.*, 1987).

oTP-1 and bTP-1 were both able to inhibit the incorporation of [^{3}H]thymidine into sheep lymphocytes that had been stimulated with Concanavalin A or phytohaemagglutinin (Roberts *et al.*, 1989; Fig. 4). They appeared to do so by reducing cell responsiveness to interleukin-2 (Niwano *et al.*, 1989). These effects were comparable to those of a recombinant human IFN-α (IFN-α_I1). Such inhibition of lymphocyte proliferation has frequently been used as a criterion for immunosuppression, and it is possible that the production of IFN by the expanding trophoblast mutes a local immune reaction mounted by the mother against the conceptus. However, it is not even clear whether the conceptus, and particularly the trophoblast, expresses major histocompatability antigens at this stage of development and constitutes, therefore, a functional allograft. Moreover, the effects of IFN on the immune system and other target cells are complex and not necessarily immunosuppressive. Natural killer cells become activated and MHC expression is usually enhanced by IFN-α treatment (Rossi, 1985; Tamm *et al.*, 1987). Conceivably, these trophoblast IFNs, possibly acting synergistically with other conceptus-derived factors, induce a locally propitious environment at the site of conceptus attachment to the uterine wall which promotes further embryonic growth and development (see Athanassakis *et al.*, 1987).

Interaction of interferons with uterine endometrium

Because recent studies have shown that at least some of the biological effects of oTP-1 and bTP-1 could be successfully mimicked by using recombinant IFN-α_Is (see below) it is perhaps not surprising that oTP-1 and IFN-α_Is compete with each other for binding to endometrial receptors (Godkin *et al.*, 1984b; Stewart *et al.*, 1987; Hansen *et al.*, 1989b; Knickerbocker & Niswender, 1989). This observation together with the reports of 'receptors' for oTP-1 in other organs, including the corpus luteum, liver and kidney, and the observation that the highest concentration of IFN receptors in ewes is in the uterus (Knickerbocker & Niswender, 1989), suggests that oTP-1 and bTP-1 might act on the endometrium as classical IFN-αs. However, it still cannot be ruled out that these trophoblast IFNs have some specialized functions that are not mediated through usual Type I, IFN-α/β receptors (Rubenstein & Orchansky, 1986). For example, studies with bifunctional cross-linking reagents have revealed that oTP-1 forms a complex with two polypeptide chains (M_r 100 000 and 70 000), while rboIFN-α_I1 becomes covalently cross-linked to only the larger chain (Hansen *et al.*, 1989b). The situation is clearly complicated as rboIFN-α_I1 in excess prevents oTP-1 from becoming cross-linked to both polypeptides. Thus, such studies must be interpreted with caution since the absence of cross-linking does not prove a lack of receptor binding.

Presumably through its interaction with specific endometrial receptors, oTP-1 altered the output of secretory proteins synthesized by cultured endometrial explants from Day 12 non-pregnant ewes (Godkin *et al.*, 1984b; Vallet *et al.*, 1987). Production of several proteins, particularly of an acidic protein of M_r 70 000 (70K protein) and one of M_r 15 000, was enhanced. These results could be mimicked by substituting either human or bovine IFN-α_I for oTP-1 in the medium (Silcox *et al.*, 1988; unpublished results). Production of the 70K protein also appeared diagnostic of endometrium that had been in local contact with the conceptus (Sharif *et al.*, 1989). It may be that these proteins form part of the microenvironment in which the conceptus, attached but still not implanted, continues to develop.

Several studies have pointed to the possibility that oTP-1 and bTP-1 may be able to reduce the pulsatile output of the luteolysin, PGF-2α, by the uterus and thus spare the corpus luteum from

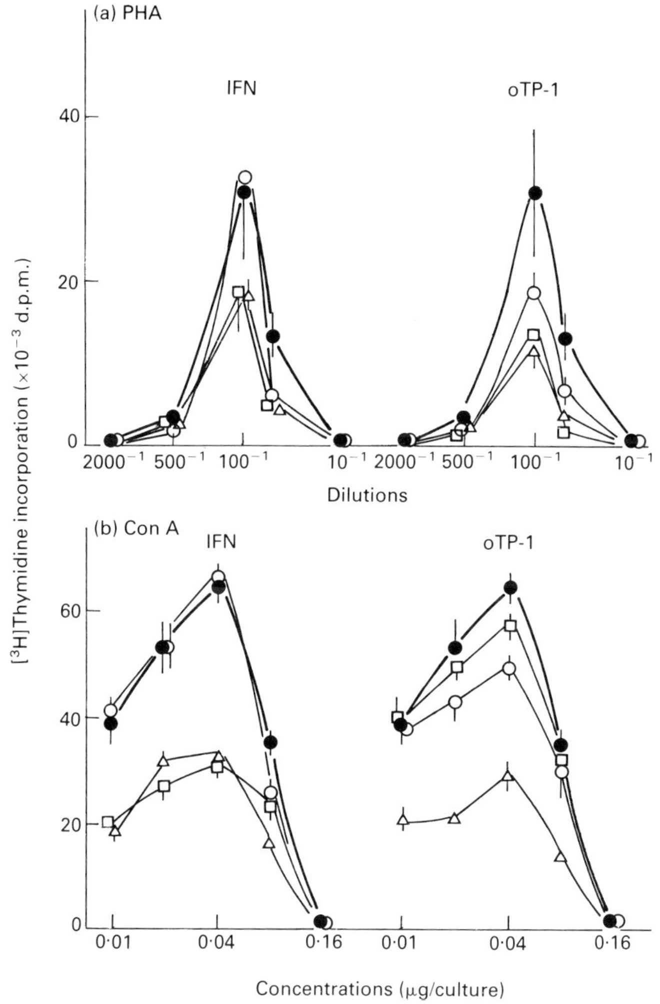

Fig. 4. Comparison of the abilities of recombinant IFN-α_I1 and oTP-1 to inhibit the mitogen-stimulated incorporation of [^{3}H]thymidine into ovine lymphocytes. Values ($\pm$s.e.m. for 3 determinations) are for controls (●), 10^{-7} M-oTP-1 (△), 10^{-9} M-oTP-1 (□), or 10^{-11} M-oTP-1 (○) in right panels and 10^{-8} M-IFN (△), 10^{-10} M-IFN (□), or 10^{-12} M-IFN (○) on left. (a) The effect of *Phaseolus* phytohaemagglutinin (PHA) on [^{3}H]thymidine incorporation was tested on dilutions of the original reagent supplied by Difco Laboratories. (b) Concanavalin A (Con A) solution was prepared by weighing the dried lectin. Reproduced from Roberts *et al.* (1989).

regression (Fincher *et al.*, 1986; Knickerbocker *et al.*, 1986b; Vallet *et al.*, 1988). Attempts have been made to confirm these results *in vitro*. Salamonsen *et al.* (1988) showed that both oTP-1 and a human IFN-α_I (IFN-α_I2) could inhibit PGF-2α and PGE synthesis by primary cultures of ovine endometrial cells. However, Thatcher *et al.* (1989) have indicated that while bTP-1 inhibited production of PGF-2α by cultured bovine endometrial explants, a recombinant bovine IFN-α_I (rboIFN-α_I1) had no effect. Further work is clearly required to resolve the issue as to whether PGF-2α synthesis is altered and whether the trophoblast interferons act uniquely in this respect. It should be emphasized that the effects of IFNs on arachidonic acid metabolism and prostaglandin production in various test systems have been by no means consistent (see Roberts *et al.*, 1990), and on present information it is difficult to understand how their actions are mediated.

Embryonic interferons and improvements of reproductive performance

The availability of a recombinant bovine IFN-α_I1 (from CIBA-GEIGY Ltd, Basle, Switzerland) has allowed a number of experiments to be conducted to determine whether a 166 amino acid IFN-α_I could substitute for the natural trophoblast product in delaying luteolysis in cattle and sheep. In our laboratory 125 μg rboIFN-α_I1 were injected into the uterine lumen of ewes twice daily between Days 12 and 15 of the oestrous cycle. No cycle length extensions were noted (H. Francis, J. C. Cross & R. M. Roberts, unpublished results), though similar amounts of oTP-1 were clearly effective in this regard (Godkin *et al.*, 1984a). A comparable study was done by Stewart *et al.* (1989b) except much higher doses of the IFN were used (2 mg daily). In this case length of the oestrous cycle and CL function were significantly extended. Plante *et al.* (1988) introduced rboIFN-α_I1 into the uteri of cyclic cows from Days 15 to 21 of the cycle. Again oestrous cycle length was extended and concentrations of progesterone in serum higher than 1·5 ng/ml were noted for about 4 additional days in the treated cows. Such experiments strongly suggested that the IFN-α_Is had comparable action to oTP-1 and bTP-1 *in utero* even though higher amounts of the recombinant homologues had to be used for them to be effective.

Of particular interest are the recent observations that intramuscular injections of rboIFN-α_I1 (rather than infusion directly into the uterine lumen) during the period of maternal recognition of pregnancy extends luteal function in non-pregnant cows (Plante *et al.*, 1989). This experiment has raised the possibility that simple systemic injection of IFN might be used to manipulate oestrous cycle length and possibly fertility in cattle.

However, two aspects concerning the ability of IFNs to maintain luteal function should be noted. First, there is no evidence that rboIFN-α_I1 or purified bTP-1 or oTP-1 can extend luteal function for more than a few days even when the product is introduced directly into the uterine lumen at a time when the conceptus would be secreting the trophoblast IFNs maximally. Therefore, other factors produced by the conceptus probably play a role in the longer-term maintenance of the CL and may even act synergistically with oTP-1 or bTP-1 to mediate proper recognition of the conceptus. Second, rboIFN-α_I, even though it appears to bind effectively to oTP-1 receptors and exhibits antiviral activity as high or higher than oTP-1 or bTP-1, must be used at much higher concentrations than the natural trophoblast IFNs if it is to be effective. It is therefore possible that the 'natural' trophoblast IFNs have additional activities relating to maternal recognition of pregnancy that are mediated only poorly by IFN-α_Is.

The question also arises as to whether IFNs play any role in maternal recognition of pregnancy in species other than domestic ungulates. There are many references indicating that IFNs are associated with placental tissues of humans and rodents (reviewed by Roberts *et al.*, 1990). In many cases these IFNs have been poorly defined structurally and serologically, and their sites of localization have not in general been shown. The pig also produces an IFN at about the time of blastocyst expansion (Cross & Roberts, 1989), but the protein has not been purified; nor has its cDNA been cloned. Therefore, its relationship to oTP-1 remains unclear. Certainly, at present no specific function for IFNs in pregnancy of non-ruminant species can be inferred, and there is no compelling evidence to suggest they are involved in rescue of the corpus luteum. Nevertheless, it is tempting to believe that these substances may play some important role in conceptus–maternal interactions of eutherian mammals in general.

We thank Gail Foristal for typing this manuscript; and Jim Bixby, Harriet Frances, Tammie Schalue-Francis, Dr P. V. Malathy, Dr M. Kazemi and Dr Duane Keisler for their involvement in many of the studies. The research was supported by NIH Grant 21896. This is paper 10 842 of the Missouri Agricultural Experimental Station.

References

Anthony, R.V., Helmer, S.D., Sharif, S.F., Roberts, R.M., Hansen, P.J., Thatcher, W.W. & Bazer, F.W. (1988) Synthesis and processing of ovine trophoblast protein-1 and bovine trophoblast protein-1, conceptus secretory proteins involved in the maternal recognition of pregnancy. *Endocrinology* **123,** 1274–1280.

Ashworth, C.J. & Bazer, F.W. (1989) Changes in ovine conceptus and endometrial function following asynchronous embryo transfer or administration of progesterone. *Biol. Reprod.* **40,** 425–434.

Athanassakis, I., Bleackley, R.C., Paetkau, V., Guilbert, L., Barr, P.J. & Wegmann, T.G. (1987) The immunostimulatory effect of T cells and T cell lymphokines on murine fetally-derived placental cells. *J. Immunol.* **138,** 37–44.

Bartol, F.F., Roberts, R.M., Bazer, F.W., Lewis, G.S., Godkin, J.D. & Thatcher, W.W. (1985) Characterization of proteins produced in vitro by peri-attachment bovine conceptuses. *Biol. Reprod.* **32,** 681–694.

Bazer, F.W., Vallet, J.L., Roberts, R.M., Sharp, D.C. & Thatcher, W.W. (1986) Role of conceptus secretory products in establishment of pregnancy. *J. Reprod. Fert.* **76,** 841–850.

Capon, D.J., Shepard, H.M. & Goeddel, D.V. (1985) Two distinct families of human and bovine interferon-α genes are coordinately expressed and encode functional polypeptides. *Molec. cell. Biol.* **5,** 768–799.

Cross, J.C. & Roberts R.M. (1989) Porcine conceptuses secrete an interferon during the pre-attachment period of early pregnancy. *Biol. Reprod.* **40,** 1109–1118.

DeMaeyer, E. & DeMaeyer-Guignard, J. (1988) *Interferons and Other Regulatory Molecules.* John Wiley & Sons, New York, NY.

Farin, C.E., Imakawa, K. & Roberts, R.M. (1989) In situ localization of mRNA for the interferons, ovine trophoblast protein-1 during early embryonic development of the sheep. *Molec. Endocrinol.* **3,** 1099–1107.

Fincher, K.B., Bazer, F.W., Hansen, P.J., Thatcher, W.W. & Roberts, R.M. (1986) Ovine conceptus secretory proteins suppress induction of uterine prostaglandin F-2α release by oestradiol and oxytocin. *J. Reprod. Fert.* **76,** 425–433.

Godkin, J.D., Bazer, F.W., Lewis, G.S., Geisert, R.D. & Roberts, R.M. (1982) Synthesis and release of polypeptides by pig conceptuses during the period of blastocyst elongation and attachment. *Biol. Reprod.* **27,** 977–987.

Godkin, J.D., Bazer, F.W., Thatcher, W.W. & Roberts, R.M. (1984a) Proteins released by cultured Day 15–16 conceptuses prolong luteal maintenance when introduced into the uterine lumen of cyclic ewes. *J. Reprod. Fert.* **71,** 57–64.

Godkin, J.D., Bazer, F.W. & Roberts, R.M. (1984b) Ovine trophoblast protein 1, an early secreted blastocyst protein, binds specifically to uterine endometrium and affects protein synthesis. *Endocrinology* **114,** 120–130.

Godkin, J.D., Lifsey, B.J., Fujii, D.K. & Baumbach, G.A. (1988) Bovine trophoblast protein-1: purification, antibody production, uterine cell interaction and antiviral activity. *Biol. Reprod.* **38** (Suppl. 1), 79, abstr.

Hansen, P.J., Bazer, F.W. & Roberts, R.M. (1985) Appearance of β-hexosaminidase and other lysosomal-like enzymes in the uterine lumen of gilts, ewes and mares in response to progesterone and oestrogens. *J. Reprod. Fert.* **73,** 411–424.

Hansen, T.R., Imakawa, K., Polites, H.G., Marotti, K.R., Anthony, R.V. & Roberts, R.M. (1988) Interferon RNA of embryonic origin is expressed transiently during early pregnancy in the ewe. *J. biol. Chem.* **263,** 12801–12804.

Hansen, T.R., Cross, J.C. & Farin, C.E. (1989a) Rates of transcription, mRNA degradation and synthesis of oTP-1. *Biol Reprod.* **40** (Suppl. 1), 86 abstr.

Hansen, T.R., Kazemi, M., Keisler, D.H., Malathy, P.V., Imakawa, K. & Roberts, R.M. (1989b) Complex binding of the embryonic interferon, ovine trophoblast protein-1, to endometrial receptors. *J. Interferon Res.* **9,** 215–225.

Hauptmann, R. & Swetly, P. (1985) A novel class of human type 1 interferons. *Nucl. Acids Res.* **13,** 4739–4749.

Helmer, S.D., Hansen, P.J., Anthony, R.V., Thatcher, W.W., Bazer, F.W. & Roberts, R.M. (1987) Identification of bovine trophoblast protein-1, a secretory protein immunologically related to ovine trophoblast protein-1. *J. Reprod. Fert.* **79,** 83–91.

Himmler, A., Hauptmann, R., Adolf, G.R. & Swetly, P. (1987) Structure and expression in E. coli of canine interferon-a genes. *J. Interferon Res.* **7,** 173–183.

Imakawa, K., Anthony, R.V., Kazemi, M., Marotti, K.R., Polites, H.G. & Roberts, R.M. (1987a) Interferon-like sequence of ovine trophoblast protein secreted in embryonic trophectoderm. *Nature, Lond.* **330,** 377–379.

Imakawa, K., Anthony, R.V., Niwano, Y., Hansen, T.R., Kazemi, M., Polites, H.G., Marotti, K.R. & Roberts, R.M. (1987b) Ovine trophoblast protein-1 (oTP-1), a polypeptide implicated in mediating maternal recognition of pregnancy in sheep, is an interferon of the alpha class. *J. Cell. Biol.* **105,** 11a, abstr.

Imakawa, K., Hansen, T.R., Malathy, P-V., Anthony, R.V., Polites, H.G., Marotti, K.R. & Roberts, R.M. (1989) Molecular cloning and characterization of complementary deoxyribonucleic acids corresponding to bovine trophoblast protein-1: a comparison with ovine trophoblast protein-1 and bovine interferon-α_{II}. *Molec. Endocrinol.* **3,** 127–139.

Knickerbocker, J.J. & Niswender, G.D. (1989) Characterization of endometrial receptors for ovine trophoblast protein-1 during the estrous cycle and early pregnancy in sheep. *J. Reprod. Fert.* **40,** 361–369.

Knickerbocker, J.J., Thatcher, W.W., Bazer, F.W., Drost, M., Barron, D.H., Fincher, K.B. & Roberts, R.M. (1986a) Proteins secreted by Day 16 to 18 bovine conceptuses extend corpus luteum function in cows. *J. Reprod. Fert.* **77,** 381–391.

Knickerbocker, J.J., Thatcher, W.W., Bazer, F.W., Barron, D.H. & Roberts, R.M. (1986b) Inhibition of uterine prostaglandin F2α production by bovine conceptus secretory proteins. *Prostaglandins* **31,** 777–793.

Lifsey, B.J., Baumbach, G.A. & Godkin, J.D. (1989) Isolation, characterization and immunocytochemical localization of bovine trophoblast protein-1. *Biol. Reprod.* **40,** 343–352.

Niwano, Y., Hansen, T.R., Malathy, P.V., Johnson, H.D., Roberts, R.M. & Imakawa, K. (1989) Suppression of T-lymphocyte blastogenesis by ovine trophoblast protein-1 and human interferon-alpha may be independent of interleukin-2 production. *Am. J. Reprod. Immunol. Microbiol.* **20,** 21–26.

Pestka, S., Langer, J.A., Zoon, K.C. & Samuel, C.E. (1987) Interferons and their actions. *Ann. Rev. Biochem.* **56,** 727–777.

Plante, C., Hansen, P.J. & Thatcher, W.W. (1988) Prolongation of luteal lifespan in cows by intrauterine infusion of recombinant bovine alpha-interferon. *J. Endocr.* **122,** 2342–2344.

Plante, C., Hansen, P.J., Martinod, S., Siegenthaler, B., Thatcher, W.W. & Leslie, M.V. (1989) Intrauterine and intramuscular administration of recombinant bovine interferon-alpha-I prolongs luteal lifespan in cattle. *J. Dairy. Sci.* **72,** 1859–1865.

Pontzer, C.H., Torres, B.A., Vallet, J.L., Bazer, F.W. & Johnson, H.M. (1988) Antiviral activity of the pregnancy recognition hormone ovine trophoblast protein-1. *Biochem. Biophys. Res. Commun.* **152,** 801–807.

Roberts, R.M., Imakawa, K., Niwano, Y., Kazemi, M., Malathy, P.V., Hansen, T.R., Glass, A.A. & Kronenberg, L.H. (1989) Interferon Production by the preimplantation sheep embryo. *J. Interferon Res.* **9,** 175–187.

Roberts, R.M., Farin, C.E. & Cross, J.C. (1990) Trophoblast proteins and maternal recognition of pregnancy. *Oxford Rev. Reprod. Biol.* **12.**

Rossi, G.B. (1985) Interferons and Cell Differentiation. In *Interferon*, vol. 6, pp. 31–68. Ed. I. Gresser. Academic Press, London.

Rubenstein, M. & Orchansky, R. (1986) The interferon receptors. *CRC Crit. Rev. Biochem.* **21,** 249–274.

Salamonsen, L.A., Stuchbery, S.J., O'Grady, C.H., Godkin, J.D. & Findlay, J.K. (1988) Interferon-alpha mimics effects of ovine trophoblast protein-1 on prostaglandin and protein secretion by ovine endometrial cells *in vitro*. *J. Endocr.* **117,** R1–R4.

Sharif, S.F., Francis, H., Keisler, D.H. & Roberts, R.M. (1989) Correlation between the release of ovine trophoblast protein-1 by the conceptus and the production of polypeptides by the maternal endometrium of ewes. *J. Reprod. Fert.* **85,** 471–476.

Silcox, R.W., Francis, H. & Roberts, R.M. (1988) The effects of ovine trophoblast protein-1 (oTP-1) on ovine endometrial protein synthesis are mimicked by human alpha-1 interferon. *J. Anim. Sci.* **66** (Suppl. 1), 153–154.

Stewart, H.J., McCann, S.H.E., Barker, P.J., Lee, K.E., Lamming, G.E. & Flint, A.P.F. (1987) Interferon sequence homology and receptor binding activity of ovine trophoblast antiluteolytic protein. *J. Endocr.* **115,** R13–R15.

Stewart, H.J., McCann, S.H.E., Northrop, A.J., Lamming, G.E. & Flint, A.P.F. (1989a) Sheep antiluteolytic interferon: cDNA sequence and analysis of mRNA levels. *J. molec. Endocr.* **2,** 65–70.

Stewart, H.J., McCann, S.H.E., Lamming, G.E. & Flint, A.P.F. (1989b) Evidence for a role for interferon in the maternal recognition of pregnancy. *J. Reprod. Fert., Suppl.* **37,** 127–138.

Tamm, I., Lin, S.L., Pffeffer, L.M. & Sehgal, P.B. (1987) Interferons α and β as cellular regulatory molecules. In *Interferon*, vol. 9, pp. 13–73. Ed. I. Gresser. Academic Press, Inc., London.

Thatcher, W.W., Hansen, P.J., Gross, T.S., Helmer, S.D., Plante, C. & Bazer, F.W. (1989). Antiluteolytic effects of bovine trophoblast protein-1. *J. Reprod. Fert., Suppl.* **37,** 91–99.

Vallet, J.L., Bazer, F.W. & Roberts, R.M. (1987) The effect of ovine trophoblast protein-1 on endometrial protein secretion and cyclic nucleotides. *Biol. Reprod.* **37,** 1307–1315.

Vallet, J.L., Bazer, F.W., Fliss, M.F.V. & Thatcher, W.W. (1988) Effect of ovine conceptus secretory proteins and purified ovine trophoblast protein-1 on interoestrous interval and plasma concentrations of prostaglandin F-2α and E and of 13,14-dihydro-15-keto prostaglandin F-2 in cyclic ewes. *J. Reprod. Fert.* **84,** 493–504.

von Heijne, G. (1986) A new method for predicting signal sequence cleavage sites. *Nucleic Acids Res.* **14,** 4683–4690.

Wetzel, R. (1981) Assignment of the disulphide bonds of leukocyte interferon. *Nature, Lond.* **289,** 606–607.

TRANSGENIC PIGS

Chairman
D. Beermann

J. Reprod. Fert., Suppl. **41** (1990), 77–87
Printed in Great Britain

Integration, expression and germ-line transmission of growth-related genes in pigs

V. G. Pursel*, R. E. Hammer†§, D. J. Bolt*, R. D. Palmiter‡ and R. L. Brinster†

**US Department of Agriculture, Agricultural Research Service, Beltsville, Maryland 20705, USA; †School of Veterinary Medicine, University of Pennsylvania, Philadelphia, Pennsylvania 19104, USA; and ‡Howard Hughes Medical Institute Laboratory and Department of Biochemistry, University of Washington, Seattle, Washington 98195, USA*

Summary. We have produced transgenic pigs that harbour structural genes for bovine and human growth hormone (bGH and hGH) ligated to a mouse metallothionein-I (MT) promoter, human growth hormone-releasing factor (hGRF) ligated to the MT or mouse albumin (ALB) promoter, and human insulin-like growth factor-I (hIGF-I) ligated to MT promoter. From 0·31 to 1·03% of microinjected ova developed into transgenic pigs with the various fusion genes. Foreign GH was present in plasma of 61% of the MT–hGH and 89% of the MT–bGH transgenic pigs. Two of 7 pigs with MT–hGRF and all 3 ALB–hGRF transgenic pigs had high concentrations of GRF in their plasma, but plasma concentrations of porcine GH (pGH) were not higher in GRF transgenic pigs than in littermate control pigs. In contrast, plasma concentrations at birth ranged from 3 to 949 ng hGH/ml for MT–hGH transgenic pigs and 5 to 944 ng bGH/ml for MT–bGH transgenic pigs. Presence of the foreign GH depressed endogenous pGH to non-detectable levels. In MT–bGH transgenic pigs, plasma IGF-I was elevated more than 2-fold, plasma glucose was elevated about 30 mg/dl, and plasma insulin was 20-fold higher than in littermate or sibling control pigs. Two lines of pigs expressing the MT–bGH transgene gained 11·1% and 13·7% faster, and were 18% more efficient in converting feed to body weight gain than were sibling control pigs. Expression of the MT–bGH transgene caused a marked repartitioning of nutrients from subcutaneous fat into other carcass components, including muscle, skin, bone and certain organs. The persistent excess hGH or bGH in transgenic pigs was detrimental to general health; lameness, lethargy and gastric ulcers were the most prevalent problems. Gilts that expressed the hGH or bGH transgenes were anoestrous. Germ-line transmission was obtained in 4 of 5 expressing transgenic boars and 4 of 5 non-expressing transgenic boars and gilts. From 2% to 73% of progeny inherited a transgene from founder transgenics. All transgenic progeny of MT–hGH, MT–bGH and MT–hGRF founder males expressed the transgene if their sire also expressed the gene. The concentration of bGH or hGH in plasma of transgenic progeny was similar to the concentration present in the founder transgenic.

Keywords: pig; transgenic; growth; gene transfer; gene expression

Introduction

The dramatic achievements in molecular biology during the past decade and the development of micromanipulation for early-stage embryos provided the combined capabilities for introducing

§Present address: Howard Hughes Medical Institute Laboratory and Department of Cell Biology and Anatomy, University of Texas Southwestern Medical Center, Dallas, TX 75235, USA.

cloned genes into the mouse genome in 1980 (see review by Brinster & Palmiter, 1986; Palmiter & Brinster, 1986). The transfer of genes was immediately recognized as an important scientific achievement, but the subsequent creation of the 'super' mouse by the transfer of a rat growth hormone (rGH) gene provided the convincing evidence that demonstrated the potential offered by gene transfer (Palmiter *et al.*, 1982).

The general strategy used by Palmiter *et al.* (1982) with the mouse, which we subsequently applied to pigs (Hammer *et al.*, 1985c; Pursel *et al.*, 1987, 1989b), was to insert foreign growth-regulating genes under control of a heterologous promoter to direct the production of peptides in ectopic tissues and evade the usual regulatory system for these peptides.

The purpose of this report is to review our recent progress on the gene integration rate, the frequency, level, and consequences of transgene expression, and the subsequent germ-line transmission of several growth-regulating genes in pigs.

Transfer of fusion genes

Fusion genes

Growth-related fusion genes that we introduced into the genome of pigs are shown in Fig. 1. A mouse metallothionein (MT) promoter was used to direct each of the four structural genes. In addition, a mouse albumin (ALB) promoter was used with human growth hormone-releasing factor (hGRF). The human and bovine growth hormone (hGH and bGH) structural genes were cloned from genomic DNA libraries and contained intron sequences (Hammer *et al.*, 1985a). The hGRF structural gene was a fusion of genomic and cDNA sequences and contained only one intron (Hammer *et al.*, 1985b). The human insulin-like growth factor-I (hIGF-I) structural gene was composed of a cDNA sequence and a signal peptide from a rat somatostatin gene.

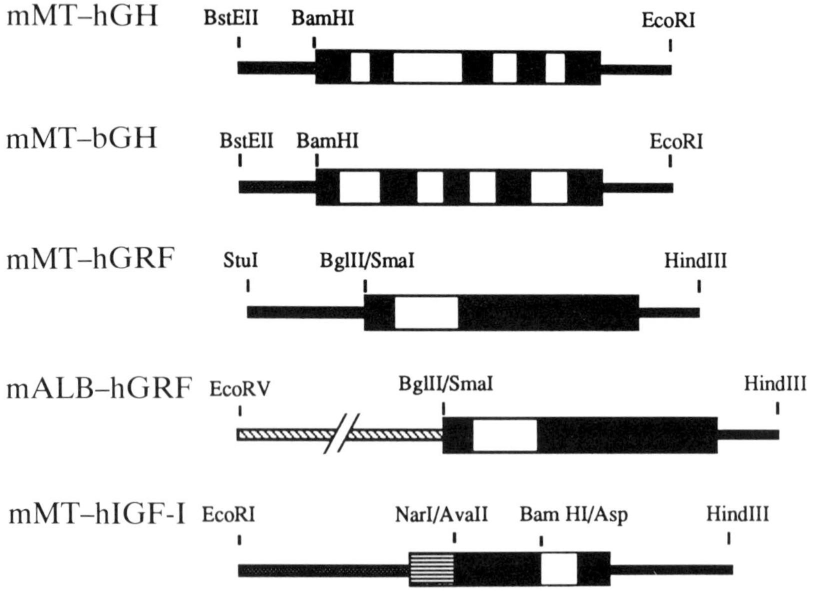

Fig. 1. Diagram of fusion genes that were excised from plasmids and microinjected into pig ova. Selected restriction sites for each construct are shown. The promoter and flanking sequences are a stippled box, exons are solid, and introns are open boxes. The horizontal-lined box in mMT–hIGF-I represents a signal peptide from a rat somatostatin gene.

Microinjection

The only technique that has so far produced transgenic pigs is the microinjection of genes into a pronucleus of 1-cell ova or nuclei of 2-cell ova. Our initial attempts to microinject genes into the pronuclei of pig ova were impeded because the opacity of the cytoplasm of pig ova precluded visualization of pronuclei. However, the lipid granules in the cytoplasm can be displaced to one end of pig ova by centrifugation at 15 000 ***g*** for 5 min, leaving pronuclei in the equatorial segment of the cytoplasm (Wall *et al.*, 1985). The pronuclei are then visible for microinjection with the aid of interference-contrast microscopy.

Synchrony of donors and recipients

Oestrous cycles of all donor and many recipient gilts or sows were hormonally synchronized by treatment with altrenogest or methallibure. Donor gilts also were injected with pregnant mares' serum gonadotrophin (PMSG) to induce superovulation and subsequently were injected with human chorionic gonadotrophin (hCG) 78–82 h after the PMSG injection to induce ovulation. Hormonally synchronized recipients received only the hCG injection.

The synchrony of the oestrous cycles of the donor and recipient females influenced the farrowing rate obtained after the transfer of microinjected ova. During one experiment, ova were either transferred into recipients that received hCG synchronous with donor gilts or transferred into recipients that received hCG 11–17 h after the donors (delayed). The farrowing rate of recipient gilts with delayed ovulation was significantly higher than the farrowing rate of recipients ovulating in synchrony with the donor gilts (Table 1).

Table 1. Pregnancy rate and litter size of recipient gilts ovulating in synchrony with or 11–17 h after donors of microinjected ova

Ovulation in recipient	No. of recipients	Pregnant		Pigs or fetuses	
		No.	%	Mean	s.e.m.
Synchronous	21	9	43*	8·6	0·83
Delayed	25	18	72*	8·2	0·59

$^{*}P < 0{\cdot}05$ ($\chi^2 = 3{\cdot}99$).

Whenever possible thereafter, we attempted to use recipient gilts that ovulated 11–17 h after the donors. However, during the course of our studies, if insufficient numbers of gilts with delayed ovulation were available, or if recipients that were hormonally treated did not exhibit oestrus or did not ovulate, then naturally cyclic gilts or sows were used as recipients if they had begun oestrus on the morning after the donors received an injection of hCG and were still in oestrus on the following morning. The results that are summarized in Table 2 show that the farrowing rate of the naturally cyclic gilts and sows was similar to that of synchronous recipients (Table 1) and was significantly lower than for the recipients with delayed ovulation. Possibly, these results are influenced by the 8–14-h period of in-vitro incubation to which our microinjected ova were subjected to accommodate their transport from Beltsville after recovery to Philadelphia for microinjection and subsequent return to Beltsville for transfer into recipients.

Gene integration

Integration efficiency

The overall efficiency of integrating growth-related transgenes into the pig genome is quite low. Only 8·3% of all transferred microinjected ova were represented by offspring at birth (Table 3).

Table 2. Pregnancy rate and litter size of recipient gilts ovulating naturally or 11–17 h after donors of microinjected ova

Ovulation in recipient	No. of recipients	Pregnant		Pigs or fetuses	
		No.	%	Mean	s.e.m.
Delayed	138	93	67*	6·9	0·30
Natural†	100	43	43*	6·5	0·44

*$P < 0{\cdot}001$ ($\chi^2 = 14{\cdot}09$).
†In oestrus on morning after donors received injection of hCG.

Integration efficiency varied from 0·31% to 1·03% when expressed as the percentage of injected ova that resulted in transgenic pigs (Table 3). Our efficiencies of gene integration are similar to those reported by Brem *et al.* (1985, 1988) and Ebert *et al.* (1988) but are slightly lower than efficiencies of 1·42% reported by Vize *et al.* (1988) and 1·73% reported by Polge *et al.* (1989). In contrast, transfer of the same fusion genes into mice resulted in an efficiency averaging in excess of 3% (Brinster *et al.*, 1985).

Table 3. Efficiency of transferring growth-regulating genes into pigs (after Pursel *et al.*, 1989a, b)

Fusion gene	No. of ova injected	No. of recipient gilts	Gilts pregnant		Offspring		Transgenic		Expressing transgene	
			No.	%	No.	%	No.	%	No.	%
MT–hGH	2035	64	37	58	192	9·4*	20	0·98†	11/18	61‡
MT–bGH	2330	49	24	49	150	6·4	9	0·41	8	89
MT–hGRF	2236	66	35	53	177	7·9	7	0·31	2	29
ALB–hGRF	968	32	20	62	108	11·2	5	0·52	3/3	100
MT–hIGF–I	387	13	5	38	34	8·8	4	1·03	1/2	50

*Percentage of injected ova resulting in offspring.
†Percentage of injected ova resulting in a pig with gene integration.
‡Percentage of transgenic pigs expressing the fusion gene.

The number of transgene copies integrated per cell in MT–hGH transgenic pigs varied from 1 to 490 (Hammer *et al.*, 1985c), <1 to 28 for MT–bGH (Miller *et al.*, 1989), <1 to 15 for MT–pGH (Vize *et al.*, 1988), and <1 to >10 for prolactin–bGH (Polge *et al.*, 1989) with most integrations probably occurring at a single locus. Southern blot analysis showed that transgenic pigs with MT–hGH contained intact copies of the fusion gene with many orientated in tandem head-to-tail arrays and some in a head-to-head configuration (Hammer *et al.*, 1985c). Only one of four MT–pGH transgenic pigs studied contained the transgene organized in the head-to-tail array (Vize *et al.*, 1988).

Expression of integrated transgenes

Incidence of expression

Integrated genes were expressed in 11 of 18 (61%) MT–hGH transgenic pigs and 8 of 9 (89%) MT–bGH transgenic pigs (Table 3). Only 2 of 7 (29%) MT–hGRF pigs expressed the transgene, which was similar to the incidence of expression of 1 of 7 (14%) transgenic lambs with the same gene (Rexroad *et al.*, 1989). In contrast, 11 of 14 (79%) transgenic mice expressed MT–hGRF (Hammer *et al.*, 1985b). Reasons for this discrepancy in incidence of expression among species are unknown.

Failure to express the transgene may be the result of integration in an inactive chromosomal locus or alteration of gene sequences during the integration process. In some cases, the presence of introns in the structural region may greatly influence whether the gene is expressed. Brinster *et al.* (1988) found in mice that genes without introns were expressed with a lower frequency and at a lower level than the same genes with introns.

Site of transgene expression

Amounts of bGH mRNA in several organs of two lines of MT–bGH transgenic pigs were determined by solution hybridization (Table 4). The MT promoter directed expression of fusion genes to the expected tissues in transgenic pigs. However, the concentrations of GH mRNA in pig tissues were considerably lower than in tissues of transgenic mice harbouring the same fusion gene (Hammer *et al.*, 1985a). While the quantity of bGH mRNA varied in both lines, pigs in the 31-04 line had high levels of bGH mRNA in liver, kidney, adrenal and pancreas, which was consistent with the high concentration of bGH found in their plasma (Table 4). In contrast, founder transgenic pig 37-06 and his transgenic descendants had very low but detectable levels of bGH mRNA in 1–4 of the organs and low concentrations of bGH were present in their plasma.

Table 4. Plasma bGH and bGH mRNA values for several organs of two lines of pigs expressing a MT–bGH transgene

		Plasma	bGH mRNA* (molecules/cell)						
Pig line	Generation	bGH (ng/ml)	Liver	Kidney	Heart	Gonad	Lung	Adrenal	Pancreas
Line 37-06									
37-06	G0	45	0	150	360	50	0	0	0
67-06	G1	111	45	0	0	0	0	14	45
67-07	G1	70	31	0	0	55	0	13	285
136-01	G2	183	56	0	0	NA†	70	0	65
136-02	G2	38	43	23	0	NA	0	14	75
136-03	G2	29	0	0	0	NA	12	0	55
137-02	G2	39	0	0	0	21	20	5	12
140-02	G2	53	0	0	0	0	14	0	0
Line 31-04									
115-14	G1	1089	1720	2320	18	500	0	975	70
182-07	G2	822	690	110	NA	203	140	965	510
185—03	G2	2138	870	1170	15	300	57	2250	2150

*Total nucleic acids were extracted from homogenized tissues and mRNA was measured by solution hybridization using a ^{32}P-labelled oligonucleotide. Molecules per cell were calculated by assuming 6·4 pg DNA per cell. 0 = less than twice background.

†NA = not assayed.

Level of gene expression

The level of gene expression varied greatly among transgenic pigs that had integrated the same fusion gene. Plasma concentrations at birth ranged from 3 to 949 ng hGH/ml and 5 to 944 ng bGH/ml in MT–hGH and MT–bGH transgenic pigs, respectively (Hammer *et al.*, 1985c; Miller *et al.*, 1989). Plasma concentrations of hGH were unrelated to the number of gene copies per cell in MT–hGH transgenic pigs (Hammer *et al.*, 1985c). However, Miller *et al.* (1989) found that, in transgenic pigs with MT–bGH, the concentration of bGH in plasma both at birth or at 30–180 days of age was positively correlated ($P < 0.05$) with the number of gene copies per cell. This relationship is evident in Fig. 2.

The concentration of GRF in plasma of transgenic pigs with MT–hGRF or Alb–hGRF was 130–380 pg/ml and 400–8000 pg/ml, respectively (Pursel *et al.*, 1989a, b). These values are 10- to 500-fold higher than concentrations of GRF in plasma of littermate control pigs. However, most of the assayable GRF in plasma of the MT–hGRF pigs was 3–44 metabolite rather than the native peptide (1–44), which may explain why the concentration of pGH in plasma was not elevated in the transgenic compared to the littermate controls (Pursel *et al.*, 1989a).

In transgenic mice harbouring fusion genes with the MT promoter, concentrations of hGH and bGH in plasma were frequently elevated more than 10-fold after zinc was added to their drinking water (Palmiter *et al.*, 1983; Hammer *et al.*, 1985a). In contrast, addition of 1000–3000 p.p.m. zinc to the feed of transgenic pigs resulted in little more than a doubling of the bGH concentration in plasma (Pursel *et al.*, 1990). Several pigs with a prolactin (PRL)–bGH transgene released bGH in an episodic fashion after pigs were injected with thyrotrophin-releasing hormone (TSH) or after infusion of a dopamine antagonist, sulpiride (Polge *et al.*, 1989). Both studies with pigs demonstrated that the promoter sequences of fusion genes can respond to induction.

Physiological consequences of transgene expression

The rapid rate of growth and large body size of mice that expressed GH transgenes (Palmiter *et al.*, 1982, 1983) created the expectation that the rate of growth might be greatly enhanced if the same GH transgenes were transferred into swine. The founder population of MT–hGH and MT–bGH transgenic pigs did not meet this expectation even though hGH and bGH were biologically active (Hammer *et al.*, 1986; Pursel *et al.*, 1987, 1989b). Pigs expressing the hGH transgene rarely had detectable concentrations of plasma pGH (Miller *et al.*, 1989), which indicates that the negative feedback mechanism was functioning. Furthermore, insulin-like growth factor-I (IGF-I) concentrations in hGH and bGH expressing transgenic pigs were considerably higher than in littermate control pigs (Fig. 3).

The effects of bGH gene expression on concentrations of glucose and several metabolic hormones in the plasma are summarized in Table 5. In comparison to littermate controls, pigs expressing MT–bGH had significantly elevated concentrations of glucose and insulin, significantly lower values of thyroxin and prolactin, and concentrations of cortisol and triiodothyronine were similar. These results are comparable to those reported for pigs injected daily with exogenous pGH, which had average increases in serum glucose ranging from 8% to 48% and concentrations of serum insulin that were 2- to 7-fold higher than in control pigs (Campbell *et al.*, 1988, 1989; Evock *et al.*, 1988). Additionally, Ebert *et al.* (1988) reported that a transgenic pig expressing rat GH had glucosuria and consistently had serum glucose concentrations more than 3-fold higher than control pigs.

The founder MT–hGH and MT–bGH transgenic pigs were fed a diet containing only 16% protein that may not have provided sufficient protein for them to gain weight faster than their littermate controls. Recent studies using pigs injected with exogenous pGH indicate that maximal growth rate is attained only if the diet contains adequate protein and, particularly, lysine (Goodband *et al.*, 1988; Newcomb *et al.*, 1988). Subsequently, we increased the levels of dietary protein and lysine during the 30–90-kg growth period and found that the G2 and G3 progeny of MT–bGH transgenic founder 37-06 gained weight 11·1% faster, and G2 progeny of 31-04 founder gained weight 13·7% faster than did sibling control pigs (Table 6).

In comparison to littermate or sibling control pigs, food intake was depressed 20% in MT–bGH founders (Pursel *et al.*, 1989b) and 17% in G2 progeny of MT–bGH 37-06 line of transgenic pigs fed *ad libitum* (Table 6). These results are comparable to a 14% and 17% depression in food intake reported for pigs injected with pGH (Campbell *et al.*, 1988; Evock *et al.*, 1988). The appetite depression that accompanies elevated GH in pigs is probably a major factor that inhibits pigs from

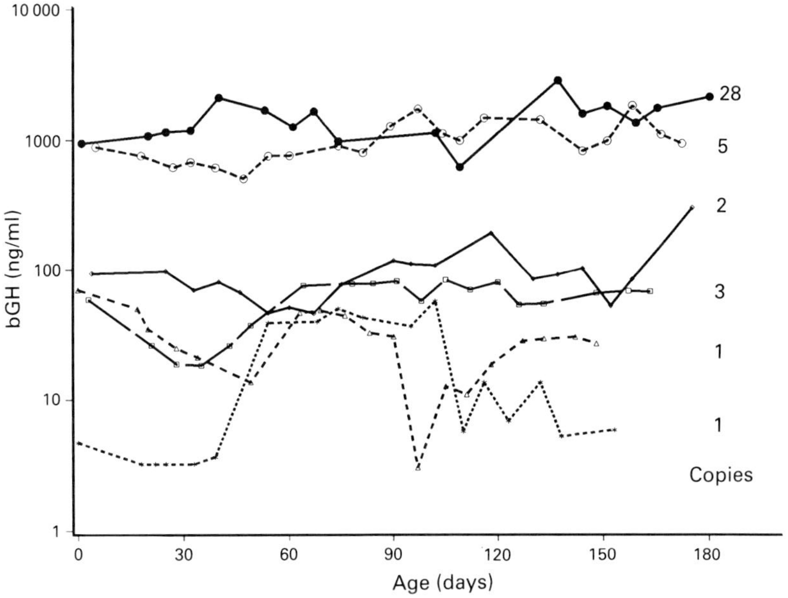

Fig. 2. Each line identifies the concentration of bovine growth hormone (bGH) in plasma for a single MT–bGH founder transgenic pig. The number of integrated gene copies per cell for each pig is indicated at the right (after Miller *et al.*, 1989).

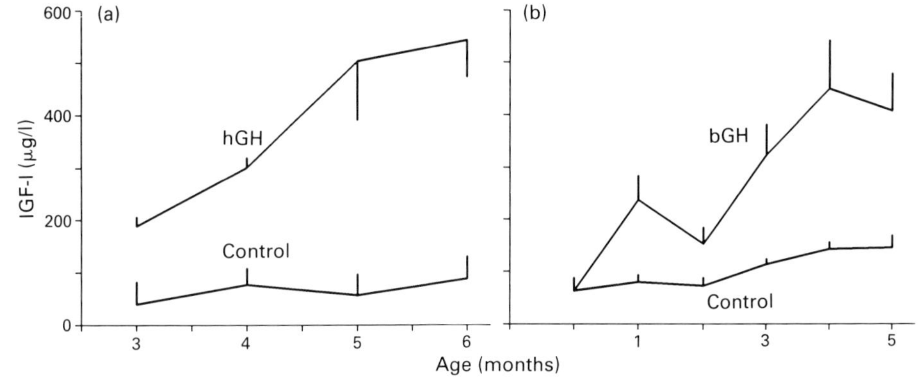

Fig. 3. Concentrations of insulin-like growth factor-I (IGF-I) in (a) plasma from control (N = 3) and MT–hGH expressing (hGH; N = 3) pigs from 3 to 6 months of age and (b) plasma from control (N = 6) and MT–bGH expressing (bGH; N = 6) pigs from 0 to 5 months of age. Values are means + s.e.m. (After Miller *et al.*, 1989.)

achieving greatly accelerated growth rates. In contrast, growth is dramatically increased in mice expressing GH transgenes or in rats with a GH-secreting tumour, which have greatly enhanced food intake (McCusker & Campion, 1986).

MT–bGH founder transgenic pigs were 16% more efficient (Pursel *et al.*, 1989b), and G2 progeny of MT–bGH 37-06 line transgenic pigs were 18% more efficient in converting food into body weight gain than were littermate or sibling controls (Table 6). These results are quite similar to improved food efficiencies of 23% (Campbell *et al.*, 1988) and 25% (Evock *et al.*, 1988) reported for pigs injected with exogenous pGH in comparison to littermate controls.

Table 5. Glucose and metabolic hormone concentrations in plasma of MT–bGH transgenic and littermate control pigs

Item	Control pigs	Transgenic pigs	Significance*
Glucose† (mg/dl)	72 ± 5 (10)	109 ± 13 (10)	$P = 0.011$
Cortisol‡ (ng/ml)	39 ± 9 (6)	37 ± 7 (8)	$P = 0.84$
Insulin† (pg/ml)	24 ± 4 (10)	480 ± 118 (10)	$P = 0.001$
Prolactin† (ng/ml)	3·9 ± 0·5 (10)	2·3 ± 0·4 (10)	$P = 0.021$
T_3‡ (ng/ml)	1·2 ± 0·3 (6)	1·3 ± 0·2 (7)	$P = 0.68$
T_4‡ (ng/ml)	49 ± 4 (6)	29 ± 3 (7)	$P = 0.003$

Values are mean ± s.e.m. for the no. of pigs indicated in parentheses. T_3 = triiodothyronine; T_4 = thyroxine.
*Comparison by least squares analysis of variance.
†Blood collected after overnight fast.
‡Blood collected from cannulated pigs.

Table 6. Average daily gain and feed efficiency of MT–bGH transgenic pigs (30–90 kg body weight) (after Pursel *et al.*, 1989b)

Line	Group*	Average daily gain (g)	Kg feed/kg gain
37-06†	Control	813 ± 17 (23)	2·99 ± 0·12 (8)
	Transgenic	903 ± 23 (13)	2·46 ± 0·16 (5)
		$P = 0.002$	$P = 0.026$
31-04‡	Control	869 ± 43 (7)	ND§
	Transgenic	988 ± 62 (7)	ND§
		$P = 0.15$	

Values are mean ± s.e.m. for the no. of animals in parentheses, and were compared by least-squares analysis of variance.
*Pigs were fed corn-soybean diet containing 18% crude protein plus 0·25% lysine.
†G2 and G3 progeny of founder 37-06.
‡G2 progeny of founder 31-04.
§Not determined because pigs were group fed.

Elevated concentrations of GH in pigs expressing MT–hGH and MT–bGH transgenes have produced marked repartitioning of nutrients away from subcutaneous fat and into other carcass components, including muscle, skin, bone and certain organs. Slaughter measurements or ultrasonic estimates of backfat thickness at the 10th rib of MT–hGH and MT–bGH transgenic pigs at ~90 kg body weight averaged 7·0 mm and 7·9 mm, respectively, while littermate control pigs averaged 18·5 and 20·5 mm, respectively ($P < 0.01$ in each case; Hammer *et al.*, 1986; Pursel *et al.*, 1989b). Additionally, backfat measurements do not adequately reflect the lack of subcutaneous fat in transgenic pigs because the skin over the 10th rib was about 1 mm thicker for transgenic pigs than for control pigs (V. G. Pursel & M. B. Solomon, unpublished data).

Pigs expressing either the MT–hGH or MT–bGH transgene exhibited several notable health problems, including lameness, susceptibility to stress, gastric ulcers, parakeratosis, lethargy, anoestrus in gilts, and lack of libido in boars (Pursel *et al.*, 1987, 1989b). Pathology in joints, characteristic of osteochondritis dissecans, was also reported in a pig expressing an rGH transgene (Ebert *et al.*, 1988) and in some pigs treated with exogenous pGH for 57 days (Evock *et al.*, 1988). In contrast, no increase in the incidence of these pathological conditions was observed in non-expressing MT–hGH or MT–bGH transgenic pigs (Pursel *et al.*, 1987) or in transgenic pigs that expressed only low levels of bGH (Polge *et al.*, 1989).

Many of the health problems observed in pigs exposed to high concentrations of GH are quite prevalent in the general pig population but at a lower incidence and with less severity. Several necropsy surveys indicate that gastric ulcers were present in 10–30% of market hogs at slaughter (O'Brien, 1986), and up to 90% of rapidly growing pigs have lesions of osteochondrosis, which leads to degenerative joint disease in certain pigs and is the major cause of lameness in swine (Reiland *et al.*, 1978; Carlson *et al.*, 1988). Additional investigation is required to determine whether these ailments would be less prevalent in pigs expressing GH transgenes if the genetic base was rigidly selected for a low incidence of these conditions.

Reproductive capacity was seriously impaired in pigs expressing either the MT–hGH or MT–bGH transgene. Gilts failed to exhibit oestrus, and their ovaries were devoid of corpora lutea or corpora albicantia when examined at necropsy (V. G. Pursel, unpublished data). Boars totally lacked libido; therefore, spermatozoa were recovered by electroejaculation or were flushed from the epididymis at necropsy to use for artificial insemination to obtain germ-line transmission of the transgene (Pursel *et al.*, 1987).

A major difference between a transgenic pig with elevated GH and a normal pig injected daily with exogenous GH is that, in the latter case, GH is elevated episodically, while in the transgenic pig, GH is elevated continuously (Miller *et al.*, 1989). We suggest that continuous exposure to elevated concentrations of GH contributes to the multiple health problems observed in our MT–hGH and MT–bGH transgenic pigs and also may prevent them from growing to their full potential. There is evidence for the latter in the rat since Robinson & Clark (1989) reported that rats infused continuously with a high concentration of GH do not attain the maximal rate of growth achieved in rats injected once daily. Use of promoters that permit expression of GH fusion genes only during the rapid growth phase or promoters that can induce the release of large episodic doses of GH may be essential to achieve only the positive aspects of elevated GH for pigs. Future research on growth regulation with transgenic pigs will most certainly be directed toward this goal, along with investigation of the potentials of IGF-I, GH receptors, and other structural genes involved in growth.

Transmission of transgenes

If economically important transgenic livestock become available, it will be essential that these animals have the ability to transmit the transgene to progeny in a predictable manner. We have tested transgenic founder pigs harbouring three different transgenes for transmission to progeny. So far, 8 of 10 transgenic pigs successfully transmitted the transgene to one or more progeny (Table 7). Many of the pigs transmitted their transgene to about 50% of the progeny as expected with simple Mendelian inheritance. The MT–hGH sow and MT–bGH transgenic boar that failed to transmit the transgene to their progeny were probably mosaic for the gene, with integration only in somatic cells (Table 7). In another MT–hGH boar, integration was evidently mosaic in the germ line since the transgene was only transmitted to 1 of 52 progeny (Pursel *et al.*, 1987). Mosaicism in the germ line was found in 25–36% of transgenic mice produced by microinjection (Wilkie *et al.*, 1986).

We obtained germ-line transmission from expressing and non-expressing transgenic pigs. All transgenic progeny sired by boars that expressed MT–hGH, MT–bGH or MT–hGRF transgenes have also expressed the transgene (Pursel *et al.*, 1987; Table 7). One of 4 MT–hGH non-expressing transgenic founders sired 17 transgenic progeny of which two expressed the transgene as was evident from concentrations of less than 10 ng hGH/ml plasma (V. G. Pursel, K. F. Miller & D. J. Bolt, unpublished data). The transgenic progeny of the other 3 non-expressing founder pigs did not express the transgene.

Table 7. Transmission of growth-regulating transgenes from transgenic founder pigs to progeny

Fusion gene	Founder	Sex	No. of gene copies	Expressing	No. of litters	Progeny No. born	Transgenic No.	Transgenic %	Expressing No.	Expressing %
MT–hGH	3-02	M	330	No	6	52	17	33	2	12
	10-04	F	23	No	1	12	5	42	0	0
	11-02	M	1	Yes	1	13	3	23	3	100
	16-03	F	3	No	3	33	0	0	0	0
	21-04	M	1	No	6	52	1	2	0	0
	25-04	F	2	No	1	8	5	63	0	0
MT–bGH	29-01	M	5	Yes	4	36	0	0	0	0
	31-04	M	28	Yes	3	19	6	32	6	100
	37-06	M	3	Yes	1	11	8	73	8	100
MT–hGRF	86-04	M	100	Yes	3	20	11	55	10/10	100

While concentrations of hGH or bGH in plasma varied widely from one founder to the next (Miller *et al.*, 1989; Fig. 2), successive generations of two lines of MT–bGH transgenic pigs maintained quite consistent levels of expression (Pursel *et al.*, 1989b). Transgenic pigs in the MT–bGH 31-04 line maintained plasma concentrations of about 1300 ng bGH/ml plasma over 3 generations while pigs in the MT–bGH 37-06 line maintained concentrations of about 85 ng bGH/ml plasma over 4 generations (Pursel *et al.*, 1989b). These findings provide evidence that fusion genes can become stably integrated into the pig genome and usually function in descendants in the same manner as they did in the founder transgenic.

We thank K. E. Mayo and L. A. Frohman for help analysing the GRF pigs; A. D. Mitchell, R. G. Campbell and M. B. Solomon for evaluating growth performance and carcass composition; and K. F. Miller, M. C. Garcia and J. P. McMurtry for analysing plasma hormones and metabolites. This work was supported in part by research grants from NIH (HD-19018) to R.L.B. and (HD-09172) to R.D.P.

References

Brem, G., Brenig, B., Goodman, H.M., Selden, R.C., Graf, F., Kruff, B., Springman, K., Hondele, J., Meyer, J., Winnaker, E-L. & Krausslich, H. (1985) Production of transgenic mice, rabbits and pigs by microinjection into pronuclei. *Zuchthygiene* **20,** 251–252.

Brem, G., Brenig, B., Muller, M., Krausslich, H. & Winnaker, E.-L. (1988) Production of transgenic pigs and possible application to pig breeding. *Occasional Publ. Br. Soc. Anim. Prod.* **12,** 15–31.

Brinster, R.L. & Palmiter, R.D. (1986) Introduction of genes into the germ line of animals. *Harvey Lect.* **80,** 1–38.

Brinster, R.L., Chen, H.Y,, Trumbauer, M.E., Yagle, M.K. & Palmiter, R.D. (1985) Factors affecting the efficiency of introducing foreign DNA into mice by microinjecting eggs. *Proc. natn. Acad. Sci. USA* **82,** 4438–4442.

Brinster, R.L., Allen, J.M., Behringer, R.R., Gelinas, R.E. & Palmiter, R.D. (1988) Introns increase transcriptional efficiency in transgenic mice. *Proc. natn. Acad. Sci. USA* **85,** 836–840.

Campbell, R.G., Steele, N.C., Caperna, T.J., McMurtry, J.P., Solomon, M.B. & Mitchell, A.D. (1988) Interrelationships between energy intake and endogenous porcine growth hormone administration on the performance, body composition, and protein and energy metabolism of growing pigs weighing 25 to 55 kilograms live weight. *J. Anim. Sci.* **66,** 1643–1655.

Campbell, R.G., Steele, N.C., Caperna, T.J., McMurtry, J.P., Solomon, M.B. & Mitchell, A.D. (1989) Interrelationships between sex and exogenous growth hormone administration on performance, body composition and protein and fat accretion of growing pigs. *J. Anim. Sci.* **67,** 177–186.

Carlson, C.S., Hilley, H.D., Meuten, D.J., Hagan, J.M. & Moser, R.L. (1988) Effect of reduced growth rate on the prevalence and severity of osteochondrosis in gilts. *Am. J. vet. Res.* **49,** 396–402.

Ebert, K.M., Low, M.J., Overstrom, E.W., Buonomo, F.C., Baile, C.A., Roberts, T.M., Lee, A., Mandel, G. & Goodman, R.H. (1988) A Moloney MLV-rat somatotropin fusion gene produces biologically active somatotropin in a transgenic pig. *Molec. Endocrinol.* **2,** 277–283.

Evock, C.M., Etherton, T.D., Chung, C.S. & Ivy, R.E. (1988) Pituitary porcine growth hormone (pGH) and a recombinant pGH analog stimulate pig growth performance in a similar manner. *J. Anim. Sci.* **66,** 1928–1941.

Goodband, R.D., Nelssen, J.L., Hines, R.H., Kropf, D.H., Thaler, R.C., Schricker, B.R. & Fitzner, G.E. (1988) The effect of porcine somatotropin (PST) and dietary lysine level on growth performance and carcass characteristics of finishing swine. *J. Anim. Sci.* **66** (Suppl. 1), 95, abstr.

Hammer, R.E., Brinster, R.L. & Palmiter, R.D. (1985a) Use of gene transfer to increase animal growth. *Cold Spring Harb. Symp. Quant. Biol.* **50,** 379–387.

Hammer, R.E., Brinster, R.L., Rosenfeld, M.G., Evans, R.M. & Mayo, K.E. (1985b) Expression of human growth hormone releasing factor in transgenic mice results in increased somatic growth. *Nature, Lond.* **315,** 413–416.

Hammer, R.E., Pursel, V.G., Rexroad, C.E., Jr, Wall, R.J., Bolt, D.J., Ebert, K.M., Palmiter, R.D. & Brinster, R.L. (1985c) Production of transgenic rabbits, sheep and pigs by microinjection. *Nature, Lond.* **315,** 680–683.

Hammer, R.E., Pursel, V.G., Rexroad, C.E., Jr, Wall, R.J., Bolt, D.J., Palmiter, R.D. & Brinster, R.L. (1986) Genetic engineering of mammalian embryos. *J. Anim. Sci.* **63,** 269–278.

McCusker, R.H. & Campion, D.R. (1986) Effect of growth hormone-secreting tumors on body composition and feed intake in young female Wistar-Furth rats. *J. Anim. Sci.* **63,** 1126–1133.

Miller, K.F., Bolt, D.J., Pursel, V.G., Hammer, R.E., Pinkert, C.A., Palmiter, R.D. & Brinster, R.L. (1989) Expression of human or bovine growth hormone gene with a mouse metallothionein I promoter in transgenic swine alters the secretion of porcine growth hormone and insulin-like growth factor-1. *J. Endocr.* **120,** 481–488.

Newcomb, M.D., Grebner, G.L., Bechtel, P.J., McKeith, F.K., Novakofski, J., McLaren, D.G. & Easter, R.A. (1988) Response of 60 to 100 kg pigs treated with porcine somatotropin to different levels of dietary crude protein. *J. Anim. Sci.* **66** (Suppl. 1), 281, abstr.

O'Brien, J.J. (1986) Gastric ulcers. In *Diseases of Swine*, 6th edn, pp. 725–737. Eds A. D. Leman, B. Straw, R. D. Glock, W. L. Mengeling, R. H. C. Scholl & E. Scholl. Iowa State University Press, Ames.

Palmiter, R.D. & Brinster, R.L. (1986) Germ line transformation of mice. *Ann. Rev. Genet.* **20,** 465–499.

Palmiter, R.D., Brinster, R.L., Hammer, R.E., Trumbauer, M.E., Rosenfeld, M.G., Birnberg, N.C. & Evans, R.M. (1982) Dramatic growth of mice that develop from eggs microinjected with metallothionein-growth hormone fusion genes. *Nature, Lond.* **300,** 611–615.

Palmiter, R.D., Norstedt, G., Gelinas, R.E., Hammer, R.E. & Brinster, R.L. (1983) Metallothionein-human GH fusion genes stimulate growth of mice. *Science, NY* **222,** 809–814.

Polge, E.J.C., Barton, S.C., Surani, M.A.H., Miller, J.R., Wagner, T., Elsome, K., Davis, A.J., Goode, J.A., Foxcroft, G.R. & Heap, R.B. (1989) Induced expression of a bovine growth hormone construct in transgenic pigs. In *Biotechnology of Growth Regulation*, pp. 189–199. Eds R. B. Heap, C. G. Prosser & G. E. Lamming. Butterworth Scientific, London.

Pursel, V.G., Rexroad, C.E., Jr, Bolt, D.J., Miller, K.F., Wall, R.J., Hammer, R.E., Pinkert, C.A., Palmiter, R.D. & Brinster, R.L. (1987) Progress on gene transfer in farm animals. *Vet. Immunol. Immunopathol.* **17,** 303–312.

Pursel, V.G., Miller, K.F., Bolt, D.J., Pinkert, C.A., Hammer, R.E., Palmiter, R.D. & Brinster, R.L. (1989a) Insertion of growth hormone genes into pig embryos. In *Biotechnology of Growth Regulation*, pp. 181–188. Eds R. B. Heap, C. G. Prosser & G. E. Lamming. Butterworth Scientific, London.

Pursel, V.G., Pinkert, C.A., Miller, K.F., Bolt, D.J., Campbell, R.G., Palmiter, R.D., Brinster, R.L. & Hammer, R.E. (1989b) Genetic engineering of livestock. *Science, NY* **244,** 1281–1288.

Pursel, V.G., Bolt, D.J., Miller, K.F., Pinkert, C.A., Hammer, R.E., Palmiter, R.D. & Brinster, R.L. (1990) Expression and performance in transgenic pigs. *J. Reprod. Fert., Suppl.* **40,** 235–245.

Reiland, S., Ordell, N., Lundeheim, N. & Olsson, S. (1978) Heredity of osteochondrosis, body constitution and leg weakness in the pig. *Acta radiol., Suppl.* **358,** 123–137.

Rexroad, C.E., Jr, Hammer, R.E., Bolt, D.J., Mayo, K.E., Frohman, L.A., Palmiter, R.D. & Brinster, R.L. (1989) Production of transgenic sheep with growth regulating genes. *Molec. Reprod. Dev.* **1,** 164–169.

Robinson, I.C.A.F. & Clark, R.G. (1989) Growth promotion by growth hormone and insulin-like growth factor-I in the rat. In *Biotechnology of Growth Regulation*, pp. 129–140. Eds R. B. Heap, C. G. Prosser & G. E. Lamming. Butterworth Scientific, London.

Vize, P.D., Michalska, A.E., Ashman, R., Lloyd, B., Stone, B.A., Quinn, P., Wells, J.R.E. & Seamark, R.F. (1988) Introduction of a porcine growth hormone fusion gene into transgenic pigs promotes growth. *J. Cell Sci.* **90,** 295–300.

Wall, R.J., Pursel, V.G., Hammer, R.E. & Brinster, R.L. (1985) Development of porcine ova that were centrifuged to permit visualization of pronuclei and nuclei. *Biol. Reprod.* **32,** 645–651.

Wilkie, T.M., Brinster, R.L. & Palmiter, R.D. (1986) Germline and somatic mosaicism in transgenic mice. *Devl Biol.* **118,** 9–18.

J. Reprod. Fert., Suppl. **41** (1990), 89–96
Printed in Great Britain

Production of transgenic pigs harbouring a rat phosphoenolpyruvate carboxykinase–bovine growth hormone fusion gene‡

M. Wieghart*¶, J. L. Hoover*, M. M. McGrane†, R. W. Hanson†, F. M. Rottman‡, S. H. Holtzman*, T. E. Wagner§ and C. A. Pinkert*§

**DNX, Animal Biology Research Center, Athens, Ohio 45701, USA; Departments of †Biochemistry & ‡Molecular Biology, Case Western Reserve University, School of Medicine, Cleveland, Ohio 44106, USA; and §Edison Animal Biotechnology Center, Ohio University, Athens, Ohio 45701, USA*

Summary. Over the past 5 years, reports detailing the production of transgenic pigs have focussed on enhanced growth performance. Phenotypic side-effects observed in pigs harbouring chimaeric constructs containing metallothionein or Moloney murine leukaemia virus transcriptional activators fused to growth hormone (GH) structural genes have been attributed to chronic overexpression of GH. In an effort to regulate a transgene product more effectively, a liver specific 460 bp 5′ flanking sequence of the rat phosphoenolpyruvate carboxykinase (PEPCK) gene was ligated to a BamHI site of the first exon of the genomic bovine GH (bGH) structural gene. Following micro-injection of the PEPCK/bGH construct into 1- and 2-cell pig zygotes, 124 offspring were produced of which 7 pigs were determined to be transgenic by dot-blot and Southern analysis. The PEPCK gene expression, in terms of tissue and developmental specificity, appears similar to that observed in PEPCK/bGH transgenic mice. Germ-line transmission was identified in 1 of 3 mated founders. Dramatic influences on backfat thickness were observed including a 41% reduction in backfat depth when compared to non-transgenic sex-matched littermate control pigs. Both the regulation and characterization of gene expression in PEPCK/bGH transgenic pigs are under investigation.

Keywords: gene transfer; transgenic; pigs; phosphoenolpyruvate carboxykinase; growth hormone

Introduction

Between 1980 and 1981, the first genetically engineered transgenic mammals were produced that expressed a transgene product (reviewed by Palmiter & Brinster, 1986). The gene transfer into mice was primarily accomplished by microinjecting DNA into the nuclei of 1- or 2-cell zygotes, then transferring the zygotes into recipient animals. This technique has tremendous potential for advancing research into many fields very rapidly; for example, we can now construct physiologically appropriate laboratory animal models for the study of human diseases and disorders, and we are now better able to evaluate and address embryological and developmental events which heretofore were poorly understood.

Gene transfer also holds great promise for the genetic improvement of farm animals. Although various groups have produced transgenic pigs, sheep and cattle (reviewed by Pinkert, 1987; Pinkert *et al.*, 1989b; Pursel *et al.*, 1989), the progress with livestock has not been as rapid as with mouse

¶Present address: University of Wisconsin, Dept. of Animal and Food Sciences, River Falls, WI 54022, USA.
‡Corresponding author: Dr C. A. Pinkert, DNX, Animal Biology Research Center, Athens, OH 45701, USA.

experiments for several reasons: (1) few groups have access to the facilities needed to house large animals, (2) generation intervals for livestock are substantially longer than that of mice, and (3) initial work with transgenic livestock containing growth hormone fusion genes yielded, in addition to the desired changes in production characteristics (e.g. leaner animals; Pursel *et al.*, 1989), animals with deleterious side effects.

The mouse metallothionein-1 (mMT-1) promoter was used as the regulatory sequence for a human growth hormone (hGH) structural gene in initial efforts to produce transgenic livestock (Brem *et al.*, 1985; Hammer *et al.*, 1985). The transgenic pigs carrying mMT-1/growth hormone fusion genes exhibited a variety of health problems ranging from lack of libido in boars and anoestrus in gilts to arthritis and greater stress susceptibility. In humans, the secretion of higher than normal levels of GH on a continuing basis leads to gigantism in juveniles and acromegaly in adults. Furthermore, many of the symptoms noted in transgenic pigs can be observed in humans exhibiting gigantism or acromegaly (Machlin, 1972). Injection of high levels of GH (≥ 220 µg/kg body weight/day) may be lethal to pigs, producing liver and kidney degeneration, oedema or severe stomach haemorrhages within 30 days of onset of hormone administration (Machlin, 1972). Injection of more moderate doses of GH (≥ 70 µg/kg body weight) for longer than 60 days led to impaired mobility of pigs, apparently due to osteochondritis (Evock *et al.*, 1988). It would therefore appear that prolonged exposure to pharmacological levels of GH is detrimental, if not lethal, in pigs.

We postulated that transgenic pigs carrying a GH structural gene under the regulation of a tissue-specific and tightly controlled or regulatable promoter would be physiologically more representative of a high yielding production pig than transgenic pigs previously produced. The gene for phosphoenolpyruvate carboxykinase (PEPCK) is expressed primarily in liver and the level of PEPCK mRNA can be regulated by cAMP administration or alteration in dietary carbohydrate and protein concentrations (McGrane *et al.*, 1988). We report here the transfer into pig zygotes of a PEPCK/bGH fusion gene which contains the PEPCK promoter–regulatory domain directing expression of the genomic bGH structural gene.

Materials and Methods

DNA for microinjection. The BamHI–EcoRI fragment from pPCbGH was prepared for microinjection as detailed by McGrane *et al.* (1988). A final concentration of 4 ng/pPCbGH/ml in Tris-EDTA buffer was used for microinjection of 1- and 2-cell pig eggs (Brinster *et al.*, 1985).

Zygote collection, microinjection and transfer. Commercial cross-bred sows and boars were used; parent stock came from Landrace, Yorkshire, Duroc and Hampshire breeds. Approximately 24 h after previous litters were weaned, sows used as zygote donors were induced to ovulate with 400 i.u. PMSG (i.m.) (Ayerst Laboratories, Montreal, Canada) and 200 i.u. hCG (i.m.) (Sigma) or allowed to ovulate naturally. Thereafter, boars were used to check for oestrus daily at 10:00 h and 16:00 h. Sows were artificially inseminated with 120 ml of fresh, extended semen 24 and 30 h after the onset of oestrus. A mid-ventral laparotomy was performed on these donor animals; zygotes were flushed from the oviducts with warm (37°C) modified BMOC-3 medium containing Hepes (Pinkert, 1987).

Once eggs were recovered from the oviduct, the zonae pellucidae were examined for the presence of spermatozoa. One- and two-cell zygotes were centrifuged at 10 000 ***g*** to facilitate visualization of pronuclear and nuclear structures. The zygotes were placed in cover-slip chambers in microdrops of modified BMOC-3 covered in silicone oil. Microinjection of approximately 10 pl pPCbGH-containing solution followed the procedure used by Wagner *et al.* (1981) in mouse experiments. Recipient sows were synchronized in oestrus with the donor sows, but were not inseminated. A mid-ventral laparotomy was performed on the recipients and at least 30 zygotes were inserted about 10 cm into one oviduct of those animals identified by ovarian morphology as having ovulated.

Nucleic acid analysis. Samples of DNA extracted from tail biopsies were screened by dot-blot analysis (McGrane *et al.*, 1988). Samples testing positive for the presence of the fusion gene were then confirmed by Southern analysis (Southern, 1975). Briefly, 20 µg Pst1-digested DNA were separated by electrophoresis and transferred to a gene screen plus nitrocellulose membrane filter. The filter was hybridized with the BamHI–BglII fragment from −547 to +73 of the PEPCK gene, which was 'random prime' labelled (Bio Rad, Richmond, CA, USA) to a specific activity of $1\cdot0 \times 10^9$/µg DNA. Endogenous PEPCK in mouse liver produces a band of 2·3 kb. A band at 1·9 kb is predicted after digestion with Pst1 and hybridization to the BamHI–BglII fragment.

Total RNA was extracted from tissues essentially as described by Chirgwin *et al.* (1979) with minor modifications (Lamers *et al.*, 1982). RNA from pig tissue was extracted and Northern analysis performed as described by McGrane *et al.* (1988). The RNA was hybridized to a 'random primed' bGH cDNA probe. S1 nuclease analysis was conducted by the method of Berk & Sharp (1977).

Results

Production of transgenic animals

Transgenic pigs were produced by microinjection of the PEPCK/bGH transgene into fertilized 1-cell and 2-cell eggs. Transgenic animals have been produced using other methods; however, transgenic pigs harbouring transcriptionally active transgenes have been produced only by microinjection techniques (Brinster & Palmiter, 1986; Pinkert, 1987).

A total of 1057 zygotes were microinjected with the PEPCK/bGH fusion gene and 124 piglets were subsequently born: 7 of the animals were identified by dot- and Southern-blot analysis as having incorporated from < 1 to 100 copies of the PEPCK/bGH fusion gene per cell (Table 1). In addition, 5 of these 7 founder animals expressed the bGH gene product.

Table 1. Production of PEPCK/bGH transgenic founder pigs*

	Transgene						
		mRNA expression*					
Pig (sex)	Integration (copies/cell)	L	I	P	K	Germ-line transmission	Life-span†
11 (♂)	2	+	–	–	–	4/9	12 months
27 (♂)	< 1	–	–	nd	–	0/13	17 months
44 (♀)	100	+ + +	–	nd	–	nd	1 week
81 (♀)	60	+ + +	+	+	+	nd	9 months
89 (♂)	10	+ + +	–	nd	–	nd	10 months
95 (♂)	1	+ +	–	nd	–	nd	10 months
100 (♂)	2	–	–	–	–	0/50	11 months

L, liver; I, intestine; P, pituitary; K, kidney; nd, not determined.
*Qualitative designation from no detectable mRNA expression (–) to high expression (+ + +).
†Pig 44 was crushed by the sow during lactation; pigs 27 and 100 were killed for tissue biopsies at 17 and 11 months of age, respectively.

DNA and RNA analyses

Southern and Northern analyses from tissues obtained from 2 representative transgenic pigs (No. 11, male, and No. 81, female) are shown in Figs 1 and 2. Southern analysis of genomic DNA from liver, kidney, spleen, intestine and adipose tissue indicated that the transgene was present in all tissues. Additionally, the DNA blot from Pig 81 indicated integration of a much higher copy number than in Pig 11. The bands at 1·9 and 3·1 kb in the DNA blot for Pig 81 demonstrate that some of the transgene copies integrated in head-to-tail and head-to-head arrays. Northern analysis was performed on RNA from liver, kidney, heart, adipose, spleen, lymph node, brain, lung, intestine, muscle, pancreas, thymus, pituitary, ovary or testis, and mammary tissues. Messenger RNA for the transgene was detected only in liver of 4 founder pigs and in the liver, small intestine, and pituitary of Pig 81 (Table 1, Fig. 2). Densitometric scans of Northern blots from Pig 11 and 3 transgenic daughter progeny (Pigs 11-3, 11-4 and 11-5) indicated that the mRNA levels in liver were 2·2-fold higher in the first generation progeny (data not shown). Founder Pig 81 had a 5·7-fold higher mRNA level in liver when compared to founder Pig 11. The mRNA levels in small intestine and pituitary of Pig 81 were 18% and 9% of liver mRNA levels, respectively.

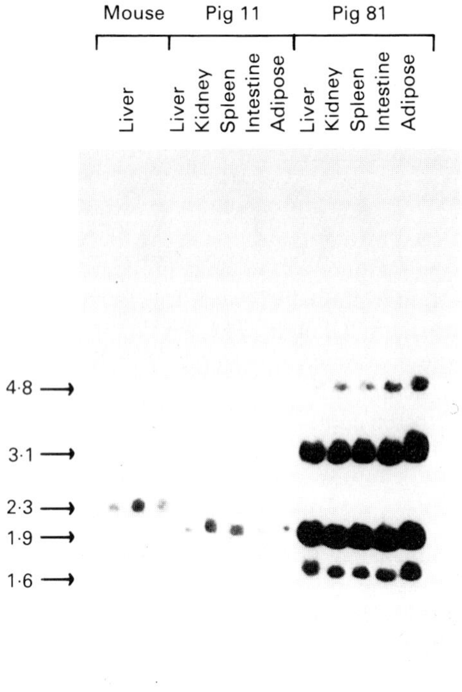

Fig. 1. Southern analysis of genomic DNA from tissues of Pigs 11 and 81. Endogenous PEPCK in mouse liver produces a band of 2·3 kb (lanes 1–3). A band at 1·9 kb was observed in tissues from Pig 11 (lanes 4–8). For Pig 81 (lanes 9–13) bands at 1·9 and 3·1 are consistent with head-to-head and head-to-tail integration arrays, with 1·7 and 4·8 kb junction-fragment bands also present.

An S1 nuclease protection analysis was performed on tissues from a young transgenic pig (Pig 44, female) and those of a transgenic mouse carrying the same PEPCK/bGH fusion gene and expressing bGH at levels in excess of 500 ng/ml serum (Fig. 3). A 133 bp fragment of the BamHI–Pst1 probe was protected from nuclease digestion by RNA in the liver of both animals.

Transmission of the PEPCK/bGH fusion gene

Of the 5 male transgenic pigs, 3 were bred to hybrid gilts. One male (No. 11) which had detectable expression of the transgene transmitted the transgene to his offspring in Mendelian fashion (Table 1). However, two mosaic founder boars, Nos 27 and 100, did not transmit the fusion gene to any of their progeny.

Health status and production traits of transgenic pigs

The transgenic pigs which expressed bGH (except Pig 44, crushed soon after birth) had life-spans of 9–12 months. The two mosaic, expression-negative pigs (Nos 27 and 100) were killed and tissue samples taken at 17 and 11 months of age, respectively. The PEPCK/bGH transgene was

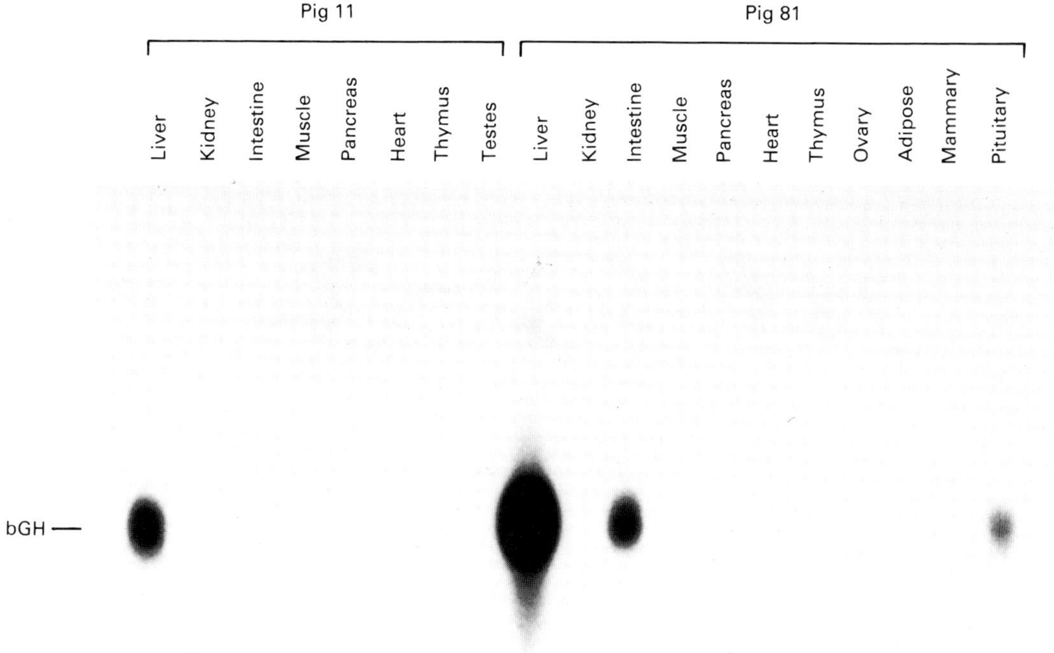

Fig. 2. Northern analysis of total mRNA from tissues of Pig 11 (lanes 1–8) yielded PEPCK/bGH transcripts in liver only. Total mRNA analysed from Pig 81 (lanes 9–19) indicated transgene message in liver, intestine and pituitary. (Tissues analysed but not represented were spleen, adipose tissue and lung for Pig 11 and spleen, lymph node, brain and lung for Pig 81. A more sensitive S1 analysis detected transgene mRNA in small intestine tissue from Pig 81.)

detected in liver tissue of Pig 27. However, of 13 tissues analysed from Pig 100, the transgene was only detected within lymphocytes.

The expression of the PEPCK/bGH transgene may have affected reproductive function in founder transgenic swine: 2 of the 3 boars (Pigs 89 and 95) exhibited poor libido and the first oestrus of Pig 81 was induced at 8 months of age. Libido in the 2 boars may have been indirectly influenced by joint pathology that was evident at sexual maturity.

A growth hormone gene has been the gene of choice for those interested in producing transgenic livestock for several reasons, one of which is its ability to enhance production characteristics, thereby yielding leaner, faster growing and more feed-efficient pigs. Backfat thickness of the transgenic pigs expressing the PEPCK/bGH transgene was measured ultrasonically at the 10th rib and last lumbar vertebra when the pigs reached market weight. The three transgenic boars (Nos 11, 89 & 95) had on average 38% less backfat ($10{\cdot}3 \pm 1{\cdot}0$ mm) than did their controls ($16{\cdot}6 \pm 2{\cdot}7$ mm) at 100 kg body weight. The one transgenic gilt (No. 81) was similarly lean (49% backfat reduction), 10·0 mm, compared with her control (19·5 mm) at 105 kg body weight.

Discussion

The efficiency of transgenic pig production can be examined at a number of levels. The purpose of analysing each stage of the process is clear—the production of transgenic pigs (or other livestock) is an extensive undertaking, in terms of animal usage, labour, facilities and expense.

The efficiencies of production of transgenic pigs at various stages in the production process have been reported (Brem *et al.*, 1985; Hammer *et al.*, 1985; Ebert *et al.*, 1988; Vize *et al.*, 1988; Pursel *et al.*, 1989). Because there is a degree of zygote lethality associated with the microinjection process, not all of the zygotes microinjected continue to develop (Pinkert, 1987). Other groups reported that 4–9·4% of the microinjected pig zygotes result in livebirths. We observed a slightly higher survivability rate with 11·7% of the injected ova resulting in livebirths. This may be due to differences in microinjection technique, sow management and care after the zygotes are transferred to recipients, or the relatively small sample size. Additionally, use of mature gilts and sows may enhance egg survivability (Pinkert *et al.*, 1989a).

Only a fraction of pigs resulting from microinjected eggs were transgenic. Other researchers have reported that 4–12% of the founder pigs resulting from zygote microinjection and transfer were transgenic. Vize *et al.* (1988) observed a much higher frequency, reporting 6 transgenic offspring from a total of 17 born; however, this appears to be an exception. Overall, our work resulted in a 6% transgenic birth rate.

There are further stages in the production process which can be examined, but the percentage of microinjected zygotes that will become expression-positive transgenic animals is critical. Our experimental efficiency is 0·47%, which is within the range and comparable to the rates obtained by other groups (Pinkert *et al.*, 1989b; Pursel *et al.*, 1989). The production of transgenic pigs is a highly

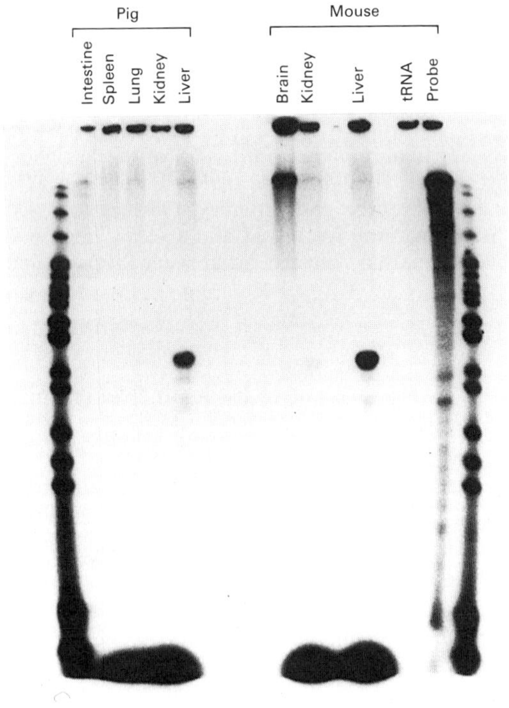

Fig. 3. An S1 nuclease protection assay for the bGH mRNA demonstrates liver specificity of the transgene product in Pig 44 (lanes 2–6) and in a transgenic PEPCK/bGH mouse (lanes 6–8).

inefficient process. Investigation into ways to enhance efficiencies is needed before production and research in this field can move rapidly.

Phosphoenolpyruvate carboxykinase catalyses the key regulatory step in gluconeogenesis and high levels of this enzyme are present in those tissues, most notably liver, with an active gluconeogenic cycle. Previous work with mice (McGrane *et al.*, 1988) had shown that the 460 bp 5′ flanking sequence of the PEPCK gene, when ligated to the genomic bovine GH structural gene, targets expression of bGH predominantly to the liver of mice at a level of expression comparable to that of endogenous PEPCK. Expression of the fusion gene in these mice was also detected in the kidney, but at a much lower level than in the liver. In transgenic PEPCK/bGH pigs, gene expression was tissue-specific and limited to the liver in all but one pig. In Pig 81, mRNA transgene expression was also detected at a much lower level in the small intestine and pituitary, and, very faintly within kidney tissue (data not shown). Overall, liver-specific mRNA expression was similar to results obtained from PEPCK/bGH transgenic mice (McGrane *et al.*, 1988). In view of work with mice and the transgene tissue expression pattern seen in Pig 81, it is possible that the aberrant expression pattern was reflective of the very high gene integration rate.

Transmission of the fusion gene to offspring was demonstrated by transgenic boar No. 11. The mRNA analysis of his 3 surviving transgenic female progeny demonstrated a higher level of gene expression than observed in boar No. 11. Correspondingly, an earlier onset phenotype, characterized by respiratory distress and joint pathology, precluded their maturation to reproductive capability. These females were either killed or died in response to stress associated with ovulation induction and laparotomy at 185–210 days of age. No ovulations were inducible, as was possible with founder female No. 81. Although few animals are represented, we suggest that the differences in chronic PEPCK-driven bGH gene overexpression between Pig 11 and his female transgenic offspring may represent a sex-limited response similar to fertility differences first observed in mMT-1/GH transgenic mice (Palmiter & Brinster, 1986).

The regulatory sequence from the PEPCK enzyme was chosen as the promoter for the bGH structural gene with the anticipation that the expression of bGH would be targeted to the liver and controllable by dietary manipulation as previously demonstrated by McGrane *et al.* (1988). However, the PEPCK promoter did provide tissue-specificity; all transgenic pigs which were expression-positive had basal plasma bGH concentrations greater than 100 ng/ml when receiving an enhanced pig production ration suitable for pigs injected with recombinantly produced pig growth hormone (18% crude protein, 1·2% total lysine).

Although expression-positive founder transgenic animals and transgenic progeny of Pig 11 exhibited enhanced production traits, including increased feed efficiency (data not shown) and decreased backfat thickness in comparison to littermate controls, as previously reported in MT/GH transgenic pigs (Pursel *et al.*, 1989), negative performance characteristics were also present. The PEPCK/bGH transgenic pigs appeared more susceptible to stress than control animals and, at 70–140 kg body weight, deleterious side-effects were initially observed. Curiously, the onset of joint pathology, stress susceptibility, and respiratory distress was observed uniformly later in development than in other trangenic pigs carrying a GH structural gene regulated by an mMT-1 or MMLV transcriptional regulatory sequence. We suggest that the substitution of the PEPCK promoter was responsible for this difference. Hence, the developmental delay in the onset of gene expression induced by the PEPCK promoter (activation at birth rather than during gestation) may be important in limiting or delaying the onset of the deleterious developmental phenotype.

We have produced transgenic bGH pigs harbouring a genomic structural gene regulated by a promoter shown to respond to dietary alterations in transgenic mice. These pigs and their offspring will provide greater insight into basic biological phenomena associated with developmental performance of pigs. However, a regulatable enhancer/promoter sequence driving structural genes is obligatory for those genes coding for qualitative traits (e.g. from the growth hormone family); one which is tightly regulated and can be precisely timed, activated or deactivated. Ultimately, such models should influence the production efficiencies associated with commercial pig production and provide a more healthy food animal product.

We thank C. Akers, J. Merriman, G. Twehues, J. Keirns, S. Choe, C. Gemma and K. Twehues for technical assistance. This work was supported by funds from DNX and The Edison Animal Biotechnology Program and from grant DK-24451 (R.W.H.) from the National Institutes of Health.

References

Berk, A.J. & Sharp, P.A. (1977) Sizing and mapping of early adenovirus mRNA's by gel electrophoresis of S1 endonuclease digested hybrids. *Cell* **12,** 721–732.

Brem, G., Brenig, B., Goodman, H.M., Selden, R.C., Graf, F., Kruff, B., Springman, K., Hondele, J., Meyer, J., Winnacker, E.L. & Krausslich, H. (1985) Production of transgenic mice, rabbits and pigs by microinjection into pronuclei. *Zuchthygiene* **20,** 251–252.

Brinster, R.L. & Palmiter, R.D. (1986) Introduction of genes into the germ line of animals. *Harvey Lect.* **80,** 1–38.

Brinster, R.L., Chen, H.Y., Trumbauer, M.E., Yagle, M.K. & Palmiter, R.D. (1985) Factors affecting the efficiency of introducing foreign DNA into mice by microinjecting eggs. *Proc. natn. Acad. Sci. USA* **82,** 4438–4442.

Chirgwin, J.M., Przybyla, A.E., MacDonald, R.J. & Rutter, W.J. (1979) Isolation of biologically active ribonucleic acid from sources enriched in ribonuclease. *Biochemistry, NY* **18,** 5294–5299.

Ebert, K.M., Low, M.J., Overstrom, E.W., Buonomo, F.C., Baile, C.A., Roberts, T.M., Lee, A., Mandel, G. & Goodman, R.H. (1988) A Moloney MLV-rat somatotropin fusion gene produces biologically active somatotropin in a transgenic pig. *Molec. Endocrinol.* **2,** 277–283.

Evock, C.M., Etherton, T.D., Chung, C.S. & Ivy, R.E. (1988) Pituitary porcine growth hormone (pGH) and a recombinant pGH analog stimulate pig growth performance in a similar manner. *J. Anim. Sci.* **66,** 1928–1941.

Hammer, R.E., Pursel, V.G., Rexroad, C.E., Jr, Wall, R.J., Bolt, D.J., Ebert, K.M., Palmiter, R.D. & Brinster, R.L. (1985) Production of transgenic rabbits, sheep and pigs by microinjection. *Nature, Lond.* **315,** 680–683.

Lamers, W.H., Hanson, R.W. & Meisner, H.M. (1982) cAMP stimulates transcription of the gene for cytosolic phosphoenolpyruvate carboxykinase in rat liver nuclei. *Proc. natn. Acad. Sci. USA* **79,** 5137–5141.

Machlin, L.J. (1972) Effect of porcine growth hormone on growth and carcass composition of the pig. *J. Anim. Sci.* **35,** 794–799.

McGrane, M.M., deVente, J., Yun, J., Bloom, J., Park, E., Wynshaw-Boris, A., Wagner, T., Rottman, F.M. & Hanson, R.W. (1988) Tissue-specific expression and dietary regulation of a chimeric phosphoenolpyruvate carboxykinase/bovine growth hormone gene. *J. biol. Chem.* **263,** 11443–11451.

Palmiter, R.D. & Brinster, R.L. (1986) Germ-line transformation of mice. *Ann. Rev. Genet.* **20,** 465–499.

Pinkert, C.A. (1987) Gene transfer and the production of transgenic livestock. *Proc. U.S. Anim. Health Assn* **91,** 129–141.

Pinkert, C.A., Kooyman, D.L., Baumgartner, A. & Keisler, D.H. (1989a) In-vitro development of zygotes from superovulated prepubertal and mature gilts. *J. Reprod. Fert.* **87,** 63–66.

Pinkert, C.A., Kooyman, D.L. & Dyer, T.J. (1989b) Enhanced growth performance in transgenic swine. In *Transgenic Technology in Medicine and Agriculture* Eds N. First & F. Hazeltine. Butterworth Scientific, London.

Pursel, V.G., Pinkert, C.A., Miller, K.F., Bolt, D.J., Campbell, R.G., Palmiter, R.D., Brinster, R.L. & Hammer, R.E. (1989) Genetic engineering of livestock. *Science, NY* **244,** 1281–1287.

Southern, E.M. (1975) Detection of specific sequences among DNA fragments separated by gel electrophoresis. *J. molec. Biol.* **98,** 503–517.

Vize, P.D., Michalska, A.E., Ashman, R., Lloyd, B., Stone, B.A., Quinn, P., Wells, J.R.E. & Seamark, R.F. (1988) Introduction of a porcine growth hormone fusion gene into transgenic pigs promotes growth. *J. Cell Sci.* **90,** 295–300.

Wagner, T.E., Hoppe, P.C., Jollick, J.D., Scholl, D.R., Hodinka, R.L. & Gault, J.B. (1981) Microinjection of a rabbit β-globin gene into zygotes and its subsequent expression in adult mice and their offspring. *Proc. natn. Acad. Sci. USA* **78,** 6376–6380.

J. Reprod. Fert., Suppl. **41** (1990), 97–104

Genetic engineering of the pseudorabies virus genome to construct live vaccines

L. E. Post, D. R. Thomsen, E. A. Petrovskis, A. L. Meyer, P. J. Berlinski and R. C. Wardley

The Upjohn Company, Kalamazoo, Michigan 49007, USA

Summary. Pseudorabies virus (PRV) is a herpesvirus of pigs. Homologous recombination with plasmids offers a method to engineer precise changes in the PRV genome to produce advantageous live vaccines. Safety can be ensured by using a non-reverting deletion to inactivate the thymidine kinase gene. One particularly important feature of new PRV vaccines is deletion of an antigen, so that vaccinated pigs are serologically distinguishable from infected pigs. We have constructed a live vaccine strain with deletions in the thymidine kinase gene and in the gene for a glycoprotein, gX. Molecular engineering techniques made it possible to choose deletion of gX, which has no known immunological significance, over deletion of other glycoproteins that contribute to protective immunity. Extensive experiments in pigs with isogenic virus pairs show that deletion of gX does not compromise efficacy of a vaccine as gI deletions do. Deletion of gX also suggests a site for replacement with antigens from other pathogens. In addition to molecular engineering of a live vaccine strain, research on PRV glycoproteins has led to the discovery that expression of the glycoprotein gp50 makes cells resistant to PRV infection. Perhaps this observation could be extrapolated to the level of a whole animal to allow engineering of pigs to become an alternative to engineered vaccines.

Keywords: Aujeszky's disease; viral glycoproteins; vaccines

Introduction

Pseudorabies virus (PRV), also known as Aujeszky's disease virus, is an alphaherpesvirus of pigs. Most strains of PRV cause minor disease in adult pigs, but cause severe and often fatal infection of the central nervous system in young piglets (reviewed by Gustafson, 1981). A feature of herpesvirus biology that has confounded attempts to control PRV is the ability of the virus to survive for long periods of time in a latent state in an asymptomatic host. Because previously infected but recovered pigs are potential sources of infection, PRV control programmes usually seek to identify and eliminate even healthy pigs that have recovered from a PRV infection. Since the only way to identify pigs that have recovered from a PRV infection is serologically, the use of vaccines has usually been considered incompatible with eradication. Since vaccines make pigs seropositive for PRV, it has traditionally been impossible to distinguish vaccinated from infected pigs.

A few years ago, the solution to this problem seemed to be production of subunit vaccines for PRV. Since it was known that PRV had multiple glycoproteins (there are now known to be at least 7), the goal was to produce one or more viral glycoproteins via recombinant DNA that would be administered as a vaccine. These would be the glycoproteins responsible for raising protective immunity. Antibodies to some viral protein not in the vaccine could then be measured to distinguish infected from vaccinated pigs. Because of research directed toward subunit vaccines, the glycoprotein genes are the most extensively characterized in the PRV genome. One of the viral glycoproteins, gp50, was shown to function as a protective subunit vaccine (Marchioli *et al.*,

1987a). During the course of this research on PRV glycoproteins, two other possible strategies for control of PRV emerged: (1) construction of a live vaccine via precise engineering of the PRV genome, using the techniques devised for herpes simplex virus (HSV) (Post & Roizman, 1981): and (2) use of the gp50 gene in a scheme to make cells resistant to PRV infection. This paper will describe how the first strategy was carried out, with the design of a vaccine to optimize safety, efficacy, and serological distinction, and present data on the second strategy to suggest that an avenue of research on transgenic animals might be to make pigs resistant to PRV.

Materials and Methods

All methods used in experiments described in this paper have been previously described by the authors (Thomsen *et al.*, 1987; Marchioli *et al.*, 1987b; Petrovskis *et al.*, 1988a).

PRV gX

When the work on PRV subunit vaccines was begun, no gene for a PRV glycoprotein had been identified. Using a scheme for mapping a transcript encoding an unidentified glycoprotein gene (Rea *et al.*, 1985), the first glycoprotein gene mapped encoded the protein that had been discovered by Ben-Porat & Kaplan (1970) to accumulate in relatively large amounts in the medium of infected cells. We called this protein gX, because of its uncertain relationship to cell-associated glycoproteins. It is now known that gX is synthesized as a higher molecular weight cell-associated glycoprotein that is released into the medium, presumably by proteolytic release of the extracellular domain(s) (Bennett *et al.*, 1986). Although the name gX has stuck with this protein, it would now be more appropriately called gG, because of its similar map location to herpes simplex virus (HSV) glycoprotein G, and its sequence homology to HSV type 2 gG.

The first evaluation of gX as a subunit vaccine candidate was made using a protein produced in *E. coli*, as a fusion protein between *E. coli trpLE*, *lacZ*, and a portion of the gX gene (protein described by Rea *et al.*, 1985). Experiments were done in mice and pigs, and in Freund's and alum adjuvants. The results were always the same: the vaccinated animals developed antibodies against gX, as measured by an ELISA assay, but no PRV-neutralizing antibodies were ever raised by the vaccine. Upon challenge with virulent PRV, the vaccinated animals showed little if any protection against the challenge.

The failure of gX produced in *E. coli* to function as an adequate vaccine left open the possibility that this material did not display proper immunogenicity when made in *E. coli*. The gX fusion protein accumulated in inclusion bodies in the *E. coli*, requiring strong denaturing conditions for solubilization and purification. Furthermore, none of the post-translational modifications known for gX (N-linked glycosylation, O-linked glycosylation, and sulphation) is added by *E. coli*. Intact gX with complete post-translational modification was produced by introducing the gene into Chinese hamster ovary cells (Bennett *et al.*, 1986). The result was the same as with the *E. coli*-produced gX: antibody response, no neutralizing antibodies, and little if any protection from challenge. As one final step to make certain that improper formulation was not responsible for the failure of gX as a vaccine, the gX gene was inserted into the thymidine kinase gene of HSV. When this recombinant HSV was used to vaccinate mice, the same result occurred: production of anti-gX antibodies, no PRV neutralizing response, and little if any protection. Since all of these experiments gave negative results, we concluded that gX was unlikely to be an adequate subunit vaccine.

Because PRV was able to infect and kill animals in the presence of a high titre of anti-gX antibodies, the question arose as to what function gX could be playing in the infection cycle. Previous work on HSV suggested a straightforward way to attempt to produce a virus with a non-functional gX gene. The scheme is illustrated in Fig. 1(a). (1) A plasmid was constructed with a thymidine kinase (tk) gene inserted to disrupt the gX coding sequence. (2) This plasmid was

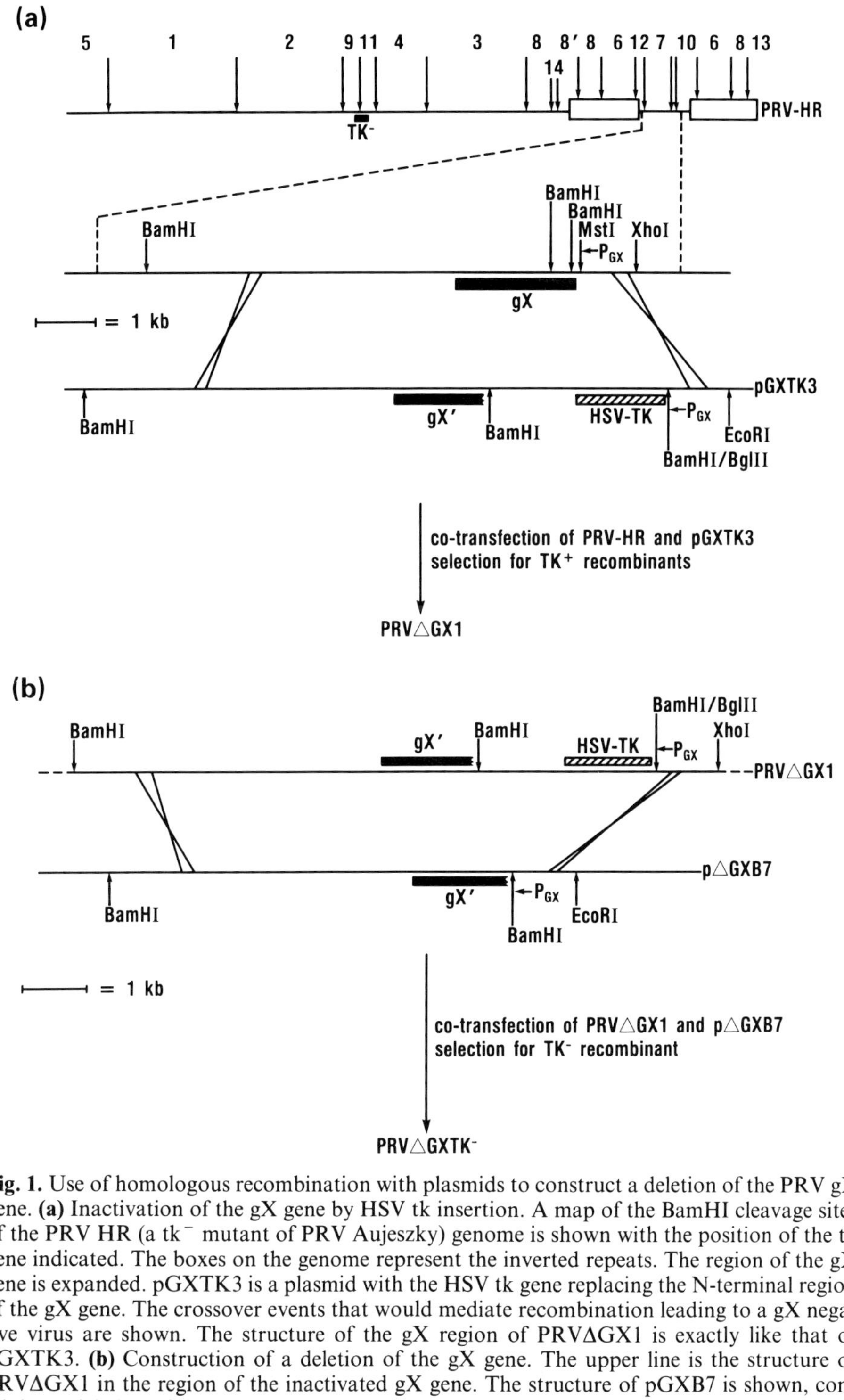

Fig. 1. Use of homologous recombination with plasmids to construct a deletion of the PRV gX gene. **(a)** Inactivation of the gX gene by HSV tk insertion. A map of the BamHI cleavage sites of the PRV HR (a tk^- mutant of PRV Aujeszky) genome is shown with the position of the tk gene indicated. The boxes on the genome represent the inverted repeats. The region of the gX gene is expanded. pGXTK3 is a plasmid with the HSV tk gene replacing the N-terminal region of the gX gene. The crossover events that would mediate recombination leading to a gX negative virus are shown. The structure of the gX region of PRVΔGX1 is exactly like that of pGXTK3. **(b)** Construction of a deletion of the gX gene. The upper line is the structure of PRVΔGX1 in the region of the inactivated gX gene. The structure of pGXB7 is shown, containing a deletion of the N-terminal amino acids of gX with no insertion of any foreign DNA. The crossover events that would mediate recombination leading to a deletion of the HSV tk gene are shown. The structure of the gX region of $\text{PRV}\Delta\text{GXTK}^-$ is exactly like that in pΔGXB7. The complete removal of the HSV sequences has been verified by Southern blots. kb, kilobase. (From Thomsen *et al.*, 1987.)

co-transfected into cells along with PRV DNA, from a tk negative mutant of PRV. After transfection, homologous recombination between the input plasmid and the virus can occur, which results in insertion of the tk gene into the gX coding region. (3) Tk positive viruses are selected by HAT medium. If viruses with a disrupted gX gene were non-viable, no tk-positive recombinant viruses containing the HSV tk gene inserted into the gX gene would be obtained. When this experiment was done, such recombinants with a disrupted gX gene, which of course produced no gX, were obtained (Thomsen *et al.*, 1987). These gX-negative viruses grew normally in cell culture, and were able to kill mice with an LD_{50} score no greater than that for gX-positive viruses.

The function of gX is still not known, because no phenotype of the gX-minus viruses has yet been identified. However, gX presumably does have a function because no naturally occurring gX-negative virus has ever been identified. What is known about gX is the following. (1) It is a very abundant, and probably the most abundant, glycoprotein synthesized during a PRV infection, and it is released into the medium of infected cells. (2) Although gX is immunogenic, the immune response to gX is not significantly protective. (3) It is non-essential for replication of PRV, either in cell culture or in animals.

A designed live vaccine

As described above, using the homologous recombination of viral DNA with plasmid DNAs, it is possible to make very precisely designed modifications in herpesvirus genomes. This raises an alternative to the subunit approach for making a serologically distinguishable PRV vaccine: a live vaccine could be constructed with a deletion of a particular immunogen gene, and be diagnostic for infected pigs based on detection of antibodies to the deleted gene product. For PRV, it is well established how to make such a live vaccine avirulent. The pioneering work of Tatarov (1968) showed that tk-minus mutants of PRV are avirulent.

To construct such a live vaccine, the choice of an immunogen for deletion focusses on the 7 glycoprotein genes of PRV. From a theoretical point of view, the choice of which glycoprotein to be deleted should be guided by the following principles: (1) The glycoprotein gene must be non-essential, and deletion should leave a virus with unimpaired ability to replicate. (2) The glycoprotein deleted should not be one that makes a significant contribution to the protective immunity, to preserve maximum efficacy. (3) The glycoprotein should be an abundant one, so that infected pigs are likely to produce a strong immune response against it that can be detected in a sensitive serological test.

These three criteria are precisely what was learned about gX after the studies exploring its use as a subunit vaccine, described above. By contrast, each of the other 6 PRV glycoproteins fails in one of these criteria, as summarized in Fig. 2, and as follows. (1) Glycoproteins gp50, gH and gII are presumably essential, and therefore cannot be deleted. Each of these glycoproteins has a counterpart in HSV which has been proved to be essential (e.g. Ligas & Johnson, 1988). We have tried to delete gp50 and gH without obtaining any viable virus (Petrovskis *et al.*, 1988b). (2) gp63 is a very minor glycoprotein. A serological assay measuring gp63 antibodies might be expected to have less than optimal sensitivity for detecting infected pigs. (3) gIII is the major target of neutralizing antibodies (Ben-Porat *et al.*, 1986) and is a target of cell-mediated immunity (Zuckerman *et al.*, 1989a). Although the precise nature of the protective immune response to PRV is not defined, it would seem desirable to have gIII in a vaccine. In addition, gIII is known to have a role in entry of virus into cells (Zuckerman *et al.*, 1989b). (4) gI is also the target of neutralizing antibodies (Mettenleiter *et al.*, 1987). The effect of gI deletion in vaccines will be examined further below.

The initial gX-negative virus described above was not suitable for use as a vaccine. First of all, the HSV tk gene inserted into the gX gene was removed, resulting in a simple deletion of the N-terminal amino acids of the gX coding sequences. Second, the original point mutation in the tk gene of the parent virus was replaced by a deletion of 276 base pairs. The latter was to ensure that the

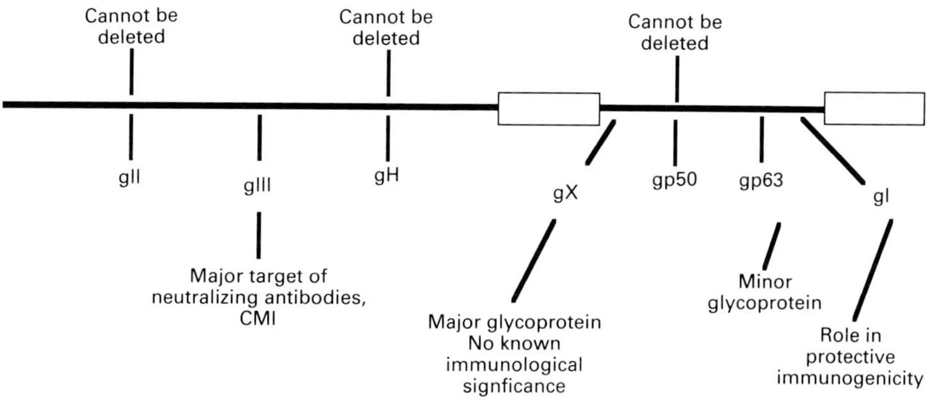

Fig. 2. Pseudorabies virus glycoproteins as potential serological markers in live vaccines. The diagram shows the long and short unique components of the PRV genome, with the short unique component flanked by boxes representing inverted repeats in the genome. The positions of the glycoproteins are not drawn to scale on the genome. See text for further discussion.

tk-negative phenotype could not revert, and therefore that the vaccine could not revert to virulent phenotype. The resulting virus strain was a wild-type PRV-strain Aujeszky with two modifications: deletion of tk for safety, and deletion of gX to provide a serological marker. This virus was shown to function as an effective vaccine (Marchioli *et al.*, 1987b) and was named Tolvid. Tolvid is now approved by USDA for marketing, and an application for approval is pending for the diagnostic kit to detect gX antibodies in infected pigs.

By a variety of criteria, Tolvid was shown to be a particularly efficacious vaccine, comparing very favourably with the existing marketed PRV vaccines (see, e.g. Wardley & Post, 1989). The conventional live PRV vaccines have been attenuated by a long series of passages in cell culture, and are known to have accumulated multiple mutations. We undertook a series of experiments to test whether the rationale described above for the design of Tolvid was in fact correct, particularly with respect to choice of gX deletions versus gI deletions. This issue had more than theoretical importance, since many existing PRV vaccines fortuitously have gI deletions (Mettenleiter *et al.*, 1985, 1988), and new recombinant PRV vaccines are being developed with tk and gI deletions (e.g. Gielkens *et al.*, 1989).

As the first test of the role of gI in live vaccines, a gI-negative derivative of Tolvid (Tolvid Δ-gI) was constructed, via the homologous recombination of Tolvid DNA with plasmids constructed to disrupt the gI gene as described above for the gX deletion. Tolvid Δ-gI has deletions in tk, gX, and gI. In a comparative vaccine trial, 62 4–6-week-old piglets were vaccinated with Tolvid, and 79 piglets were vaccinated with Tolvid Δ-gI. After 3 weeks, the pigs were challenged intranasally with the Rice strain of PRV. All of the piglets were monitored for clinical signs and the performance of the vaccines was assessed by several criteria: (i) the pigs receiving the gI-deleted vaccine showed significantly greater pyrexic response on challenge ($P = 0{\cdot}006$); (ii) the pigs receiving the gI-deleted vaccine showed longer arrest of growth after challenge (3·8 days for Δ-gI, 1·8 days for Tolvid, $P = 0{\cdot}01$). The average weight loss after challenge was significantly greater ($P = 0{\cdot}006$) in the group receiving the Δ-gI virus (1·14 kg) than in those receiving Tolvid (0·19 kg); (iii) the pigs receiving the Δ-gI virus shed challenge virus for a significantly greater time ($P = 0{\cdot}006$) than did the pigs vaccinated with Tolvid. The shedding of challenge virus is a significant epidemiological issue.

By all criteria measured, the removal of gI from Tolvid significantly compromised the efficacy of the vaccine. It was tentatively concluded, therefore, that gI does play a role in the protective immunogenicity of a live vaccine. This conclusion was tentative because in the experiment described above the gI-negative virus also had a gX deletion, i.e. two glycoproteins deleted from wild-type virus. The experiment was repeated using a set of 3 viruses: a tk-deletion virus containing no other deletions; a Δ-tk Δ-gI virus, containing deletions in gI and tk, but a functional gX gene; and Tolvid, containing gX and tk deletions but a functional gI gene. Groups of pigs were vaccinated with each of the 3 viruses, challenged, and the parameters of the previous experiment were measured. The conclusions were that the Δ-tk, gX-positive, gI-positive virus and Tolvid showed no statistically significant difference in performance. However, the gI-negative, gX-positive virus gave poorer protection against weight loss, time of growth arrest, and shedding of challenge virus.

These experiments show that what was known about gX was information that could be used in a predictive way to design a live vaccine. The recombinant DNA tools for precise manipulation of viral genomes are well-developed to take advantage of basic knowledge of viruses. Basic immunological and functional information on components of pathogens will undoubtably be used to design future vaccines.

Cells resistant to PRV

During the efforts to make a subunit vaccine for PRV, it was found that the PRV glycoprotein gp50 could function as a subunit vaccine (Marchioli *et al.*, 1987a). To study the function of this protein, we made several unsuccessful attempts to disrupt the gp50 gene, never obtaining a viable virus (Petrovskis *et al.*, 1988b). In attempting to establish that gp50 is an essential glycoprotein, we made cell lines expressing gp50 with the idea that these lines could support growth of gp50-negative viruses, as in the experiments of Ligas & Johnson (1988). Cell lines Vero, HeLa and MVPK (a swine kidney line) were each transfected with a gp50 expression vector and clones were isolated that expressed gp50. Each of these lines supported replication of PRV very poorly, as reflected by delayed and lower yield of progeny virus (Petrovskis *et al.*, 1988b) (see Table 1). Table 1 also shows that the gp50 cell lines were resistant to HSV, but not to the unrelated vaccinia virus. An HSV glycoprotein, gD, is homologous to gp50 (Petrovskis *et al.*, 1986). Campadelli-Fiume *et al.* (1988) demonstrated a similar resistance to HSV infection of cells expressing HSV gD. These workers further showed that, in the gD-producing cells, the infecting HSV undergoes some aberrant entry process that results in unsuccessful infection by most of the input virus. Apparently gD, and presumably gp50, produced by the cells binds to a site normally used by entering virus for the fusion of virions to the cell membrane.

Table 1. Growth of viruses on Vero cell lines that express glycoprotein gp50 (from Petrovskis *et al.*, 1988b)

Cell line	PRV	HSV	Vaccinia
Vero	$1{\cdot}4 \times 10^8$	$1{\cdot}3 \times 10^8$	$5{\cdot}3 \times 10^6$
Vero-tPA*	$1{\cdot}0 \times 10^8$	$1{\cdot}4 \times 10^8$	$7{\cdot}4 \times 10^6$
A-8†	$1{\cdot}0 \times 10^7$	$1{\cdot}5 \times 10^6$	$5{\cdot}8 \times 10^6$
B-4†	$1{\cdot}5 \times 10^7$	$1{\cdot}6 \times 10^7$	$6{\cdot}6 \times 10^6$

Number express yield of progeny virus in plaque-forming units/ml culture medium. Confluent monolayers of the cell lines were infected with PRV(Rice), HSV(F^+), or vaccinia (WR) at a multiplicity of infection of 0·01. The virus was harvested 29 h after infection by freezing, thawing, and sonication. Virus yields were determined by titration on Vero cells.

*Vero-tPA is a line of Vero cells transfected to express human tissue plasminogen activator.

†Lines of Vero cells transfected to express gp50.

The resistance of cells containing gp50 or gD to infecting virus is reminiscent of the phenomenon of interference long known in retroviruses (see Steck & Rubin, 1966), whereby a cell producing a retrovirus glycoprotein is resistant to superinfection by a similar retrovirus. Sanford & Johnston (1985) have coined the term "pathogen-derived resistance" to describe the general principle that production of a component of a pathogen might interfere with normal development of that pathogen.

It is exciting to speculate that pathogen-derived resistance to a virus can be extrapolated to the level of a transgenic animal. The case of retrovirus interference has been shown to operate at the level of intact animals (see Robinson *et al.*, 1981). A transgenic pig producing gp50 in most of its cells, or particular target cells, might well be resistant to PRV. There are other suggestions of ways to make pathogen-derived resistance to herpesviruses. Orberg & Schaffer (1987) found that cell lines overproducing ICP8, the HSV DNA-binding protein, in response to infection showed inhibited HSV growth. Also, Friedman *et al.* (1988) made cell lines producing a truncated version of the HSV α-transactivating protein, and these also had impaired ability to support HSV growth.

In summary, as in the case of design of live vaccines, a knowledge of the function of various viral proteins offers opportunities for engineering animals resistant to infection by particular viral pathogens.

References

Bennett, L.M., Timmins, J.G., Thomsen, D.R. & Post, L.E. (1986) The processing of pseudorabies virus glycoprotein gX in infected cells and in an uninfected cell line. *Virology* **155,** 707–715.

Ben-Porat, T. & Kaplan, A.S. (1970) Synthesis of proteins in cells infected with herpesvirus. V. Viral glycoproteins. *Virology* **41,** 265–273.

Ben-Porat, T., DeMarchi, J.M., Lomniczi, B. & Kaplan, A.S. (1986) Role of glycoproteins of pseudorabies virus in eliciting neutralizing antibodies. *Virology* **154,** 325–334.

Campadelli-Fiume, G., Arsenakis, M., Farabegoli, F. & Roizman, B. (1988) Entry of herpes simplex virus 1 in BJ cells that constitutively express viral glycoprotein D is by endocytosis and results in degradation of the virus. *J. Virol.* **62,** 159–167.

Friedman, A.D., Triezenberg, S.J. & McKnight, S.L. (1988) Expression of a truncated viral trans-activator selectively impedes lytic infection by its cognate virus. *Nature, Lond.* **22,** 452–454.

Gielkens, A.L.J., Moorman, R.J.M., van Oirschot, J.T. & Berns, A.J.M. (1989) Vaccine efficacy and innocuity of strain 783 of Aujeszky's disease virus. In *Vaccination and Control of Aujeszky's Disease*, pp. 27–35. Ed. J. T. van Oirschot. Kluwer, London.

Gustafson, D.P. (1981) Pseudorabies. In *Diseases of Swine*, 5th edn, pp. 209–223. Eds A. D. Leman, R. D. Gloack, W. L. Mengeling, R. H. Penny, E. H. Scholl & B. Straw. Iowa State University Press, Ames.

Ligas, M.W. & Johnson, D.C. (1988) A herpes simplex virus mutant in which glycoprotein D sequences are replaced by β-galactosidase sequences binds to but is unable to penetrate into cells. *J. Virol.* **62,** 1486–1494.

Marchioli, C.C., Yancey, R.J., Petrovskis, E.A., Timmins, J.G. & Post, L.E. (1987a) Evaluation of pseudorabies virus glycoprotein gp50 as a vaccine for Aujeszky's disease in mice and swine: expression by vaccinia virus and Chinese hamster ovary cells. *J. Virol.* **61,** 3977–3982.

Marchioli, C.C., Yancey, R.J., Wardley, R.C., Thomsen, D.R. & Post, L.E. (1987b) A vaccine strain of pseudorabies virus with deletions in the thymidine kinase and glycoprotein X genes. *Am. J. vet. Res.* **48,** 1577–1583.

Mettenleiter, T.C., Lukacs, N. & Rziha, H.J. (1985) Pseudorabies virus avirulent strains fail to express a major glycoprotein. *J. Virol.* **56,** 307–311.

Mettenletier, T.C., Schreurs, C., Thiel, H.-J. & Rziha, H.-J. (1987) Variability of pseudorabies virus glycoprotein I expression. *Virology* **158,** 141–146.

Mettenletier, T.C., Lomniczi, B., Sugg, N., Schreurs, C. & Ben-Porat, T. (1988) Host cell-specific growth advantage of pseudorabies virus with a deletion in the genome sequences encoding a structural glycoprotein. *J. Virol.* **62,** 12–19.

Orberg, P.K. & Schaffer, P.A. (1987) Expression of herpes simplex virus type 1 major DNA-binding protein, ICP8, in transformed cell lines: complementation of deletion mutants and inhibition of wild-type virus. *J. Virol.* **61,** 1136–1146.

Petrovskis, E.A., Timmins, J.G., Armentrout, M.A., Marchioli, C.C., Yancey, R.J. & Post, L.E. (1986) DNA sequence of the gene for pseudorabies virus gp50, a glycoprotein without N-linked glycosylation. *J. Virol.* **59,** 216–223.

Petrovskis, E.A., Meyer, A.L. & Post, L.E. (1988a) Reduced yield of infectious pseudorabies virus and herpes simplex virus from cell lines producing viral glycoprotein gp50. *J. Virol.* **62,** 2196–2199.

Petrovskis, E.A., Meyer, A.L., Thomsen, D.R., Berlinski, P.J., Wardley, R.C. & Post, L.E. (1988b) Pseudorabies virus gp50: an effective subunit vaccine and interference with virus replication in cell lines. In *Technological Advances in Vaccine Development*, pp. 147–156. Ed. L. Laskey. Alan R. Liss, New York.

Post, L.E. & Roizman, B. (1981) A generalized technique for deletion of specific genes in large genomes: α gene 22 of herpes simplex virus is not essential for growth. *Cell* **25,** 227–232.

Rea, T.J., Timmins, J.G., Long, G.W. & Post, L.E. (1985) Mapping and sequence of the gene for the pseudorabies virus glycoprotein which accumulates in the medium of infected cells. *J. Virol.* **54,** 21–29.

Robinson, H.L., Astrin, S.M., Senior, A.M. & Salazar, F.H. (1981) Host susceptibility to endogenous viruses: defective glycoprotein-expressing proviruses interfere with infections. *J. Virol.* **40,** 745–751.

Sanford, J.C. & Johnston, S.A. (1985) The concept of parasite-derived resistance—deriving resistance genes from the parasites own genome. *J. theor. Biol.* **113,** 395–405.

Steck, F.T. & Rubin, H. (1966) The mechanism of interference between an avian leukosis virus and Rous sarcoma virus. II. Early steps of infection by RSV of cells under conditions of interference. *Virology* **29,** 642–653.

Tatarov, G. (1968) Apathogenic mutant of the Aujeszky virus induced by 5-iodo-2-deoxyuridine (IUDR). *Zentbl. VetMed.* **15,** 847–853.

Thomsen, D.R., Marchioli, C.C., Yancey, R.J. & Post, L.E. (1987) Replication and virulence of pseudorabies virus mutants lacking glycoprotein gX. *J. Virol.* **61,** 229–232.

Wardley, R.C. & Post, L.E. (1989) The use of the gX deleted vaccine PRV τ-tk τ-gX 1 in the control of Aujeszky's disease. In *Vaccination and Control of Aujeszky's Disease*, pp. 13–25. Ed. J. van Oirschot. Kluwer, London.

Zuckerman, F., Mettenleiter, T.C. & Ben-Porat, T. (1989a) Role of pseudorabies virus glycoproteins in immune response. In *Vaccination and Control of Aujeszky's Disease*, pp. 107–117. Ed. J. van Oirschot. Kluwer, London.

Zuckerman, F., Zsak, L., Reilly, L., Sugg, N. & Ben-Porat, T. (1989b) Early interactions of pseudorabies virus with host cells: functions of glycoprotein gIII. *J. Virol.* **63,** 3323–3329.

TRANSGENIC FISH

Chairman
P. Bruns

J. Reprod. Fert., Suppl. **41** (1990), 107–116
Printed in Great Britain

Strategies for introducing foreign DNA into the germ line of fish

J. G. Cloud

Department of Biological Sciences, and the University of Idaho Aquaculture Program, University of Idaho, Moscow, ID 83843, USA

Keywords: gene transfer; germ line; fish; transgenic

Introduction

The ability to add specific genes into the genome of fish has potential applications in both the basic sciences and in the aquaculture industry. In support of research efforts in the basic sciences, transgenic fish have been and will continue to be produced to provide specific model systems with which to study and evaluate the genetic control of developmental and physiological events. Likewise, in order to provide more efficient strains for commercial aquaculture, many investigators are presently manipulating the genome of food fish for the purpose of increasing growth rate and temperature tolerance. It is predictable that these more practical concerns will be expanded in the near future to include improvement in feed efficiency and disease resistance.

As described by Andreason & Evans (1988), a successful programme in gene transfer involves (1) the isolation and purification of specific genes, (2) modifying the gene or the elements that normally control its expression and (3) reintroducing these modified genes into cultured cells or the germ line of intact animals. This paper will review the progress that has been made to date in transferring genes into fish, indicate some of the efforts that are being made to develop alternative strategies for the introduction of foreign DNA into the germ line of fish and will develop suggestions as to needs for additional research to further utilize this technology.

Methods of gene transfer

Two methods have been used successfully to transfer genes into fish. The most common method has been to microinject the exogenous DNA into fertilized eggs. The second method, chromosome-mediated gene transfer, involves the use of irradiated spermatozoa at fertilization to introduce DNA from another species; although this method may have limited use in developing new genetic lines, it could become an important means to help identify genes.

Microinjection into fertilized eggs

Transfer of new or additional copies of genes into the vertebrate genome was developed in the mouse (Gordon *et al.*, 1980; Brinster *et al.*, 1981; Costantini & Lacy, 1981; Wagner, E. F. *et al.*, 1981; Wagner, T. E. *et al.*, 1981). From these and subsequent studies, it has been clearly demonstrated that isolated genes injected into the pronucleus of the mouse zygote can integrate into the genome, be expressed and enter the germ line. Using this same basic microinjection technique, gene transfer was extended to rabbits, sheep and pigs (Hammer *et al.*, 1985). Development of this methodology in the mammalian system clearly indicated the need to inject the exogenous DNA into a pronucleus (Brinster *et al.*, 1985). Because pronuclei of fertilized fish eggs cannot be visualized and since DNA injected into the cytoplasm of fertilized eggs of the toad, *Xenopus laevis*, was successfully integrated and expressed (Etkin *et al.*, 1984), the initial studies in the production of transgenic fish for the various species were conducted primarily by injecting exogenous DNA into the zygote cytoplasm.

Microinjection is a fairly straightforward procedure; it only requires a good quality microscope, a micromanipulator, a pipette-puller and a thumb-screw syringe or equivalent component to inject the DNA-containing solution. For most species, the difficult part of injecting the fish zygote is penetrating the outer surface coat or chorion. In general 3 different approaches have been used to deal with this barrier. One approach is to remove it by manual dissection (Stuart *et al.*. 1988) or by enzymic digestion (Zhu *et al.*, 1985; Hallerman *et al.*, 1988). In species for which this approach is utilized, the chorion is thin, and the perivitelline space is relatively large. Additionally, the zygote must have sufficient structural integrity such that it can withstand its own weight in the absence of the mechanical support provided by the chorion. A second approach is to penetrate the chorion at the micropyle, the opening through which the spermatozoa gain access to the oocyte (Brem *et al.*, 1988; Fletcher *et al.*, 1988). This method requires that the micropyle be easily visualized and that the egg be positioned properly with respect to the angle of the micropipette. The third approach, which was developed by Chourrout *et al.* (1986b) for rainbow trout, is to cut a hole in the chorion. This procedure was originally described using a broken end of a Pasteur pipette to make a hole approximately 100 µm in size. Although our laboratory has found this procedure to be satisfactory, we use a microsurgery probe (Micromanipulator Microscope Company, Inc., Carson City, NV, USA) to make a small cut in the chorion; in our experience these microinstruments are easier to use.

The data obtained from the initial studies of microinjecting exogenous DNA into the fertilized fish egg are summarized in Table 1. A major conclusion that can be drawn from this information is that the fish zygote or early embryo will tolerate the delivery of exogenous DNA to the cytoplasm by microinjection. Even though there are no specific numbers estimating the reduction in embryo survival in these studies as a result of a sham injection alone, impalement of the blastodisc does not appear to be a significant detrimental factor for any of the species investigated, as indicated by the absence of a large and consistent difference between the injected and non-injected control embryos.

The optimal amount of DNA to inject into the fertilized egg has not yet been determined for the various species but appears to be of considerable importance. Stuart *et al.* (1988) demonstrated that survival of the zebrafish embryo at 10 days of age decreased as the amount of DNA injected into the eggs at fertilization increased. Conversely, these same investigators reported that the amplification of injected DNA during gastrulation was followed by a period of degradation and that the level of exogenous genes in the embryo during somatogenesis was directly related to the amount injected. The conclusion drawn from these data is that the amount of DNA injected needs to be optimized between providing an adequate amount of integration and compromising embryonic survival with too much.

From Table 1, it is evident that exogenous DNA injected into the cytoplasm of fertilized eggs is integrated into the genome. However, because of the use of different promoters/genes, species of fish and criteria for integration of the injected DNA, comparison of these results provides little insight into the optimal conditions for microinjection. What is not apparent in this table but is discussed by many of the investigators is the probable variation in the time of integration of the injected DNA in cells within an embryo and the resultant mosaicism of the transgenic fish. This mosaicism is especially evident in the germ line (Stuart *et al.*, 1988). This presumed delay in integration may result from the fact that the DNA is not being injected into one of the pronuclei.

An alternative method of microinjection was used by Ozato *et al.* (1986); these investigators injected exogenous DNA into the germinal vesicle of the egg before the reinitiation of meiosis. The relatively high rate of integration (15%) and the subsequent expression of the transgene (chicken delta-crystallin) suggests that this was an effective means of gene transfer in fish. This particular microinjection procedure does not appear to have been tested in any other species, even though the hormonal events controlling oocyte maturation and ovulation have been defined for a number of different species of fish (reviewed by Goetz, 1983) and rainbow trout eggs capable of fertilization

Table 1. Transfer of DNA into the genome of various species of fish by microinjection

Species	Time of injection (after fertilization	Amount of DNA injected	Gene transferred	Promoter	% Survival	% Integration§	Expression of transgene(s)	Germ-line transmission	Reference
Goldfish (*Carassius auratus*)	Before first cleavage	7×10^6 copies	hGH	MT	—†	50 (author's calc.)	—†	—†	Zhu *et al.* (1985)
Rainbow trout (*Salmo gairdneri*)	2·25–5·75 h	200 pg	hGH	SV-40	50–100 (at hatching)	—† (embryos pooled)	—†	—†	Chourrout *et al.* (1986b)
Medaka (*Oryzias latipes*)	GV*	5×10^3–10^4 copies	cryst	SV-40	48 (stage 32)	15	15%	—†	Ozato *et al.* (1986)
Channel catfish (*Ictalurus punctatus*)	5 min	10^6 copies	hGH	MT	13 (3 weeks)	2·5	—†	—†	Dunham *et al.* (1987)
Tilapia (*Oreochromis niloticus*)	10 min 21–23 h 40–43 h	10^6 copies	hGH	MT	25 51 98 (90 days)	0 7 2	—†	—†	Brem *et al.* (1988)
Atlantic salmon (*Salmo salar*)	3–8 h	200 pg	β-gal	MT	7 (80 days)	—‡	1% (6 of 600)	—†	McEvoy *et al.* (1988)
Zebrafish (*Brachydanio rerio*)	<1 h	4·5 pg 15 pg 30 pg 90 pg	hygro	SV-40	43 24 16 3 (10 days)	0·8	—†	+	Stuart *et al.* (1988)
Atlantic salmon (*Salmo salar*)	2–3 h	10^6 copies	AFP		80 (at hatching)	5·3	—†	—†	Fletcher *et al.* (1988)

hGH = human growth hormone; cryst = chicken delta-crystallin; β-gal = beta-galactosidase; hygro = hygromycin resistance; AFP = antifreeze peptides; MT = mouse metallothionein.
*Germinal vesicle stage/prefertilization.
†Information not available.
‡Data inconclusive.
§% Integration based on number injected.

have been produced by incubating isolated follicles with the exogenous hormones (Jalabert, 1978). One of the disadvantages of this method is that at least a portion of the ovary must be removed and the isolated follicles need to be cultured for a period of time.

Chromosome-mediated gene transfer

Several transgenic rainbow trout (*Salmo gairdneri*) have been produced by using an in-vivo modification of the chromosome-mediated gene transfer method (Thorgaard *et al.*, 1985; Disney *et al.*, 1987). In these studies paternal DNA from normally pigmented rainbow or brook trout was fragmented by gamma irradiation of spermatozoa before fertilization. Eggs from albino females were fertilized with treated spermatozoa and heat shocked (29°C for 10 min after fertilization) to induce the retention of the second polar body with the resultant production of gynogenic diploid fish. Of these resultant embryos, 2–13% survived through early development; their pigmentation pattern supported the conclusion that they were mosaics. Analysis of cells from these embryos revealed that they contained chromosomal fragments in addition to their regular number of diploid chromosomes (Thorgaard *et al.*, 1985; Disney *et al.*, 1987); the individual size of the chromosomal fragments decreased with increasing levels of irradiation. Disney *et al.* (1988) estimated that these transgenic embryos contained an average of 7% additional DNA. The data supported the conclusion that the paternally derived genes were replicated, retained in the adults (Disney *et al.*, 1988) and expressed (Thorgaard *et al.*, 1985; Disney *et al.*, 1987, 1988). Furthermore, the mosaicism of these transgenic fish suggests that the fragments were initially unstable or were partitioned unequally during cleavage. Interestingly, the genes that were retained did not have an apparent relationship with their distance from the centromere (Disney *et al.*, 1987).

The additional chromosome fragments of the transgenic parents were inherited by the offspring (Disney *et al.*, 1988), but the number of fragments in the progeny from a single cross was variable. At present it is unclear as to whether the individual fragments will remain stable over a number of generations. However, as indicated by Disney *et al.* (1987), the transgenic fish produced by this procedure may be utilized to identify and isolate genes of interest such as those involved in disease resistance or temperature tolerance.

Possible strategies for gene transfer

In addition to the two methods of gene transfer already discussed, there are several possible strategies currently being investigated. Each of these possibilities is being pursued because of presumed or realized advantages over present methods.

Electroporation of embryos

Electroporation has been used successfully to transfer genes into many different kinds of isolated cells (Andreason & Evans, 1988). Since the chorion can be easily removed from the fertilized egg in some species and the resultant embryo will complete development in the absence of this extracellular surface coat, electroporation of the dechorionated zygote is a potential means of transferring genes into large numbers of embryos very efficiently. Initial attempts at using this method to transfer the chloramphenicol acetyltransferase (CAT) gene into fertilized goldfish (*Carassius auratus*) eggs were unsuccessful (Hallerman *et al.*, 1989), but improvements or changes in the methodology are presently being pursued. Although this technique, if successful, will probably be limited to those species in which the chorion can be easily removed, it should be an effective means by which genes can be transferred into large batches of fertilized eggs.

Embryonic stem cells

Embryonic stem cells, pluripotent cells that remain undifferentiated under proper culture conditions, have been established from embryos of mice (Evans & Kaufman, 1981; Martin, 1981) and hamsters (Doetschman *et al.*, 1988). Because embryonic stem cells contribute to the germ cell population when injected into early embryos, these cells have been utilized for gene transfer in the mouse (Gossler *et al.*, 1986; Robertson *et al.*, 1986). One of the main advantages of this approach is that transfected embryonic stem cells can be screened before being introduced into the germ line. Using this method of gene transfer coupled with homologous recombination to target the gene insertion (Smithies *et al.*, 1985), Doetschman *et al.* (1987) have demonstrated that genes can be inserted at a specific locus in a germ line.

Although this method appears to provide a number of advantages in gene transfer, it has not been utilized in fish. The primary reason for this absence is probably that the methodology to re-establish pluripotent cells or to produce fish chimaeras has not previously been available. Therefore as a first step in adapting this procedure to fish, Nilsson & Cloud (1989) recently demonstrated that chimaeric rainbow trout embryos can be produced by microinjecting isolated blastomeres from stage 6C embryos (Ballard, 1973) into recipient embryos of the same age. Chimaerism was detected by the incorporation of fluorescein-labelled cells into the recipient embryo. These labelled blastomeres were produced by injecting fluorescein isothiocyanate–Dextran (FITC-D) into zygotes (Kimmel & Law, 1985) at 8–12 h after fertilization and allowing the resultant embryos to develop to stage 6C. At this time, the FITC-D-labelled embryos were manually dissected free of the yolk sac and dispersed into single cells by incubating the blastoderm in a Ca^{2+} + Mg^{2+}-free Niu-Twitty solution (King, 1966). Approximately 100 isolated blastomeres were microinjected into the blastoderm of unlabelled recipient embryos (Fig. 1). The variation in the location of labelled cells in embryos 6 days after introduction of donor cells suggests that incorporation of the injected cells was random. The resultant chimaeric embryos, observed at the time of retinal pigmentation, appeared to have developed normally. Although we have suggested that the injected blastomeres are pluripotent and will enter the germ line, we have not yet investigated these assumptions.

Piggybacking exogenous DNA on spermatozoa

Gametes and the fertilization process in fish are unlike those of mammals. The spermatozoa are immotile in seminal plasma, are only motile for 30–60 sec following activation and do not have an acrosome or undergo the acrosomal reaction. Similarly, fish eggs remain inactive in ovarian fluid and can be stored in this fluid for days after spawning with little reduction in fertility. When spermatozoa and eggs are combined at fertilization, activation and the incorporation of the spermatozoon into the egg is initiated with the addition of water or some other appropriate solution.

Because of the relative simplicity of the fertilization process in fish, the spermatozoa may be capable of piggybacking sufficient exogenous DNA encapsulated in immunoliposomes into the egg for gene transfer to occur. To examine the feasibility of this hypothesis, the surface requirements of fish spermatozoa were first defined. Several monoclonal antibodies (MAbs) to surface determinants of rainbow trout spermatozoa have been produced (J. C. Beck, K. D. Fulcher, C. F. Beck & J. G. Cloud, unpublished). Although most of these MAbs cross-reacted with trout somatic cells, two were specific for spermatozoa. One of the MAbs specific for spermatozoa reduced fertility in a dose-dependent manner when incubated with spermatozoa before fertilization (Fig. 2). Based on the distribution of FITC-labelled secondary antibody, the cell surface component recognized by the MAb that reduced fertility appeared to be restricted to the sperm head. From these investigations, two interrelated conclusions are possible. Firstly, rainbow trout spermatozoa have a specific surface component that is required for fertilization. And secondly, the addition of a number of antibody molecules to spermatozoa at non-essential sites is compatible with fertilization. Whether spermatozoa can carry exogenous DNA into the egg at fertilization in quantities sufficient for gene transfer, as reported for the mouse (Lavitrano *et al.*, 1989) or as we have suggested in immunoliposomes, has yet to be completely tested.

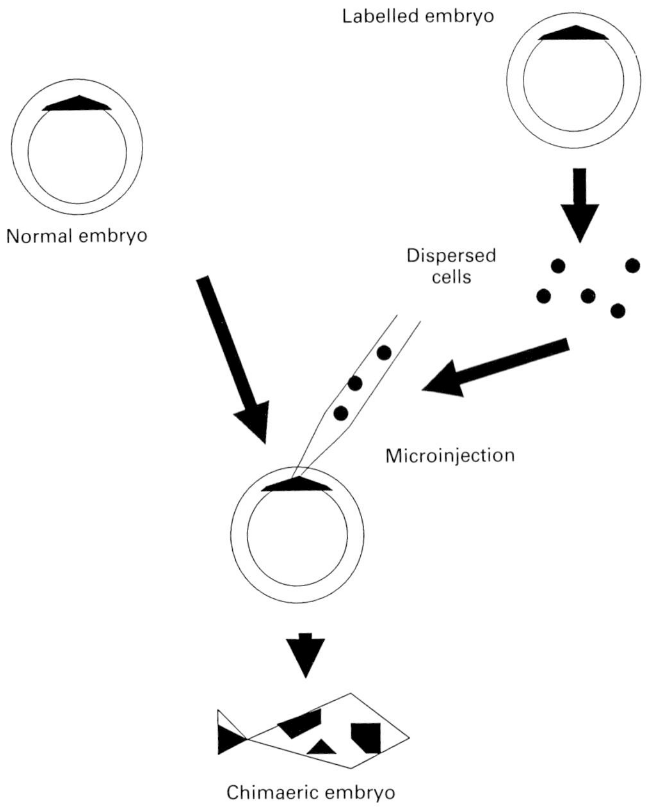

Fig. 1. Chimaeric embryos were produced by injecting blastomeres from labelled blastulae into unlabelled embryos at the same development stage.

Muscle precursor cells

Growth of vertebrate skeletal muscle continues after embryological development; the increase in muscle mass occurs as a result of both myofibre hypertrophy and hyperplasia. During skeletal muscle growth there is an increase in the number of myonuclei (Enesco & Puddy, 1964). Myogenic stem cells (satellite cells) are considered to be the source of these myonuclei since these cells proliferate and subsequently fuse with pre-existing myofibres during myofibre hypertrophy (Moss & Leblond, 1971; Cardasis & Cooper, 1975; Lipton & Schultz, 1979) and fuse with themselves to produce new myofibres during myofibre hyperplasia (Chiakulas & Pauly, 1965; Lipton & Schultz, 1979).

Transplantation of exogenous myogenic stem cells into skeletal muscle has been used to introduce new nuclei into myofibres in an effort to provide genes for normal muscle components to alleviate certain muscle pathologies (Watt *et al.*, 1984). Alternatively, it would appear that this same methodology can be used to introduce nuclei with specific transgenes into muscle cells.

The present hypothesis is that transplantation of isolated myogenic cells into skeletal muscle of recipient animals can be used to transfer genes into fish. Testing this hypothesis will require the isolation of satellite cells from muscle of donor animals, transfection of these myogenic stem cells in culture and transplantation of selected transfected cells into skeletal muscle of recipient animals. From a number of studies, satellite cells have been identified and described in muscle tissue from many different species of fish (Nag & Nursall, 1972; Kryvi, 1975; Kryvi & Eide, 1977; Sandset & Korneliussen, 1978). Powell *et al.* (1989) harvested mononucleated cells from enzymically dispersed skeletal muscle of rainbow trout and maintained them in culture. The harvested cells appeared to

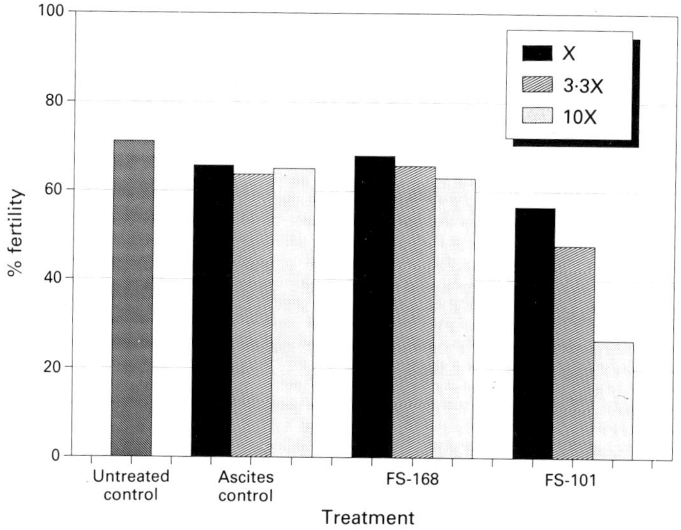

Fig. 2. Fertility profiles of spermatozoa treated with the same relative concentration of antibodies. Each bar represents the average response of 9 replicates (spermatozoa from 3 different males was used to fertilize eggs from 3 different females; ~100 eggs per cross or 900 eggs total). X equals the antibody dilution equivalent to sperm saturation at 37°C after 2 h of incubation; for the fertility study, spermatozoa were incubated with antibody for 10 h at 4–5°C. FS-169 was an antibody that cross-reacted with all somatic cells; FS-101 was an antibody that was specific for spermatozoa.

be fibroblasts or fusiform satellite cells. In culture, the satellite cells increased in number and fused together to form multinucleated structures that have been interpreted to be myotubes. Some of these myotubes in turn developed into elongated structures that contained striations. Taken together, these data support the conclusion that viable satellite cells can now be obtained, at least from rainbow trout, and maintained *in vitro*.

Non-transmittable transgenes

With the development of transgenic fish comes the need and responsibility to provide adequate safeguards against unwanted or unauthorized introduction of transgenes into native/wild fish populations or the inadvertent establishment of a transgenic fish population in a watershed. Although these proposed safeguards will probably take different forms for different species of fish, in some instances it may be more efficient to produce transgenic fish that cannot pass the transgenes to subsequent generations. Two obvious means of fulfilling this restriction are to produce fish that are sterile or do not have the transgene in the germ line.

Sterile fish

Although sterile fish can be produced by a variety of techniques, the method of choice needs to be simple and fail-safe. There are at least two different ways to design a security system against release of the transgenes. The first type would involve maintaining fertile broodstock in an indoor facility with multiple traps in the outflow to ensure against escape and sterilizing the resultant offspring before their removal from this facility. The induction of triploidy might be one possible means of sterilizing these offspring. In rainbow trout for example, female triploid fish have been shown to be sterile (Thorgaard & Gall, 1979; Lincoln & Scott, 1984), but the method by which triploidy is produced is an important consideration. Triploid salmonid females can be produced by fertilizing eggs with X-bearing spermatozoa from sex-reversed females followed by a heat shock to induce the retention of the second polar body. While this procedure will produce the desired triploid product, it is not fail-safe. For instance, if the heat treatment is not applied properly or for long enough periods, a large percentage of the resultant embryos may be diploid. While certification of triploidy appears to be feasible for some commercial uses of fish (Allen & Wattendorf, 1987), it would generally be cost prohibitive. An alternative method to produce triploids is by crossing diploid with tetraploid broodstock. While this method provides the requisite safeguards, the production and fertility of tetraploids may need to be improved to make this procedure functional (Chourrout *et al.*, 1986a). Another way to produce sterile offspring is to generate triploid hybrids in which the diploid is not viable (Chevassus *et al.*, 1983; Scheerer & Thorgaard, 1983). This method is attractive from the standpoint that it would provide both sterility and certification of triploidy. However, the resultant cross would also need to have desirable characteristics to be useful as an aquacultural product.

A second type of genetic security system would involve the development of genetic strains that are infertile in the wild but can be made fertile in the hatchery. One example of this approach would be to produce fish that do not have a complete and functional gonadal duct; as such they would be infertile if lost to the natural environment but could contribute gametes for in-vitro fertilization. This example is used because the variation in the presence of a functional sperm duct following the masculinization of genotypic female salmonids (Bye & Lincoln, 1986) suggests that the formation and development of the duct is susceptible to its hormonal environment. Another possible approach would be the development of a genetic strain in which gametogenesis is dependent upon an exogenous component (vitamin, nutrient, hormone) that could be supplied to the broodstock as a feed additive or as an implant to reverse temporarily the infertile condition. Regardless of the strategy, research in reproduction and embryonic development appears to be needed to support the overall programme of genetic engineering in order to produce transgenic fish that can be used commercially without undue risk.

Somatic cell transplantation

With further development and understanding of tissue rejection, transplantation of transgenic somatic cells is a potential means of introducing foreign genes into specific tissues of animals. One advantage of the approach, especially for fish, is that the transgene would be limited to the somatic cells. Whether this method is feasible and can be adapted for aquaculture is still an open question.

Conclusion

Specific genes can now be transferred into the germ line of fish. Research efforts are presently ongoing to improve the efficiency of gene transfer, to target the site of gene integration and to restrict the expression and inheritance of the resultant transgene.

I thank Dr Gary Thorgaard for valuable discussions on fish genetics and gene transfer; Dr Allen Schuetz for reviewing the manuscript; and Pauly Waldron for help in the preparation of the manuscript. Support was provided by the Idaho State Board of Education and by the National Science Foundation during the preparation of this review.

References

Allen, S.K, Jr & Wattendorf, R.J. (1987) Triploid grass carp: status and management implications. *Fisheries* **12,** 20–24.

Andreason, G.L. & Evans, G.A. (1988) Introduction and expression of DNA molecules in eukaryotic cells by electroporation. *BioTechniques* **6,** 650–660.

Ballard, W.W. (1973) Normal embryonic stages for salmonid fishes, based on *Salmo gaidneri* Richardson and *Salvelinus fontinalis* (Mitchill). *J. exp. Zool.* **184,** 7–25.

Brem, G. Brenig, B., Horstgen-Schwark, G. & Winnacker, E.-L. (1988) Gene transfer in tilapia (*Oreochromis niloticus*). *Aquaculture* **68,** 209–219.

Brinster, R.L., Chen, H.Y., Trumbauer, M.E., Senear, A.W., Warren, R. & Palmiter, R.D. (1981) Somatic expression of herpes thymidine kinase in mice following injection of a fusion gene into eggs. *Cell* **27,** 223–231.

Brinster, R.L., Chen, H.Y., Trumbauer, M.E., Yagle, M.K. & Palmiter, R.D. (1985) Factors affecting the efficiency of introducing foreign DNA into mice by microinjecting eggs. *Proc. natn. Acad. Sci. USA* **82,** 4438–4442.

Bye, V.J. & Lincoln, R.F. (1986) Commercial methods for the control of sexual maturation in rainbow trout (*Salmo gairdneri* R.). *Aquaculture* **57,** 299–309.

Cardasis, C.A. & Cooper, G.W. (1975) An analysis of nuclear numbers in individual muscle fibers during differentiation and growth: a satellite cell-muscle fiber growth unit. *J. exp. Zool.* **191,** 347–358.

Chevassus, B., Guyomard, R., Chourrout, D. & Quillet, E. (1983) Production of viable hybrids in salmonids by triploidization. *Genet. Sel. Evol.* **15,** 519–532.

Chiakulas, J.J. & Pauly, J.E. (1965) A study of postnatal growth of skeletal muscle in the rat. *Anat. Rec.* **152,** 55–62.

Chourrout, D., Chevassus, B., Krieg, F., Happe, A., Burger, G. & Renard, P. (1986a) Production of second generation triploid and tetraploid rainbow trout by mating tetraploid males and diploid females. Potential of tetraploid fish. *Theor. appl. Genet.* **72,** 193–206.

Chourrout, D., Guyomard, R. & Houdebine, L.-M. (1986b) High efficiency gene transfer in rainbow trout (*Salmo gairdneri* Rich.) by microinjection into egg cytoplasm. *Aquaculture* **51,** 143–150.

Costantini, F. & Lacy, E. (1981) Introduction of a rabbit B-globin gene into the mouse germ-line. *Nature, Lond.* **294,** 92–94.

Disney, J.E., Johnson, K.R. & Thorgaard, G.H. (1987) Intergeneric gene transfer of six isozyme loci in rainbow trout by sperm chromosome fragmentation and gynogenesis. *J. exp. Zool.* **244,** 151–158.

Disney, J.E., Johnson, K.R., Banks, D.K. & Thorgaard, G.H. (1988) Maintenance of foreign gene expression and independent chromosome fragments in adult transgenic rainbow trout and their offspring. *J. exp. Zool.* **248,** 335–344.

Doetschman, T., Gregg, R.G., Maeda, N., Hooper, M.L., Melton, D.W., Thompson, S. & Smithies, O. (1987) Targetted correction of a mutant HPRT gene in mouse embryonic stem cells. *Nature, Lond.* **330,** 576–578.

Doetschman, T., Williams, P. & Maeda, N. (1988) Establishment of hamster blastocyst-derived embryonic stem (ES) cells. *Devl Biol.* **127,** 224–227.

Dunham, R.A., Eash, J., Askins, J. & Townes, T.M. (1987) Transfer of the metallothionein-human growth hormone fusion gene into channel catfish. *Trans. Am. Fish Soc.* **116,** 87–91.

Enesco, M. & Puddy, D. (1964) Increase in the number of nuclei and weight in skeletal muscle of rats at various ages. *Am. J. Anat.* **114,** 235–244.

Etkin, L.D., Pearman, B., Roberts, M. & Bektesh, S.L. (1984) Replication, integration, and expression of exogenous DNA injected into fertilized eggs of *Xenopus laevis*. *Differentiation* **26,** 194–202.

Evans, M.J. & Kaufman, M. (1981) Establishment in culture of pluripotent cells from mouse embryos. *Nature, Lond.* **292,** 154–156.

Fletcher, G.L., Shears, M.A., King, M.J., Davies, P.L. & Hew, C.L. (1988) Evidence for antifreeze protein gene transfer in atlantic salmon (*Salmo salar*). *Can. J. Fish. Aquat. Sci.* **45,** 352–357.

Goetz, F.W. (1983) Hormonal control of oocyte final maturation and ovulation in fishes. In *Fish Physiology*, Vol. 9B, pp. 117–170. Eds W. S. Hoar, D. J. Randall & E. M. Donaldson. Academic Press, New York.

Gordon, J.W., Scangos, G.A., Plotkin, D.J., Barbosa, J.A. & Ruddle, F.H. (1980) Genetic transformation of mouse embryos by microinjection of purified DNA. *Proc. natn. Acad. Sci. USA* **77,** 7380–7384.

Gossler, A. Doetschman, T., Korn, R., Serfling, E. & Kemler, R. (1986) Transgenesis by means of blastocyst-derived embryonic stem cell lines. *Proc. natn. Acad. Sci. USA* **83,** 9065–9069.

Hallerman, E.M., Schneider, J.F., Gross, M.L., Faras, A.J., Hackett, P.B., Guise, K.S. & Kapuscinski, A.R. (1988) Enzymatic dechorionation of goldfish, walleye and northern pike eggs. *Trans. Am. Fish Soc.* **117,** 456–460.

Hallerman, E.M., Faras, A.J., Hackett, P.B., Kapuscinski, A.R. & Guise, K.S. (1989) Attempted introduction of a novel gene into goldfish through electroporation. *J. Cell. Biochem., Suppl.* **13B,** p. 170, abstr.

Hammer, R.E., Pursel, V.G., Rexroad, C.E., Jr, Wall, R.J., Bolt, D.J., Ebert, K.M., Palmiter, R.D. & Brinster, R.L. (1985) Production of transgenic rabbits, sheep and pigs by microinjection. *Nature, Lond.* **315,** 680–683.

Jalabert, B. (1978) Production of fertilizable oocytes from follicle of rainbow trout (*Salmo gairdnerii*) following *in vitro* maturation and ovulation. *Annls Biol. anim. Biochem. Biophys.* **18,** 416–470.

Kimmel, C.B. & Law, R.D. (1985) Cell lineage of zebrafish blastomeres 1. Cleavage pattern and cytoplasmic bridges between cells. *Devl Biol.* **108,** 78–85.

King, T.J. (1966) Nuclear transplantation in amphibia. *Methods Cell Physiol.* **2,** 1–36.

Kryvi, H. (1975) The structure of the myosatellite cells in axial muscles of the shark *Galeus melastomus*. *Anat. Embryol.* **147,** 35–44.

Kryvi H. & Eide, A. (1977) Morphometric and autoradiographic studies on the growth of red and white axial muscle fibres in the shark *Etmopterus spinax*. *Anat. Embryol.* **151,** 17–28.

Lavitrano, M., Camaioni, A., Fazio, V.M., Dolci, S., Farace, M.G. & Spadafora, C. (1989) Sperm cells as vectors for introducing foreign DNA into eggs: genetic transformation of mice. *Cell* **57,** 717–723.

Lincoln, R.F. & Scott, A.P. (1984) Sexual maturation in triploid rainbow trout. *Salmo gairdneri* Richardson. *J. Fish Biol.* **25,** 385–392.

Lipton, B.H. & Schultz, E. (1979) Developmental fate of skeletal muscle satellite cells. *Science, NY* **205,** 1292–1294.

Martin, G.R. (1981) Isolation of a pluripotent cell line from early mouse embryos cultured in medium conditioned by teratocarcinoma stem cells. *Proc. natn. Acad. Sci. USA* **78,** 7634–7638.

McEvoy, T., Stack, M., Keane, B., Barry, T., Sreenan, J. & Gannon, F. (1988) The expression of a foreign gene in salmon embryos. *Aquaculture* **68,** 27–37.

Moss, F.P. & Leblond, C.P. (1971) Satellite cells as the source of nuclei in muscles of growing rats. *Anat. Rec.* **170,** 421–436.

Nag, A.C. & Nursall, J.R. (1972) Histogenesis of white and red muscle fibres of trunk muscles of a fish *Salmo gairdneri. Cytobios* **6,** 227–246.

Nilsson, E. & Cloud, J.G. (1989) Production of chimeric embryos of trout (*Salmo gairdneri* by introducing isolated blastomeres into recipient blastulae. *Biol. Reprod.* **40,** (Suppl. 1), p. 109, abstr.

Ozato, K., Kondoh, H., Inohara, H., Iwamatsu, T., Wakamatsu, Y. & Okada, T.S. (1986) Production of transgenic fish: introduction and expression of chicken-crystallin gene in medaka embryos. *Cell Diff.* **19,** 237–244.

Powell, R.L., Dodson, M.V. & Cloud, J.G. (1989) Cultivation and differentiation of satellite cells from skeletal muscle of the rainbow trout *Salmo gairdneri. J. exp. Zool.* **250,** 333–338.

Robertson, E., Bradley, A., Kuehn, M. & Evans, M. (1986) Germ-line transmission of genes introduced into cultured pluripotential cells by retroviral vector. *Nature, Lond.* **323,** 445–448.

Sandset, P.M. & Korneliussen, H. (1978) Myosatellite cells associated with different muscle fibre types in the Atlantic hagfish (*Myxine glutinosa,* L.). *Cell Tissue Res.* **195,** 17–27.

Scheerer, P.D. & Thorgaard, G.H. (1983) Increased survival in salmonid hybrids by induced triploidy. *Can. J. Fish. Aquat. Sci.* **40,** 2040–2044.

Smithies, O., Gregg, R.G., Boggs, S.S., Koralewski, M.A. & Kucherlapati, R.S. (1985) Insertion of DNA sequences into the human chromosomal-globin locus by homologous recombination. *Nature, Lond.* **317,** 230–234.

Stuart, G.W., McMurray, J.V. & Westerfield, M. (1988) Replication, integration and stable germ-line transmission of foreign sequences injected into early zebrafish embryos. *Development* **103,** 403–412.

Thorgaard, G.H. & Gall, G.A.E. (1979) Adult triploids in a rainbow trout family. *Genetics Princeton* **93,** 961–973.

Thorgaard, G.H., Scheerer, P.D. & Parsons, J.E. (1985) Residual paternal inheritance in gynogenetic rainbow trout: Implications for gene transfer. *Theor. appl. Genet.* **71,** 119–121.

Wagner, E.F., Stewart, T.A. & Mintz, B. (1981) The human β-globin gene and a functional thymidine kinase gene in developing mice. *Proc. natn. Acad. Sci. USA* **78,** 5016–5020.

Wagner, T.E., Hoppe, P.C., Jollick, J.D., Scholl, D.R., Hodinka, R.L. & Gault, J.B. (1981) Microinjection of a rabbit-globin gene in zygotes and its subsequent expression in adult mice and their offspring. *Proc. natn. Acad. Sci. USA* **78,** 6376–6380.

Watt, D.J., Morgan, J.E. & Partridge, T.A. (1984) Use of mononuclear precursor cells to insert allogeneic genes into growing mouse muscles. *Muscle & Nerve* **7,** 741–750.

Zhu, Z., Li, G., He, L. & Chen, S. (1985) Novel gene transfer into the fertilized eggs of gold fish (*Carassius auratus* L. 1758). *Z. Angew. Ichthyol.* **1,** 31–34.

TRANSGENIC RUMINANTS

Chairman
C. Batt

J. Reprod. Fert., Suppl. 41 (1990), 119–124

Insertion, expression and physiology of growth-regulating genes in ruminants

C. E. Rexroad, Jr*, R. E. Hammer†§, R. R. Behringer†, R. D. Palmiter‡ and R. L. Brinster†

**U.S. Department of Agriculture, Agricultural Research Service, Beltsville Agricultural Research Center, Livestock and Poultry Sciences Institute, Reproduction Laboratory, Beltsville, Maryland 20705, USA; †Laboratory of Reproductive Physiology, University of Pennsylvania, Philadelphia, Pennsylvania 19104, USA; and ‡Howard Hughes Medical Institute, University of Washington, Seattle, Washington 98195, USA*

Summary. Transgenic sheep with elevated concentrations of circulating growth hormone (GH) were produced by microinjecting recombinant DNA into pronuclei of zygotes. The transgenes were fusion genes of non-GH promoters with coding sequences of various growth hormone genes including human, ovine or bovine. In addition, sheep transgenic with the human growth hormone releasing factor gene were produced. Non-GH promoters for fusion genes allowed novel regulation of GH production in ectopic tissues, including the kidney, liver and gut. Elevated levels of GH profoundly altered plasma IGF-1 without significantly altering rate of growth or feed efficiency. Carcass composition was altered with reduced fat. Elevated GH induced diabetes, resulting in death by 1 year of age. These studies indicate the need for improved regulation of inserted genes or investigation of alternative systems, such as GH receptors, to improve growth using the transgenic approach in ruminants.

Keywords: sheep; transgenic; growth hormone; expression; diabetes

Introduction

Mice transgenic with genes that coded for human growth hormone (hGH), rat growth hormone (rGH), or human growth hormone releasing factor (hGRF) grew in some cases to almost twice the size of control mice (Palmiter *et al.*, 1982, 1983; Hammer *et al.*, 1985a, b). The additional growth resulted from excess production of GH. The transgenes were fusion genes of the coding sequences of the GH genes or the hGRF gene ligated to the mouse metallothionein-I promoter. The use of fusion genes permitted the excess production of GH because GH was no longer under normal feedback control. Insertion of genes with novel regulatory sequences may offer the potential to increase growth rates and/or alter body composition in ruminants. Since the report of Hammer *et al.* (1985c), who inserted a hGH gene into a lamb, a number of experiments have been conducted to insert GH-related genes into ruminants, especially sheep. In this paper we will discuss those experiments with respect to the efficiency of insertion and expression of transgenes in ruminants. In addition, we will discuss the effects of excess production of GH on the physiology of lambs.

Gene insertion

Transgenic ruminants with genes for growth hormones or human growth hormone-releasing factor have been produced by microinjection of DNA into a pronucleus of fertilized eggs or into the nuclei

§Present address: Howard Hughes Medical Institute Laboratory and Department of Cell Biology, University of Texas Southwestern Medical Center, Dallas, Texas 75235, USA.

of 2-cell eggs (Table 1). Microinjection of ruminant eggs is difficult because yolk in the cytoplasm obstructs the view of the pronuclei. Poor visibility can be overcome for sheep pronuclei by using differential interference contrast (DIC) microscopy. Hammer *et al.* (1985c, 1986) used DIC and were able to observe pronuclei in about 80% of fertilized 1-cell sheep eggs. Visibility was not as good as for mouse ova. Murray *et al.* (1989) reported that, with DIC on an inverted microscope, changing to a thinner depression slide for microinjection increased the proportion of visible pronuclei in sheep eggs from 69% to 92%. DIC alone does not permit visualization of cow pronuclei. Centrifugation of cow eggs at high speed for a brief period results in visible equatorial pronuclei with yolk material at one pole of ova (Wall *et al.*, 1985).

Superovulation, collection of ova, microinjection of GH or GRF genes, and transfer of sheep embryos resulted in 5–17% of the eggs producing lambs (Table 1). Similar procedures in mice resulted in about 20% of eggs producing fetuses (Brinster *et al.*, 1985). Reduced viability of sheep embryos resulted from handling embryos for microinjection and from microinjection (Rexroad & Wall, 1987). Surprisingly, injection of buffer and buffer containing DNA did not differ greatly in degree of adverse effect. These results were similar to those reported for mice in which microinjection of buffer resulted in 28% of eggs producing fetuses and microinjection of DNA resulted in about 20% of eggs producing fetuses (Brinster *et al.*, 1985). Walton *et al.* (1987) found that reducing the largest portion of the microinjection needle that was likely to penetrate sheep eggs from 9·6 μm to 7·5 μm decreased the proportion of eggs that lysed after DNA microinjection.

Table 1. Efficiency of production of sheep containing growth-related transgenes

Transgene	No. of injected embryos transferred	Offspring		Transgenic offspring		Transgenic offspring expressing	
		No.	%*	No.	%*	No.	%
mMT-hGH[a]	1032	73	7·1	1	0·10		
mMT-bGH[b]	842	47	5·6	2	0·24	2	100
sMT-sGH5[c]	1089	83	7·4	4	0·37	0	0
sMT-sGH9[c]	409	23	5·6	3	0·74	3	100
mMT-hGRF[d]	435	63	14·5	9	2·07	1/7	14
mTF-bGH[d]	247	42	17·1	11	4·45	3	27
mAL-hGRF[e]	171	16	9·4	4	2·34	2	50
Total	4225	347	8·2	34	0·82	11	37·5

*% of embryos transferred.
[a]Hammer *et al.* (1985c).
[b]Hammer *et al.* (1986); Pursel *et al.* (1987).
[c]Murray *et al.* (1989).
[d]Rexroad *et al.* (1988).
[e]C. E. Rexroad, Jr, R. R. Behringer, D. J. Bolt, L. A. Frohman, K. E. Mayo, R. D. Palmiter & R. L. Brinster (unpublished results).

The proportion of microinjected eggs that produced transgenic lambs in a number of experiments varied from a low of 1 of 1032 eggs to a high of 11 of 247 (Table 1). Both studies were conducted at the same laboratory but utilized different genes, and the study with the higher rate was peformed 3 years after the first study. Although the role of the DNA cannot be ruled out as a source of variation between experiments in the production of transgenics, experience in microinjection is probably an important factor in determining the success of microinjection.

Production of transgenic mice was optimal for overall efficiency when 1–2 μg linear DNA/ml were injected into the male pronucleus (Brinster *et al.*, 1985). In sheep, microinjection of a sheep metallothionein–sheep GH fusion gene resulted in 0 of 20 transgenic lambs when the DNA concentration was 0·5 μg/ml but 5 of 44 when the concentration was 5 μg/ml. Transgenic lambs were also produced with concentrations of 1 and 2 µg/ml (Murray *et al.*, 1989). These concentrations are similar to those reported to be optimal in mice for overall efficiency of producing transgenics (Brinster *et al.*, 1985).

Besides DNA concentration, microinjection experience, and gene construct, a number of other factors may be important for microinjection. Brinster *et al.* (1985) found that hybrid mouse eggs (C57 × SJL) produced more transgenics (27·1%) than did C57 eggs (3·3%). The stage of egg development and quality of eggs may affect the efficiency of producing transgenics. The requirement for large numbers of embryos and transfers to study microinjection makes optimization of the microinjection technique difficult in ruminants.

Transgene expression

The proportion of transgenic sheep that expressed inserted genes varied from a low of 0 of 4 for sMT-sGH5 (sheep metallothionein–sheep GH fusion gene) to a high of 2 of 2 with mMT-bGH (mouse metallothionein-I–bovine GH fusion gene) and 3 of 3 with the sMT-sGH9 construct, and the overall proportion was 11 of 32 (Table 1). Several factors may account for cases of non-expression. The construction of the DNA used for microinjection can affect expression. In mice, the presence of introns in transgenes increases the proportion of mice that express several transgenes and the overall level of expression of each transgene (Brinster *et al.*, 1988). In addition, in mice, inclusion of vector sequences in the injected DNA reduced expression (Hammer *et al.*, 1985a). A similar phenomenon may have been observed in sheep. Murray *et al.* (1990) injected sMT-sGH5, which contained origin of replication and enhancer sequences from SV40, and obtained no expression in 4 transgenic lambs. The sMT-sGH5 construct expressed in transformed L-cells, but, as in sheep, failed to express in mice (Ward *et al.*, 1988). They then constructed sMT-sGH9 without the viral sequences and obtained expression in 3 of 3 lambs (Table 1). Other factors that may prevent expression of transgenes are gene rearrangement, as was the case for an mMT-hGH lamb (Hammer *et al.*, 1985c), or perhaps integration into a region of a chromosome that is unavailable for transcription.

Table 2. Plasma growth hormone concentrations in growth hormone transgenic lambs

Transgene	No. of lambs	GH conc. (ng/ml)	Reference
mMT-bGH	2	36–718	Rexroad *et al.* (1989)
mTF-bGH	2	42–289	Rexroad *et al.* (1988)
sMT-sGH9	3	3800–>26 000	Murray *et al.* (1989)

The promoter used in fusion genes allows novel sites and levels of expression of the coding sequences to which it is ligated and must therefore be chosen carefully. Rexroad *et al.* (1989) reported that only 1 of 7 lambs transgenic with a fusion gene of the mouse metallothionein-I promoter (mMT) ligated to the human growth hormone releasing factor minigene expressed the transgene. Similarly, only 2 of 7 pigs expressed the same transgene (Pursel *et al.*, 1989). The same minigene was expressed in 11 of 14 transgenic mice (Hammer *et al.*, 1985a, b). This observation may suggest that promoters work best in homologous species, but the same mMT promoter caused 50–100% expression with other transgenes (Table 1).

Additional support for the concept that some species homology between a promoter and the transgenic species may be beneficial for gene expression is seen in Table 2. The construct sMT-sGH9, which has a sheep metallothionein promoter, resulted in very high concentration of plasma

GH in lambs while mouse metallothionein and transferrin promoters elevated plasma GH but not to the same degree.

Different promoters may result in different plasma concentrations by defining different sites of expression. The mMT-bGH (Rexroad *et al.*, 1989) and sMT-sGH9 (Ward *et al.*, 1989) constructs were expressed in similar sets of tissues that included kidney, brain and liver. However, quantitative differences in the relative amount of mRNA in each tissue existed between lambs for mMT-bGH and may be the cause for each lamb appearing to have its own level of plasma GH. Thus, one mMT-bGH lamb had a plasma GH range of 36–84 ng/ml while another had a range of 470–720 ng/ml. One mTF-bGH lamb had a plasma GH range of 117–289 ng/ml and another had a range of 42–207 ng/ml (Rexroad *et al.*, 1988). One sMT-sGH9 lamb had plasma GH that ranged around 900 ng/ml, and 2 had values of $>$20 000 ng/ml.

Transgenes with metallothionein promoters should respond to changes in dietary heavy metal concentrations by increased production of gene product. In mMT-bGH lambs (Rexroad *et al.*, 1989) and sMT-sGH9 lambs (Murray *et al.*, 1989), plasma concentrations of GH were elevated from shortly after birth. These observations indicate that the inserted genes express constitutively or that trace amounts of metals such as zinc that are normally found in milk and in the environment were sufficient to induce the expression of the transgenes. If the integrated genes should not function until induced, then the metallothionein promoters reported in these studies are not satisfactory for use in ruminants.

Physiology of transgenic sheep with increased GH

Transgenic sheep with excess plasma GH resulting from expressing ovine GH (Ward *et al.*, 1989), from expressing bovine GH (Rexroad *et al.*, 1988), or having elevated ovine GH as the result of an mMT-hGRF transgene (Rexroad *et al.*, 1989) grew at rates similar to or slightly lower than those of control lambs. These results may be similar to those reported after injection of lambs with purified preparations of ovine or bovine GH in which non-significant increases for average daily gain were observed for lambs (20%: Wagner & Veenhuizen, 1978; 4%: Muir *et al.*, 1983) and to a report in which significant increases (22%) in average daily gain did not result in increased carcass weight (Johnson *et al.*, 1985). Small numbers and inability to propagate lines of transgenic sheep may have precluded observations of effects of transgene expression on growth. Pursel *et al.* (1989) observed little difference in growth between founder mMT-bGH transgenic pigs and controls. Lines of pigs derived from some founders had significantly higher average daily gains than did controls.

Ward *et al.* (1989) observed that three sMT-sGH9 transgenic lambs had one-half to one-fifth of the body fat of control lambs. The mTF-bGH-expressing transgenic lambs also had little or no backfat or depot fat at autopsy (C. E. Rexroad, Jr, R. R. Behringer, D. J. Bolt, L. A. Frohman, K. E. Mayo, R. D. Palmiter & R. L. Brinster, unpublished results). These findings agree with the observation that lambs injected with ovine GH had significantly less fat (Muir *et al.*, 1983). Reduced fat in lambs injected with purified bovine GH was accompanied by an increase in lean in the carcass (Johnson *et al.*, 1985). The sMT-sGH9 sheep did not have increased nitrogen retention. Even though fat deposition was reduced, feed/gain was not reduced in transgenic lambs (Rexroad *et al.*, 1988), which may reflect a large heat production increment in GH transgenic sheep (Ward *et al.*, 1989).

In spite of unchanged growth pattern, increased GH in expressing transgenics had pronounced effects on IGF-1. Plasma IGF-1 was increased in mTF-bGH (Rexroad *et al.*, 1988) and sMT-sGH9 (Ward *et al.*, 1989) lambs. Plasma insulin before 100 days of age was also elevated in mTF-bGH lambs (C. E. Rexroad, Jr, R. R. Behringer, D. J. Bolt, L. A. Frohman, K. E. Mayo, R. D. Palmiter & R. L. Brinster, unpublished results) and in sMT-sGH9 transgenic lambs (Ward *et al.*, 1989). Injections of purified bovine GH into growing lambs also increased plasma insulin concentrations (Johnson *et al.*, 1985). Increased insulin may have been secondary to a hyperglycaemia caused by elevated GH (for review, see Holly *et al.*, 1988). The mTF-bGH lambs had elevated plasma glucose and hydroxybutyrate concentrations after 100 days of age and were found to have glucosuria and abnormally low plasma insulin values at autopsy (Rexroad *et al.*, 1988). These observations suggest that excess GH secretion in transgenic lambs produced a lethal diabetes. Ward *et al.* (1989) reported that sMT-sGH9 lambs had degeneration of liver and kidneys at necropsy which may reflect diabetes associated degenerative changes.

Conclusions

Genes to increase growth hormone production, either GH genes or GRF genes, can be inserted into ruminants. Inserted genes function appropriately by directing production of their products to tissues other than the pituitary (GH) or hypothalamus (hGRF). Inducibility of inserted genes, for instance, increased production of GH by an mMT-bGH transgene in response to heavy metals, has not been demonstrated for ruminants. Lack of precise control of the transgenes with resultant over-production of GH caused lethal diabetes in sheep, with death by 1 year of age. Early death of the transgenic lambs prevented development of lines of transgenic sheep that could be thoroughly evaluated for GH effects on growth and feed efficiency. In pigs transgenic with mMT-bGH, lines could be developed from male founders (Pursel *et al.*, 1989). In these transgenic lines of pigs, it was demonstrated that the transgenes resulted in increased average daily gain, feed efficiency, and reduced backfat.

These observations suggest other options for further research to improve growth in ruminants using the transgenic approach. One option is to spend more effort developing promoters to allow precise regulation of the production of GH. Promoters might be selected that function during specific periods or that direct only induced secretion of GH. Another option to improve growth would be to look at other systems that affect growth. Other systems could include increasing receptors or the affinity of receptors for factors such as steroids, IGF-1, and growth hormone. The most significant restrictions to investigating these further applications appear to be the inefficiency (costliness) of producing transgenics and our lack of understanding of the actions of single genes in altering growth.

We thank Linda Neuenhahn for manuscript preparation.

References

Brinster, R.L., Chen, H.Y., Trumbauer, M.E., Yagle, M.K. & Palmiter, R.D. (1985) Factors affecting the efficiency of introducing foreign DNA into mice by microinjecting eggs. *Proc. natn. Acad. Sci. USA* **82,** 4438–4432.

Brinster, R.L., Allen, J.M., Behringer, R.R., Gelinas, R.E. & Palmiter, R.D. (1988) Introns increase transcriptional efficiency in transgenic mice. *Proc. natn. Acad. Sci. USA* **85,** 836–840.

Hammer, R.E., Brinster, R.L. & Palmiter, R.D. (1985a) Use of gene transfer to increase animal growth. *Cold Spring Harbor Symp.* **50,** 379–387.

Hammer, R.E., Brinster, R.L., Rosenfeld, M.G., Evans, R.M. & Mayo, K.E. (1985b) Expression of human growth hormone releasing factor in transgenic mice results in increased somatic growth. *Nature, Lond.* **315,** 413–416.

Hammer, R.E., Pursel, V.G., Rexroad, C.E., Jr, Wall, R.J., Bolt, D.J., Ebert, K.M., Palmiter, R.D. & Brinster, R.L. (1985c) Production of transgenic rabbits, sheep and pigs by microinjection. *Nature, Lond.* **315,** 680–683.

Hammer, R.E., Pursel, V.G., Rexroad, C.E., Jr, Wall, R.J., Bolt, D.J., Palmiter, R.D. & Brinster, R.L. (1986) Genetic engineering of mammalian embryos. *J. Anim. Sci.* **63,** 269–278.

Holly, J.M.P., Amiel, S.A., Sandhu, R.R., Rees, L.H. & Wass, J.A.H. (1988) The role of growth hormone in diabetes mellitus. *J. Endocr.* **118,** 353–364.

Johnson, I.D., Hart, I.C. & Butler-Hogg, B.W. (1985) The effects of exogenous bovine growth hormone and bromocriptine on growth, body development, fleece weight and plasma concentrations of growth hormone, insulin and prolactin in female lambs. *Anim. Prod.* **41,** 207–217.

Muir, L.A., Wien, S., Duquette, P.F., Rickes, E.L. & Cordes, E.H. (1983) Effects of exogenous growth hormone and diethylstilbestrol on growth and carcass composition of growing lambs. *J. Anim. Sci.* **56,** 1315–1323.

Murray, J.D., Nancarrow, C.D., Marshall, J.T., Hazelton, I.G. & Ward, K.A. (1989) The production of transgenic Merino sheep by microinjection of ovine metallothionein-ovine growth hormone fusion genes. *Reprod. Fert. Devel.* **1,** 147–155.

Palmiter, R.D., Brinster, R.L., Hammer, R.E., Trumbauer, M.E., Rosenfeld, M.G., Birnberg, N.C. & Evans, R.M. (1982) Dramatic growth of mice that develop from eggs microinjected with metallothionein-growth hormone fusion genes. *Nature, Lond.* **300,** 611–615.

Palmiter, R.D., Norstedt, G., Gelinas, R.E., Hammer, R.E. & Brinster, R.L. (1983) Metallothionein-human GH fusion genes stimulate growth of mice. *Science, NY* **222,** 809–814.

Pursel, V.G., Rexroad, C.E., Jr, Bolt, D.J., Miller, K.F., Wall, R.J., Hammer, R.E., Pinkert, C.A., Palmiter, R.D. & Brinster, R.L. (1987) Progress on gene transfer in farm animals. *Vet. Immunol. Immunopathol.* **17,** 303–312.

Pursel, V.G., Pinkert, C.A., Miller, K.F., Bolt, D.J., Campbell, R.G., Palmiter, R.D., Brinster, R.L. & Hammer, R.E. (1989) Genetic engineering of livestock. *Science, NY* **244,** 1281–1288.

Rexroad, C.E., Jr & Wall, R.A. (1987) Development of one-cell fertilized sheep ova following microinjection into pronuclei. *Theriogenology* **27,** 611–619.

Rexroad, C.E., Jr, Behringer, R.R., Bolt, D.J., Miller, K.F., Palmiter, R.D. & Brinster, R.L. (1988) Insertion and expression of a growth hormone fusion gene in sheep. *J. Anim. Sci.* **66** (Suppl. 1), 267, abstr.

Rexroad, C.E., Jr, Hammer, R.E., Bolt, D.J., Mayo, K.E., Frohman, L.A., Palmiter, R.D. & Brinster, R.L. (1989) Production of transgenic sheep with growth-regulating genes. *Molec. Reprod. Dev.* **1,** 164–169.

Wagner, J.F. & Veenhuizen, E.L. (1978) Growth performance, carcass deposition and plasma hormone levels in wether lambs when treated with growth hormone and thyroprotein. *J. Anim. Sci.* **47** (Suppl. 1), 397, abstr.

Wall, R.J., Pursel, V.G., Hammer, R.E. & Brinster, R.L. (1985) Development of porcine ova that were centrifuged to permit visualization of pronuclei and nuclei. *Biol. Reprod.* **32,** 645–651.

Walton, J.R., Murray, J.D., Marshall, J.T. & Nancarrow, C.D. (1987) Zygote viability in gene transfer experiments. *Biol. Reprod.* **37,** 957–967.

Ward, K.A., Murray, J.D., Shanahan, C.M., Rigby, N.W. & Nancarrow, C.D. (1988) The creation of transgenic sheep for increased wool production. In *The Biology of Wool and Hair*, pp. 465–477. Eds G. E. Rogers, P. J. Reis, K. A. Ward & R. C. Marshall. Chapman and Hall, London.

Ward, K.A., Nancarrow, C.D., Murray, J.D., Wynn, P.C., Speck, P. & Hales, J.R.S. (1989) The physiological consequences of growth hormone fusion gene expression in transgenic sheep. *J. cell. Biochem.* **13B,** 164, abstr.

J. Reprod. Fert., Suppl. **41,** (1990), 125–134

Cloning embryos by nuclear transfer

R. S. Prather*†‡ and N. L. First*

**Department of Meat and Animal Science, College of Agricultural and Life Sciences, University of Wisconsin, Madison, Wisconsin 53706, USA; and †Department of Animal Science, College of Agriculture, University of Missouri, Columbia, Missouri 65211, USA*

Summary. Nuclear transfer has been used to study the differentiation process in embryogenesis, as well as a method to produce multiple identical individuals. When nuclei are transferred to activated, enucleated oocytes the nuclei swell in diameter, synthesize DNA, acquire cytoplasmic proteins and release nuclear proteins. This protein exchange is thought to result in specific genomic modifications resulting in the transferred nucleus behaving as a zygotic nucleus. The limitations of development observed with relatively differentiated nuclei are thought to result from asynchronies in the length of the cell cycle between the donor cell and the recipient cytoplasm, as well as insufficient genomic modifications. This results in incomplete DNA synthesis and incomplete reprogramming before the first cell division. These nuclear modifications are discussed with data from amphibians and mammals.

Keywords: embryo; cloning; mammals; genome; reprogramming

Introduction

The transfer of individual gene constructs to host cells for propagation is standard procedure for the molecular biologist. However, transfer of an entire nuclear genome to a surrogate cell for propagation is not commonly considered. Nuclear transfer is conceptually no different from transfection experiments. In both cases the DNA is transferred to a host or surrogate cell and cell replication is encouraged. Transferring an entire nucleus to a cell from which the endogenous chromosomes have been removed requires that the transferred nucleus directs future mitotic activity. Hence, for nuclear transfer to succeed the transferred nucleus and the surrogate cytoplasm must co-operate to direct future divisions. When conducting nuclear transfer in early mammalian cells, a further restriction is placed on the nucleus and cytoplasm in that the nucleus must be reprogrammed in its developmental cascade of events. For example, a cow 32-cell stage blastomere will participate in forming a blastocyst after 1 or 2 cell divisions, and if a 32-cell stage blastomere is transferred to an enucleated meiotic metaphase II oocyte and is not reprogrammed it may attempt to cavitate at the 2- or 4-cell stage. This would be detrimental to development as too few cells would be present and a developmentally competent blastocyst would not be formed. However, if the nucleus of the 32-cell stage blastomere is reprogrammed to behave as a zygotic nucleus, then blastulation would occur at the correct cell number and the cell mass would be competent to develop to term. Therefore, nuclear transfer in early embryos requires that the transferred nucleus be both mitotically competent and developmentally reprogrammable.

The idea that nuclei change during development, i.e. differentiate, was proposed by the founders of biology and has remained a central concept of early development. Spemann (1938) is

‡Present address: Department of Animal Science, University of Missouri, Columbia, MO 65211, USA.

given credit for devising the experiment to answer the question of nuclear equivalence or 'do nuclei change as development progresses?' He described an experiment in which more and more advanced nuclei were transferred to enucleated 1-cell eggs (Spemann, 1938). The limit of development of the resulting eggs would indicate the degree of differentiation. This was designed to determine exactly when the nuclei became restricted in their developmental potential. It is unfortunate that his experiment was beyond the contemporary technology, for Spemann died in 1941 and never saw completion of his experiment. In 1952 Briggs & King completed Spemann's experiment and ushered in a new era of developmental biology. Now one could show that certain nuclei were equivalent; i.e. they could direct complete development to a sexually mature adult.

Since it is assumed, at least in early development, that all the nuclei in an individual are identical, then after nuclear transfer to an enucleated egg all nuclei from a single source should result in similar offspring. The resulting individuals would all have identical nuclear genomes and thus be clones. Therefore the practice of cloning is in actuality a spin-off of a procedure designed to answer the much more fundamental question of differentiation. To use the term clone in a more rigid sense we must prove that the cytoplasmic contributions are equal and no chromosomal rearrangements have taken place. Some possibilities of chromosomal changes include gene amplification (Tobler, 1975), gene rearrangements (Alt *et al.*, 1987), translocations (King & Linares, 1983) and diminution (Beerman, 1977). Even if these conditions are met, to be considered identical environmental factors must also be equivalent. A list of such factors that must be satisfied is given by Seidel (1983).

The advantages for having clones in research are obvious. Truly identical individuals would be perfect controls for all types of studies such as those relating to the environment, nutrition and reproduction. However, caution should be exercised when interpreting such data in an attempt to extrapolate to the entire species, because these nuclear clones should be considered similar to an inbred strain of mouse. Treatments may have different effects in different strains. Individuals with identical nuclear genomes, but made with different cytoplasmic sources would be useful for evaluating cytoplasmic inheritance and nuclear–cytoplasmic communication. On a commercial level, identical individuals would enhance marketing of livestock, as a more uniform product could be sold. Additionally, management of a group of identical individuals would be easier, as all would respond similarly to management changes. Therefore, there is both scientific and economic incentive to produce clones.

Procedures for nuclear transfer

The procedure used for nuclear transfer in amphibians has remained essentially unchanged since 1952 (Briggs & King, 1952) and has been described in detail by Elsdale *et al.* (1960) and Gurdon & Laskey (1970). The technique involves aspirating a cell containing a donor nucleus into a pipette that is just small enough to cause the cell to rupture. The pipette is then inserted into an unfertilized oocyte in metaphase II of meiosis and the ruptured cellular contents ejected. In some species this alone causes activation of the oocyte (*Xenopus* and *Rana*: Gurdon, 1986; hamster: Naish *et al.*, 1987), in other species the oocyte must then be activated (*Ambystoma* and *Pleurodeles*: Gurdon, 1986). After activation the egg attempts to elicit its second polar body. This is seen as a black speck on the surface of the oocyte and can be physically removed as in *Rana* or exposed to ultraviolet light as in *Xenopus*. The ultraviolet light effectively enucleates the egg as the exposed chromosomes do not participate in development (Gurdon, 1960).

In mammals a similar procedure has been developed for the mouse by McGrath & Solter (1983). Eggs to be manipulated are cultured in medium containing cytochalasin and colchicine. The cytochalasin acts as a microfilament inhibitor and imparts an elasticity to the plasma membranes such that a portion of a cell may be drawn up into a micropipette and removed as a cytoplast or karyoplast. The colchicine acts as an inhibitor of microtubules. Colchicine is only necessary when a large number of microtubules are present such as the pronuclear stage egg, but is not as important for the metaphase II oocyte since the majority of microtubules are present in the meiotic spindle (Schatten *et al.*, 1985). As in the amphibian the donor nucleus is aspirated into a micropipette; however, the membranes are not ruptured and the nucleus remains in the karyoplast. The karyoplast can then be transferred adjacent to a previously enucleated cell, generally within a zona pellucida. The two cells, karyoplast and cytoplast, are then fused together with the aid of Sendai virus (Graham, 1969; McGrath & Solter, 1983) or an electrical pulse (Berg, 1982; Robl *et al.*, 1987). This effectively transfers the donor nucleus into the recipient cytoplasm (Fig. 1). If electricity is used to complete the transfer this can also activate the oocyte (Prather *et al.*, 1987, 1989a).

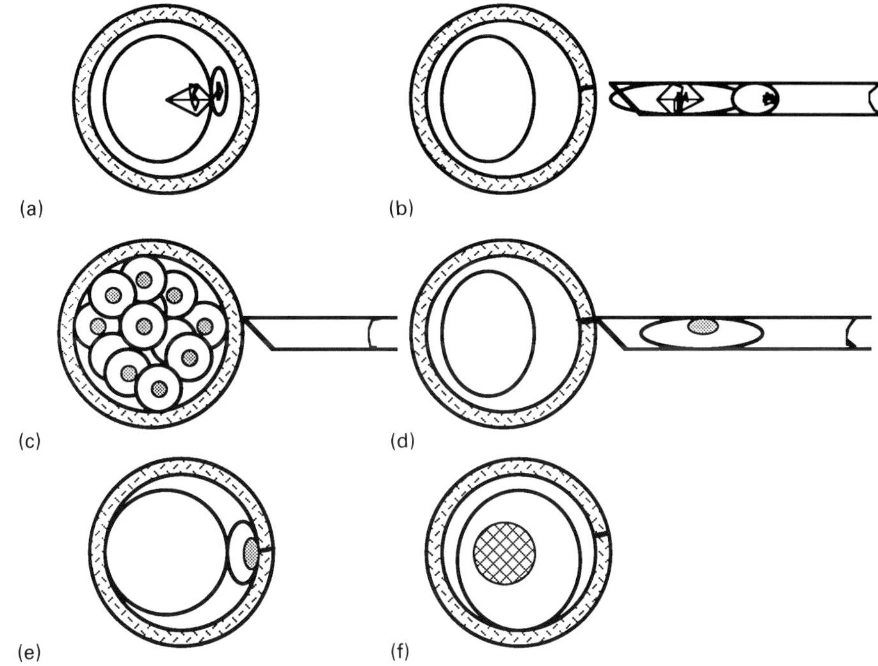

Fig. 1. Scheme of nuclear transfer procedure. (a) Meiotic metaphase II oocyte. (b) Meiotic metaphase II oocyte after removal of polar body and metaphase chromosomes. (c) Sixteen-cell stage donor embryo. (d) Before transfer of 16-cell stage blastomere to enucleated oocyte from (b). (e) After transfer of donor blastomere to enucleated oocyte. (f) After exposure to electric pulse and subsequent fusion. Note that swelling of the transferred nucleus as well as chromatin reorganization is depicted.

In both procedures described above the transfer results in transfer of some cytoplasm associated with the donor nucleus into the recipient cell. This can be a relatively small contribution as in the amphibian or a relatively large amount in mammals such as when an 8-cell stage blastomere is transferred to a one-half oocyte (Prather *et al.*, 1987). The contamination of donor cytoplasm may have effects on the developmental potential of the nucleus as well as containing mitochondria or other cytoplasmic organelles that would confound experiments attempting to distinguish between the nuclear and cytoplasmic inheritance.

The exchange of nuclei between two similar cell types can be used to validate the manipulation procedures and show that development can continue even after exposure to drugs that disrupt the cytoskeleton and actual micromanipulation. This has been achieved in mice with Sendai virus-mediated cell fusion for both the zygote (McGrath & Solter, 1983) and the 2-cell stage (Robl *et al.*, 1986), as well as with electrically mediated cell fusion in both the mouse (Tsunoda *et al.*, 1987a) and rat (Kono *et al.*, 1988). In domestic animals, pronuclear exchange (see Figs 2 and 3) has resulted in

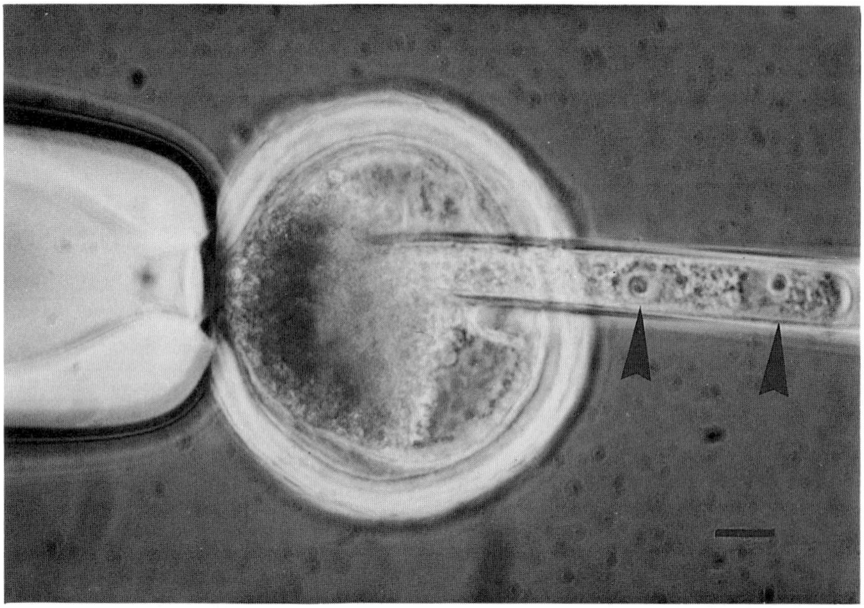

Fig. 2. Pig zygote during pronuclear exchange. The embryo is centrifuged, then treated with cytoskeletal inhibitors and micromanipulated. The two pronuclei (maternally and paternally derived) can be seen in the micropipette (arrows). Each pronucleus has a single nucleolus. After withdrawal of the pipette the membranes will bud off, thus forming a karyoplast containing 2 pronuclei and a cytoplast within the zona pellucida. Bar = 25 μm.

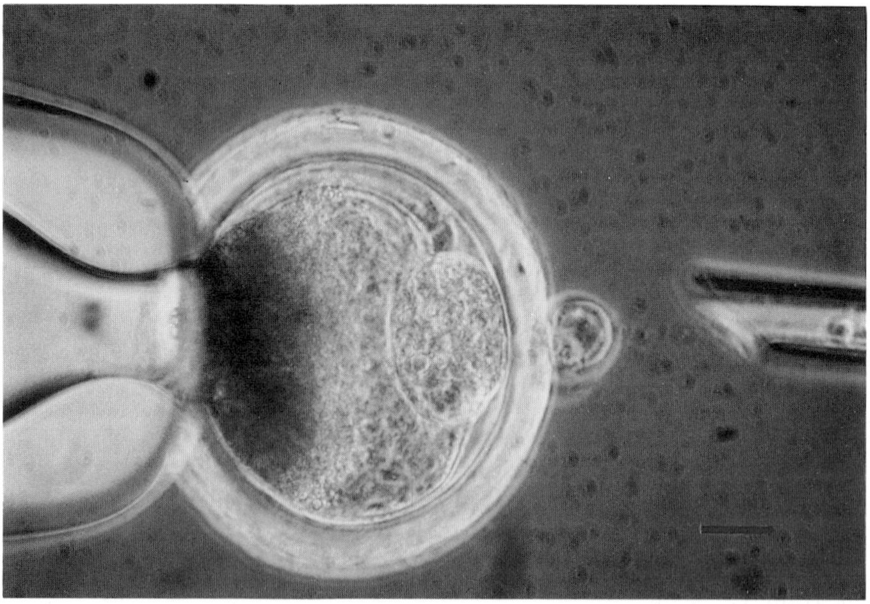

Fig. 3. Pig zygote after pronuclear exchange. This figure is after transfer of a karyoplast (Fig. 2) to another cytoplast (enucleated zygote). The embryo is now ready for electroporation and subsequent membrane fusion. Note the stratification of the cytoplasm due to the centrifugation. Bar = 25 μm.

offspring in cows (Robl *et al.*, 1987) and pigs (Prather *et al.*, 1989a). Therefore the procedures described above are compatible with development to term.

Events after nuclear transfer

After the transfer of (pro)nuclei between two similar types of cells, few major nuclear modifications would be expected. However, after the transfer of a nucleus to an enucleated, activated, oocyte various changes occur to the transferred nucleus. These changes are characterized by a remodelling of the nucleus and can be observed as a swelling of the transferred nucleus (Gurdon, 1964). This remodelling presumably results in a reprogramming of the transferred nucleus such that it behaves developmentally as if it were a 1-cell zygote. The events that occur after nuclear transfer have been well characterized in the amphibian and will be discussed in that context with information about mammals added as appropriate.

Nuclear remodelling

Nuclear remodelling is a process of changes that occur to a nucleus after transfer to an oocyte. These morphological and biochemical changes in the nuclear structure are thought to be a result of cytoplasmic components from the oocyte acting upon the transferred nucleus that result in a reprogramming of genomic expression. A morphological indication of nuclear remodelling is the disappearance of nucleoli after transfer to an oocyte, as nucleoli are present in *Xenopus* larval stage nuclei, but not in early cleavage stages (Gurdon & Brown, 1965). Another indicator is swelling of the transferred nucleus. This occurs in amphibians (Gurdon, 1964; Gurdon & Brown, 1965) as well as mammals (mouse: Czołowska *et al.*, 1984; rabbit: Stice & Robl, 1988; pig: Prather *et al.*, 1988) and is dependent upon the time of transfer in relation to the time of activation. There appears to be a 1·5-h window around the time of activation when an introduced nucleus can swell to a size similar to that of an endogenous pronucleus (Czołowska *et al.*, 1984). This is a similar time frame for events that occur during formation of the male pronucleus at fertilization (Usui & Yanagimachi, 1976). In addition, swelling does not occur if the recipient cell is an enucleated zygote, enucleated 2-cell stage blastomere (Barnes *et al.*, 1987), or an enucleated germinal vesicle oocyte (Dettlaff *et al.*, 1964).

The remodelling that does occur may be a result of the exchange of proteins between the nuclear and cytoplasmic compartments. The exchange of proteins between the nucleus and cytoplasm does not appear to be limited by the nuclear envelope, but by selective binding sites within the nucleus (Feldher & Pomerantz, 1978). Some proteins appear to migrate out of or into the nucleus at different rates than others. Specifically in *Rana*, non-histone [^{3}H]tryptophan-containing proteins leave advanced endodermal nuclei after transfer to an oocyte, but [^{3}H]lysine-containing proteins remain in the nucleus (DiBerardino & Hoffner, 1975; Leonard *et al.*, 1982). Simultaneously, cytoplasmic proteins, both basic and acidic, are acquired by the nucleus (Merriam, 1969; Gurdon, 1986). This protein uptake appears to be an inducer of the nuclear swelling and is not a consequence of nuclear swelling (Merriam, 1969).

In amphibians and mammals there are changes in the distribution of nuclear antigens during early cleavage (Dreyer, 1987; Stricker *et al.*, 1989). One class of proteins that make a similar change is the nuclear lamins. The A/C type of nuclear lamins becomes undetectable after the transition to zygotic control of development in the mouse, pig and cow (Schatten *et al.*, 1985; Prather *et al.*, 1989b). However, in the mouse (R. S. Prather, N. L. First & G. Schatten, unpublished observations) and the pig (Prather *et al.*, 1989b) a nucleus past the transition to zygotic control of development can acquire the A/C epitope if transferred to an enucleated activated oocyte. Additionally, in the mouse this epitope is only lightly acquired if the recipient cell is an intact or enucleated zygote (R. S. Prather, N. L. First & G. Schatten, unpublished observations).

Correlated with swelling and the exchange of proteins is the regulation of DNA synthesis. Nuclei in the G_1 or S phase of the cell cycle initiate or continue DNA synthesis after transfer to an oocyte in amphibians (Graham *et al.*, 1966) and mammals (Naish *et al.*, 1987), but nuclei in G_2 do not undergo DNA synthesis (DeRoeper *et al.*, 1977). The ability to cause replication of DNA occurs at germinal vesicle breakdown, and so the amphibian egg need not be activated to initiate DNA synthesis (Gurdon, 1967). The DNA synthetic activity then remains for up to 65 min after activation (Gurdon, 1967). Cells that are more differentiated, and so more slowly dividing, initiate DNA synthesis later and take longer to complete DNA synthesis than do less differentiated rapidly dividing cells (Graham *et al.*, 1966). Therefore, in a rapidly dividing population of embryonic donor cells there is no effect of stage of the cell cycle in subsequent development, unless the donor cell is actually entering mitosis (McAvoy *et al.*, 1975; Ellinger, 1978; von Beroldington, 1981). However, some cells develop at a higher rate if they are in G_2 versus G_1 (von Beroldington, 1981). Limitations in development can often be traced to chromosomal abnormalities (Gurdon, 1964; Briggs *et al.*, 1964; DiBerardino, 1979) that appear to occur in the first cell cycle after nuclear transfer (DiBerardino & Hoffner, 1970). Serial nuclear transfer reveals that these chromosomal abnormalities show stable developmental restrictions (Briggs *et al.*, 1964; DiBerardino & King, 1965). Many of these chromosomal abnormalities can be rescued by chimaerization or triploidy (Subtelny, 1965). The amount of cytoplasm transferred may be very important since adult liver proteins can cause chromosomal abnormalities in normal fertilized *Rana* eggs (Markert & Ursprung, 1963; Ursprung & Markert, 1963).

Nuclear remodelling is a physical change in the structure of the nucleus that may be a result of the exchange of proteins between the nucleus and cytoplasm. Transferring nuclei to metaphase II oocytes results in swelling of the nucleus, the disappearance of nucleoli and DNA synthesis in non-G_2 nuclei.

Nuclear reprogramming

The nuclear remodelling that occurs after nuclear transfer as discussed above presumably results in a reprogramming of the transferred nucleus. To be effective the reprogramming should modify the nucleus so that it behaves as if it were a zygote nucleus. Reprogramming can therefore be defined both morphologically and biochemically, based on the timing of the appearance of developmental stages and specific gene products, respectively.

Amphibians. Nuclear reprogramming has been well characterized both morphologically and biochemically in amphibians. For example, in *Xenopus* RNA synthesis begins shortly before the mid-blastula transition (MBT: Newport & Kirschner, 1982; Nakakura *et al.*, 1987). Cells past the MBT actively synthesize rRNA; this synthesis stops after nuclear transfer of a post-MBT nucleus to an enucleated activated metaphase II oocyte (Gurdon & Brown, 1965), but resumes when the resulting embryo reaches the MBT. Correlated with the synthesis of rRNA is the appearance of nucleoli at the MBT and their disappearance after nuclear transfer (Gurdon & Brown, 1965). The gene coding for muscle-specific actin begins producing RNA in myotome cells at the gastrula stage. After nuclear transfer of a post-gastrula myotome nucleus the actin gene stops transcription until the resulting embryo achieves the gastrula stage (Gurdon *et al.*, 1984). Similarly, the 5 S^{ooc} gene is active for only a short period of time at the late blastula stage. Neurula-stage nuclei can express this gene again at the late blastula stage if transferred to an oocyte and allowed to continue development to the late blastula stage (Wakefield & Gurdon, 1983). Even cells that are terminally differentiated and inactive in RNA and DNA synthesis can be induced to direct limited development after nuclear transfer (Orr *et al.*, 1986).

Mammals. Morphological criteria have evaluated those nuclear transfers in mammals that result in limited or complete development. In the sheep (Willadsen, 1986), cow (Prather *et al.*, 1987), rabbit (Stice & Robl, 1988) and pig (Prather *et al.*, 1989a), nuclear transfer embryos form a blastocoele at a time corresponding to a zygote. Results in the mouse, when using an enucleated

2-cell stage blastomere as a recipient, are conflicting. Tsunoda *et al.* (1987b) state that compaction occurs earlier than for 2-cell stage controls, whereas Barnes *et al.* (1987) showed that the time of blastocoele formation was the same as for control 2-cell stage blastomeres. Both Tsunoda *et al.* (1987b) and Barnes *et al.* (1987) show a delay in the timing of developmental events, suggestive of some degree of morphological reprogramming.

Developmental potential of donor nuclei

The desired result of nuclear transfer is the ability of the transferred nucleus to direct development to normal adulthood. However, most nuclear transfer embryos stop development before adulthood. This section will evaluate the limits of development for different types of nuclei and will draw upon the data discussed above to aid the explanations.

Amphibians. Blastula-stage nuclei from *Xenopus* are capable of directing development to adulthood after nuclear transfer to an enucleated activated oocyte. However, as more and more developmentally advanced nuclei are used for the nuclear transfer, the maximal development progressively decreases (reviewed by Gurdon, 1986; DiBerardino, 1987). This decrease in development sometimes occurs abruptly as when nuclei are harvested from the notochord of neurula-stage *Ambystoma* embryos. Differentiation of the notochord begins at the anterior end and progresses to the posterior end. The resulting development after nuclear transfer is greater with nuclei from the anterior region than with those from the more differentiated posterior region (Briggs *et al.*, 1964). Furthermore, when limited development was observed it could be traced to chromosomal abnormalities (Briggs *et al.*, 1964). Similar restricted development is seen in *Rana* (DiBerardino & King, 1967) and this restricted development is not advanced by serial nuclear transfer. DiBerardino & King (1967) concluded that the limited development is due to the inability of the nuclei to replicate normally as development progresses. Others have shown that serial nuclear transfers show stable developmental restrictions probably due to chromosomal damage (Briggs *et al.*, 1964) caused by the inability of the DNA to replicate completely in the first cell cycle after nuclear transfer.

Serial nuclear transfer is effective in extending the limit of development when first transferring to a metaphase I oocyte followed by development to the blastula stage and subsequent transfer of nuclei to a metaphase II oocyte (Orr *et al.*, 1986). The donor nucleus in this study was a terminally differentiated red blood cell. This type of nuclear transfer results in synthesis of RNA from a previously synthetically inactive nucleus. However, the limit of development is not extended beyond the tadpole stage (Orr *et al.*, 1986).

The developmental potential of *Rana* nuclei can be enhanced with two techniques. One is to cool the embryos to 10°C during nuclear transfer (Hennen, 1970). This presumably allows a better exchange of proteins between the nucleus and cytoplasm, as cooling may reduce the binding of proteins to one another. The second technique that can enhance development is to add the polyamine spermine to the manipulation medium. Polyamines are known to associate with and affect the spatial conformation of DNA, affect both RNA chain initiation and elongation (Karpetsky *et al.*, 1977) and affect differentiation in plants and animals (Alexandre & Gueskins, 1984; Lane & Davis, 1984; Malmberg *et al.*, 1985). Cooling and spermine both appear to have a more beneficial effect in *Rana* than *Xenopus* (Gurdon, 1986). Conversely, delaying the first cleavage increases the limits of development in *Xenopus* but not *Rana* (McAvoy *et al.*, 1975). In the axolotl the addition of protamines or poly-arginine enhances development, whereas spermine has no beneficial effect (Brothers, 1985).

Mammals. The developmental potential of nuclei in mammals has just begun to be evaluated. In mice, nuclei transferred to enucleated oocytes rarely cleave more than once (Prather, 1989); however, 8-cell nuclei transferred to enucleated 2-cell stage blastomeres develop to term (Tsunoda *et al.*, 1987b). In the cow development can continue to term from 16- or 32-cell stage embryos (Prather *et al.*, 1987; N. L. First, unpublished observation). Development in the rabbit can result from 8-cell stage nuclei (Stice & Robl, 1988). Sheep nuclei from the 8- or 120-cell stage can direct development to term (Willadsen, 1986; Smith & Wilmut 1989). The development of 4-cell pig embryos after nuclear transfer can also proceed to term (Prather *et al.*, 1989a). The mammalian equivalent to the MBT occurs at the 2-cell, 4-cell, 8-cell and 8–16-cell stage in the mouse, pig, cow and sheep, respectively (reviewed by Prather & First, 1988; First & Barnes, 1989). Therefore little genomic reprogramming occurs to nuclei transferred before their respective MBT equivalent. However, for nuclei in the sheep at the 120-cell stage significant reprogramming is likely to occur. Experiments yielding information for mammals are very limited and the subject is providing a fertile area for continued research.

Conclusions

The ability to transfer nuclei and to achieve developmental reprogramming that would permit development to continue to term have both scientific and economic merit. The procedures developed to date show that many nuclei in the development of amphibians and mammals are equivalent. However, the nuclei must be reprogrammed by cytoplasmic components found in the recipient oocyte. The reprogramming that can occur is highly specific. There is a correlation between the ability of the transferred nucleus to replicate and the subsequent ability to direct development. There is also a correlation between the length of the cell cycle of the donor nucleus and the subsequent ability to replicate after nuclear transfer. Finally, a positive correlation exists between the length of the cell cycle and the degree of differentiation of the donor nucleus. Therefore, firm conclusions about the degree of differentiation and the developmental potentials cannot be made since the degree of differentiation and cell cycle length are confounded; i.e., as development progresses, the cell cycle length increases and the ability to direct subsequent development decreases (reviewed by Gurdon, 1986; DiBerardino, 1987; Prather, 1989 Prather & First 1989). Since the early embryos of mammals have relatively longer cell cycle lengths than early amphibian embryos an answer concerning development versus cell cycle length may be readily attainable. Understanding development in the mammalian embryo is in its infancy and nuclear transfer provides an experimental tool to extract a great deal of valuable information concerning mammalian differentiation and development.

This manuscript was prepared while supported by the Cooperative State Research Service, U.S. Department of Agriculture under agreement No. 88-37240-3755, and is a contribution from the Missouri Agricultural Experiment Station Journal Series Number 10 878.

References

Alexandre, H. & Gueskins, M. (1984) Relationship between the effects of polyamine depletion on DNA transcription and on blastocyst formation in the mouse. *Archs Biol.* (Brussels) **95,** 55–70.

Alt, F.W., Blackwell, K. Yancopoulos, G.D. (1987) Development of the primary antibody repertoire. *Science, NY* **238,** 1079–1087.

Barnes, F.L., Robl, J.M. & First, N.L. (1987) Nuclear transplantation in mouse embryos: assessment of nuclear function. *Biol. Reprod.* **36,** 1267–1274.

Beerman, S. (1977) The diminution of heterochromatic chromosomal segments in Cyclops. *Chromosoma* **60,** 297–344.

Berg, H. (1982) Fusion of blastomeres and blastocysts of mouse embryos. *Bioelectricity and Bioenergetics* **9,** 223–228.

Briggs, R. & King, T.J. (1952) Transplantation of living cell nuclei from blastula cells into enucleated frogs' eggs. *Proc. natn. Acad. Sci. USA* **38,** 455–463.

Briggs, R., Signornet, J. & Humphrey, R.R. (1964) Transplantation of nuclei of various cell types from neurulae of the Mexican axolotl (*Ambystoma mexicanum*). *Devl Biol.* **10,** 233–246.

Brothers, A.J. (1985) Protamine mediated enhancement of nuclear expression. *Cell Diff., Suppl.* **16,** Abstr.

Czołowska, R., Modlinski, J.A. & Tarkowski, A.K. (1984) Behavior of thymocyte nuclei in nonactivated and activated mouse oocytes. *J. Cell Sci.* **69,** 19–34.

DeRoeper, A., Smith, J.A., Watt, R.A. & Barry, J.M. (1977) Chromatin dispersal and DNA synthesis in G1 and G2 HeLa cell nuclei injected into *Xenopus* eggs. *Nature, Lond.* **265,** 469–470.

Dettlaff, T.A., Nikitina, L.A. & Stroeva, O.G. (1964) The role of the germinal vesicle in oocyte maturation in anurans as revealed by the removal and transplantation of nuclei. *J. Embryol. exp. Morph.* **12,** 851–873.

DiBerardino, M.A. (1979) Nuclear and chromosomal behavior in amphibian nuclear transplants. *Int. Rev. Cytol., Suppl.* **9,** 129–160.

DiBerardino, M.A. (1987) Genomic potential of differentiated cells analyzed by nuclear transplantation. *Amer. Zool.* **27,** 623–644.

DiBerardino, M.A. & Hoffner, N.J. (1970) Origin of chromosomal abnormalities in nuclear transplants—a reevaluation of nuclear differentiation and nuclear equivalence in amphibians. *Devl Biol.* **23,** 185–209.

DiBerardino, M.A. & Hoffner, N.J. (1975) Nucleocytoplasmic exchange of nonhistone proteins in amphibian embryos. *Expl Cell Res.* **94,** 235–252.

DiBerardino, M.A. & King, T.J. (1965) Transplantation of nuclei from the frog adenocarcinoma. II Chromosomal and histologic analysis of tumor nuclear transplant embryos. *Devl Biol.* **11,** 217–242.

DiBerardino, M.A. & King, T.J. (1967) Development and cellular differentiation of neural nuclear-transplant embryos of known karyotype. *Devl Biol.* **15,** 102–128.

Dreyer, C. (1987) Differential accumulation of oocyte nuclear proteins by embryonic nuclei of *Xenopus. Development* **101,** 829–846.

Ellinger, M.S. (1978) The cell cycle and transplantation of blastula nuclei in *Bombina orientalis. Devl Biol.* **65,** 81–89.

Elsdale, T.R., Gurdon, J.B. & Fischberg, M. (1960) A description of the technique for nuclear transplantation in *Xenopus laevis. J. Embryol. exp. Morph.* **8,** 437–444.

Feldher, C.M. & Pomerantz, J. (1978) Mechanism for the selection of nuclear polypeptides in *Xenopus* oocytes. *J. Cell Biol.* **78,** 168–175.

First, N.L. & Barnes, F.L. (1989) Development of preimplantation mammalian embryos. In *Development of Preimplantation Embryos and Their Environment*, pp. 151–170. Eds K. Yoshinaga & T. Mori. Alan R. Liss, Inc., New York.

Graham, C.F. (1969) The fusion of cells with one- and two-cell mouse embryos. In *The Heterospecific Genome Interaction*, pp. 19–35. The Wistar Institute Press.

Graham, C.F., Arms, K. & Gurdon, J.B. (1966) The induction of DNA synthesis by frog egg cytoplasm. *Devl Biol.* **14,** 349–381.

Gurdon, J.B. (1960) The effects of ultraviolet irradiation on the uncleaved eggs of *Xenopus laevis. Q. Jl microsc. Sci.* **101,** 299–312.

Gurdon, J.B. (1964) The transplantation of living cell nuclei. *Adv. Morphol.* **4,** 1–43.

Gurdon, J.B. (1967) On the origin and persistence of a cytoplasmic state inducing nuclear DNA synthesis in frogs' eggs. *Proc. natn. Acad. Sci. USA* **58,** 545–552.

Gurdon, J.B. (1986) Nuclear transplantation in eggs and oocytes. *J. Cell Sci., Suppl.* **4,** 287–318.

Gurdon, J.B. & Brown, D.D. (1965) Cytoplasmic regulation of RNA synthesis and nucleolus formation in developing embryos of *Xenopus laevis. J. molec. Biol.* **12,** 27–35.

Gurdon, J.B. & Laskey, R.A. (1970) Methods of transplanting nuclei from single cultured cells to unfertilized frogs' eggs. *J. Embryol. exp. Morph.* **24,** 227–248.

Gurdon, J.B., Brennan, S., Fairman, S. & Mohun, T.L. (1984) Transcription of muscle specific actin genes in early *Xenopus* development: nuclear transplantation and cell dissociation. *Cell* **38,** 691–700.

Hennen, S. (1970) Influence of spermine and reduced temperature on the ability of transplanted nuclei to promote normal development in eggs of *Rana pipiens. Proc. natn. Acad. Sci. USA* **66,** 630–637.

Karpetsky, T.P., Heiter, P.A., Frank, J.J. & Levy, C.C. (1977) Polyamines, ribonucleases, and the stability of RNA. *Molec. cell. Biochem.* **17,** 89–99.

King, W.A. & Linares, T. (1983) A cytogenetic study of repeat breeder heifers and their embryos. *Can. vet. J.* **24,** 112–115.

Kono, T., Shioda, Y. & Tsunoda, Y. (1988) Nuclear transplantation of rat embryos. *J. exp. Zool.* **248,** 303–305.

Lane, S.M. & Davis, D. (1984) Regulation of embryo development in pigs: effect of inhibiting polyamine synthesis. *J. Anim. Sci., Suppl.* **59,** 122, abstr.

Leonard, R.A., Hoffner, N.J. & DiBerardino, M.A. (1982) Induction of DNA synthesis in amphibian erythroid nuclei in *Rana* eggs following conditioning in meiotic oocytes. *Devl Biol.* **92,** 343–355.

Malmberg, R.L., McIndoo, J., Hiatt, A.C. & Lowe, B.A. (1985) Genetics of polyamine synthesis in tobacco: developmental studies in the flower. *Cold Spring Harb. Symp. Quant. Biol.* **50,** 475–482.

Markert, C.L. & Ursprung, H. (1963) Production of reliable persistent changes in zygote chromosomes of *Rana pipiens* by injected proteins from adult liver nuclei. *Devl Biol.* **7,** 560–577.

McAvoy, J.W., Dixon, K.E. & Marshall, J.A. (1975) Effects in differences in mitotic activity, stage cell cycle, and degree of specialization of donor cells on nuclear transplantation in *Xenopus laevis. Devl Biol.* **45,** 330–339.

McGrath, J. & Solter, D. (1983) Nuclear transplantation in the mouse embryo by microsurgery and cell fusion. *Science, NY* **220,** 1300–1302.

Merriam, R.W. (1969) Movement of cytoplasmic proteins into nuclei induced to enlarge and initiate DNA or RNA synthesis. *J. Cell Sci.* **5,** 333–349.

Naish, S., Perreault, S.J. & Zirkin, B. (1987) DNA synthesis following microinjection of heterologous sperm and somatic cell nuclei into hamster oocytes. *Gamete Res.* **18,** 109–120.

Nakakura, N., Miura, T., Yamana, K., Ito, A. & Shiokawa, K. (1987) Synthesis of heterogenous mRNA-like RNA and low-molecular RNA before the midblastula transition in embryos of *Xenopus laevis. Devl Biol.* **123,** 421–429.

Newport, J. & Kirschner, M. (1982) A major developmental transition on *Xenopus* embryos. II. Control of the onset of transcription. *Cell* **30,** 687–696.

Orr, N.H., DiBerardino, M.A. & McKinnell, R.G. (1986) The genome of the frog displays centuplicate replications. *Proc. natn. Acad. Sci. USA* **83,** 1369–1373.

Prather, R.S. (1989) Nuclear transfer in mammals and amphibians: nuclear equivalence, species specificity? In *The Molecular Biology of Fertilization*, pp. 323–340. Eds G. Schatten & H. Schatten. Academic Press, Orlando.

Prather, R.S. & First, N.L. (1988) A review of early mouse embryogenesis and its applications to domestic species. *J. Anim. Sci.* **66,** 2626–2635.

Prather, R.S. & First, N.L. (1989) Nuclear transfer in mammalian embryos. *Int. Rev. Cytol.* **120,** 169–190.

Prather, R.S., Barnes, F.L., Sims, M.M., Robl, J.M., Eyestone, W.H. & First, N.L. (1987) Nuclear transplantation in the bovine embryo: assessment of donor nuclei and recipient oocyte. *Biol. Reprod.* **37,** 859–866.

Prather, R.S., Sims, M.M. & First, N.L. (1988) Nuclear transplantation in the porcine embryo. *Theriogenology* **29,** 290, abstr.

Prather, R.S., Sims, M.M. & First, N.L. (1989a) Nuclear transplantation in early porcine embryos. *Biol. Reprod.* **41,** 414–418.

Prather, R.S., Sims, M.M., Maul, G.G., First, N.L. & Schatten, G. (1989b) Nuclear lamin antigens: porcine and bovine early developmental regulation. *Biol. Reprod.* **41,** 123–132.

Robl, J.M., Gilligan, B., Critser, E.S. & First, N.L. (1986) Nuclear transplantation in mouse embryos: assessment of recipient cell stage. *Biol. Reprod.* **34,** 733–739.

Robl, J.M., Prather, R., Barnes, F., Eyestone, W., Northey, D., Gilligan, B. & First, N.L. (1987) Nuclear transplantation in bovine embryos. *J. Anim. Sci.* **64,** 642–647.

Schatten, G., Maul, G.G., Schatten, H., Chaly, N., Simerly, C., Balczon, R. & Brown, D.L. (1985) Nuclear lamins and peripheral nuclear antigens during fertilization and embryogenesis in mice and sea urchins. *Proc. natn. Acad. Sci. USA* **82,** 4727–4731.

Seidel, G.E., Jr (1983) Production of genetically identical sets of mammals: cloning? *J. exp. Zool.* **228,** 347–354.

Smith, L.C. & Wilmut, I. (1989) Influence of nuclear and cytoplasmic activity on the development in vivo of sheep embryos after nuclear transplantation. *Biol. Reprod.* **40,** 1027–1036.

Spemann, H. (1938) *Embryonic Development and Induction*, pp. 210–211. Hafner Publishing Company, New York.

Stice, S.L. & Robl, J.M. (1988) Nuclear reprogramming in nuclear transplant rabbit embryos. *Biol. Reprod.* **39,** 657–664.

Stricker, S., Prather, R., Simerly, C., Schatten, H. & Schatten, G. (1989) Nuclear architectural changes during fertilization and development. In *The Cell Biology of Fertilization*, pp. 225–250. Eds H. Schatten & G. Schatten. Academic Press, Orlando.

Subtelny, S. (1965) On the nature of the restricted differentiation-promoting ability of transplanted *Rana pipiens* nuclei from differentiating endoderm cells. *J. exp. Zool.* **159,** 59–92.

Tobler, H. (1975) The occurrence and developmental significance of gene amplification. In *Biochemistry of Animal Development*, vol. 3, pp. 91–123. Ed. R. Weber. Academic Press, New York.

Tsunoda, Y., Kato, Y. & Shioda, Y. (1987a) Electrofusion for the pronuclear transplantation of mouse eggs. *Gamete Res.* **17,** 15–20.

Tsunoda, Y., Yasui, T., Shioda, Y., Nakamura, K., Uchida, T. & Sugie, T. (1987b) Full term development of mouse blastomere nuclei transplanted into enucleated two-cell embryos. *J. exp. Zool.* **242,** 147–151.

Ursprung, H. & Markert, C.L. (1963) Chromosomal complement of *Rana pipiens* embryos developing from eggs injected with proteins from adult liver cells. *Devl Biol.* **8,** 309–321.

Usui, N. & Yanagimachi, R. (1976) Behavior of hamster sperm nuclei incorporated into eggs at various stages of maturation, and early development. The appearance and disappearance of factors involved in sperm chromatin decondensation in egg cytoplasm. *J. Ultrastruct. Res.* **57,** 276–288.

von Beroldington, C.H. (1981) The developmental potential of synchronized amphibian cell nuclei. *Devl Biol.* **81,** 115–126.

Wakefield, L. & Gurdon, J.B. (1983) Cytoplasmic regulation of 5S RNA genes in nuclear-transplant embryos. *EMBO J.* **2,** 1613–1619.

Willadsen, S.M. (1986) Nuclear transplantation in sheep embryos. *Nature, Lond.* **320,** 63–65.

J. Reprod. Fert., Suppl. **41** (1990), 135–146

Printed in Great Britain

Modification of milk composition

I. Wilmut, A. L. Archibald, S. Harris, M. McClenaghan, J. P. Simons, C. B. A. Whitelaw and A. J. Clark

AFRC Institute of Animal Physiology and Genetics Research, Edinburgh Research Station, Roslin, Midlothian EH25 9PS, UK

Summary. Revolutionary new opportunities for the modification of milk composition have been created by the development of methods for gene transfer and targeted mutation of genes may extend the range of opportunities still further. Exploitation of these opportunities depends upon selection and cloning of milk protein genes and identification of the sequences that govern tissue-specific hormonally induced expression in the mammary gland. Fragments of the ovine β-lactoglobulin gene fused to cDNA for the human therapeutic proteins clotting factor IX and α-1 antitrypsin have directed production of these proteins in the milk of transgenic mice and sheep. Factor IX was biologically active and co-migrated with authentic proteins, but was present at too low a concentration for commercial exploitation. Recent observations suggest that fusion genes containing genomic clones direct production of higher concentrations of protein. Mouse whey acidic protein genomic sequences also directed production of low concentrations of human tissue plasminogen activator in the milk of transgenic mice. Targeted expression of this kind may be used for the production of therapeutic and industrial proteins, to increase the concentration or modify the nature of milk proteins, reduce the concentration of lactose, change the composition of fat or direct production of bacteriocidal proteins in milk in order to combat mastitis.

Keywords: milk; transgenic, beta-lactoglobulin; Factor IX; $alpha_1$-antitrypsin

Introduction

The ability to transfer genes into the germline of livestock creates revolutionary new opportunities for the modification of animal production traits. Many of these traits are subject to complex control, and while the success of selective breeding reveals the role of genotype in these traits, the molecular and physiological basis of control of these traits is poorly understood. Successful manipulation of such traits will necessarily avoid deleterious pleiotropic effects. The composition of milk, in particular the protein content of milk, is a trait which is controlled very directly by gene expression, and alteration of which may be expected to lead to no large pleiotropic effects.

The mammary gland of a dairy animal can be envisaged as an efficient vat for the production of specific proteins, sugars and fats. In this role it has long been exploited by man as a source of food, particularly protein. In developed countries up to 20% of dietary protein is derived from milk. Changes may be envisaged either to proteins that are normally produced in milk or to direct production of novel proteins to the mammary gland (Lathe *et al.*, 1986). In addition, by changes to the enzymes of the mammary epithelial cells, it may be possible to alter the nature of the carbohydrate and lipid components of milk.

This paper will consider the structure of milk, review experiments to target gene expression to the mammary gland, and discuss potential applications of the technology. Particular emphasis will be placed upon the use of techniques for gene transfer as these are available at present, but mention will also be made of some potential applications of targeted mutation.

Gene transfer in ruminants has been achieved by direct injection of a few hundred copies of the gene into a nucleus of an early embryo, usually a pronucleus in a zygote (Ward *et al.*, 1986; Simons *et al.*, 1988; Biery *et al.*, 1988). While useful, this approach has several disadvantages. Less than 1% of injected eggs become transgenic young; the site of integration is apparently random and in some 8% of cases within an endogenous gene (Palmiter & Brinster, 1986). In addition, the neighbouring DNA affects expression of the transgene (see below). Two approaches are being followed to overcome these limitations, i.e. use of retroviral vectors and manipulation of embryo stem cells.

It may be possible to exploit the ability of retroviruses to incorporate their genome into the chromosomes of the host cell (Varmus, 1988) to increase the frequency of gene incorporation. Genes have been introduced successfully by retrovirus, but the frequency of integration in these early trials was not greater than after microinjection (Jaehner *et al.*, 1985; van der Putten *et al.*, 1985). As the embryos that were used in these experiments had cleaved, the resulting mice were mosaics, having some cells with the gene and some without. In addition, there are limits to the size of the gene that can be carried by a retrovirus. In early experiments the retroviral sequences affected expression of the transgene (see Jaehner *et al.*, 1985). However, it has now been shown that efficient tissue-specific expression can be achieved if the transgene is governed by internal promoters rather than the viral sequences (Wagner *et al.*, 1985; Soriano *et al.*, 1986).

Embryo stem cells have been isolated by culture of mouse embryos in such a way that cells derived from the inner cell mass divide, but do not differentiate (Evans & Kaufman, 1981). In some cases, if such cells are injected into the blastocoele cavity of another embryo the stem cells colonize the embryo and contribute to all of the tissues of the offspring including the germ line (Bradley *et al.*, 1984). Manipulation of embryo stem cells in culture before transfer has resulted in gene transfer (Lovell-Badge *et al.*, 1985; Stewart *et al.*, 1985; Robertson *et al.*, 1986).

A further series of opportunities will be created by the development of targeted manipulation. In principle, targeted mutation may be used to insert a gene at a particular site or to modify existing genes. Targeted insertion would avoid the occurrence of damage to an endogenous gene and with experience could permit selection of a site in which gene expression would depend exclusively upon the transferred sequences. While gene transfer simply adds additional copies of a gene to the genome and recessive alleles are without effect, changes to endogenous genes are expected to be effective when homozygous. Targeted mutation of this kind in mouse embryo stem cells has repaired a genetic defect (Doetschmann *et al.*, 1987) and introduced mutations in the homeobox genes *Hox 1-1* and *En-2* (Zimmer & Gruss, 1989; Joyner *et al.*, 1989). Recent observations suggest that, if embryo stem cells can be isolated from ruminants, it may be possible to use nuclear transfer to produce young from embryo stem cells and avoid the chimaeric generation. Normal calves and lambs have been born following the transfer of nuclei from the inner cell mass cells of blastocysts to enucleated oocytes (Marx, 1988; Smith & Wilmut, 1989).

The nature of milk

Milk is a complex mixture of specific proteins, fats, minerals and sugars adapted to meet the nutritional requirements of the young. Interactions between the components create a complex microstructure and in turn this influences the nutritional value of milk and determines the response to cooking and other forms of processing. Milk is a colloid, with much of the protein in suspension in large casein micelles and much of the fat suspended in membrane-bound globules. At the peak of lactation the mammary gland secretes substantial quantities of protein into the milk: around 0·1 and 1 kg per day in sheep and cattle respectively. Most of the protein is synthesized by the mammary epithelium, but some blood serum proteins, such as albumin, are present in milk in smaller quantities. Table 1 lists the major proteins of cow, sheep, mouse and human milks.

Caseins, a family of phosphoproteins, are the major milk proteins and contribute over 80% of milk protein. Caseins and genes encoding caseins from a number of species have been characterized

Table 1. Protein composition of milk from a number of species (data taken from Lathe *et al.*, 1986)

	Concentration in milk (g/l)			
	Cow	Sheep	Mouse	Man
Caseins				
α_{S1}-Casein	10	12	—*	0·4
α_{S2}-Casein	3·4	3·8	—*	
β-Casein	10	16	—*	3
κ-Casein	3·9	4·6	—*	1
Major whey proteins				
α-Lactalbumin	1	0·8	Trace	1·6
β-Lactoglobulin	3	2·8	None	None
Whey acidic protein	None	None	2	None
Other whey proteins				
Serum albumin	0·4	—*	—*	0·4
Lysozyme	Trace	—*	—*	0·4
Lactoferrin	0·1	—*	—*	1·4
Immunoglobulins	0·7	—*	—*	1·4

*No data available.

and these studies reveal considerable species differences in protein and DNA sequences (see Rosen, 1987); they are, however, very similar in their biophysical properties. Interactions between casein molecules and calcium phosphate lead to the formation of micelles. In the presence of physiological levels of calcium, (bovine) α_{S1}-, α_{S2}- and β-caseins aggregate. The size of the aggregates is limited by κ-casein, which forms a coat around the micelle and prevents further aggregation. This casein has only one phosphoserine, compared with 5–11 in the other caseins, and so is soluble in high concentrations of calcium. It is also the only casein that is glycosylated and polarized. The hydrophobic regions of the κ-casein molecules are believed to interact with the other caseins while the hydrophilic regions interact with the aqueous phase of milk. The structure of caseins and their role in micelle formation are reviewed by Kang *et al.* (1986) and Jimenez-Flores & Richardson (1988). Kappa-casein is the substrate for the protease rennin; cleavage of κ-casein destabilizes the micelles and results in precipitation of the other caseins, forming the curds. This property of milk is exploited in cheese making. Precipitation of caseins also results from exposure of milk to low pH. Stability of the micelles during milk processing is influenced by the relative concentrations of κ-casein and β-lactoglobulin. Addition of κ-casein to milk increases the tolerance to heating.

Large numbers of proteins are found in whey. The major whey proteins of ruminants are β-lactoglobulin and α-lactalbumin, and in rodents α-lactalbumin and whey acidic protein; β-lactoglobulin is absent from the milk of rodents and man. These proteins are synthesized in the mammary epithelium, while many others enter milk from the bloodstream. In ruminants, the most abundant whey protein is β-lactoglobulin, which is present at about 3 g/l in cows and sheep. In sheep ~5% of poly-A^+ RNA encodes β-lactoglobulin (Mercier *et al.*, 1985). The protein is a dimer of M_r 18 000 which may function in the transport of retinol to the young (Papiz *et al.*, 1986). The ovine gene has been cloned (Ali & Clark, 1988) and is the subject of detailed studies (see below).

In addition to its role in providing dietary protein, α-lactalbumin is involved in the synthesis of lactose (Kuhn, 1983). Alpha-lactalbumin modifies the action of galactosyl transferase to direct transfer of galactose to glucose, making lactose. Galactosyl transferase normally functions in the glycosylation of proteins.

Whey acidic protein, present only in rodents and rabbits, has no known function. It has a mass of 14 000 and has some structural similarities and a similar distribution of cysteine residues to a family of proteins that includes neurophysin, snake venom toxin, wheat germ agglutinin and rag weed allergen Ra5 (Hennighausen & Sippel, 1982). It was suggested that this similarity may indicate that the proteins have a similar three dimensional structure. The gene has been cloned and is the subject of studies to direct production of novel proteins to the mammary gland (see below).

There are differences between species in the nature of the iron-binding proteins found in milk (Walstra & Jenness, 1984). Lactoferrin, which is synthesized in the mammary gland, is relatively abundant in human milk. By contrast, iron-binding proteins are only present in low concentrations in ruminant milk. It has been argued that, as well as having a role in iron transport, these proteins may exert a bacteriostatic effect by depriving bacteria of iron.

Milk is involved in the passive immunization of the neonate. The nature of the immunoglobulin in milk is related to the importance of immunoglobulin transfer during prenatal development (Jenness, 1982). In species such as the cow, in which immunoglobulin is primarily transferred in milk, the major form in milk is IgG. By contrast, in animals such as man, in which most immunoglobulin is transferred before birth, the major form in milk is IgA. The immunoglobulins are derived either from blood serum or from cells located in the mammary epithelium. The pattern of changes in concentration varies between species: the concentration of IgG is maximal in colostrum, produced during the first weeks of lactation. In cattle, the concentration of IgG in colostrum may reach 50 g/l before falling to concentrations around 1 g/l. In species such as pigs and horses, the concentration of IgA increases after the intestine of the neonate ceases to absorb immunoglobulin. These IgA antibodies in milk are generally not absorbed by the neonate, but rather act at the level of the mucosal surface by mechanisms that are not understood (Solari & Kraehenbuhl, 1987). They provide the effector arm in a complex mechanism by which the young are protected from enteric organisms by the common mucosal immune system. By contrast, in ruminants, this increase in IgA concentration does not occur; this may reflect a greater need to establish a digestive flora in the gut.

Most of the fat in milk is in membrane-bound spherical particles. The membrane is similar in composition to the outer cell membrane of the secretory cell from which it is derived (Mather, 1987). The fat globules also contain lipid-soluble compounds such as vitamins and steroids. A very small proportion of milk fat is in lipoprotein particles which are not membrane bound. The main sugar in milk is lactose. Typically in cows' milk there are 46 g lactose/l and approximately 0·1 g other carbohydrate/l. Milk serum contains small amounts of many other compounds, including minerals, trace elements, organic acids and traces of other proteins.

When contemplating modifications to the composition of milk, it is important to recognize that changes to the amount or structure of one component may have profound effects upon the nature of the milk because of the complex interactions between the constituents. In particular, changes to proteins may affect micelle structure and so change the milk dramatically. When using dairy animals to synthesize specific proteins for extraction and pharmaceutical or industrial use, this is not likely to be important. However, when the milk is for human consumption, any changes that make processing more difficult would be extremely inconvenient. There is also a need to consider the well being of neonates that at some time in the future may obtain the modified milk from their mothers.

Targeting expression to the mammary gland

There are considerable differences between milk proteins in the pattern of accumulation of their individual mRNAs during gestation and lactation (Rosen, 1987). In the gland of the virgin rat there are several hundred α-casein mRNA molecules per cell, but very few whey acidic protein mRNA molecules. By mid-gestation, the level of β-casein has increased more rapidly than the other mRNA molecules and it has become the most abundant species, a position it retains throughout pregnancy and lactation. These observations establish that, in addition to the mechanisms to stimulate cell division and promote the development of the mammary gland, there are mechanisms which regulate expression of the individual milk protein genes.

Analysis of the control of gene expression in the mammary gland has been handicapped by the absence of cell lines that retain specific milk protein synthesis and respond to hormones. It has become clear that control in mammary gland is particularly complex and involves interactions between cells, between cells and extra-cellular matrix and hormonally induced factors (Rosen, 1987; Bissel & Hall, 1987). Transgenic animals allow the study of expression of transferred genes in the native mammary gland, with all the normal developmental and hormonal signals. To date, studies in transgenic animals have been published with 3 milk protein genes; ovine β-lactoglobulin, murine whey acidic protein and rat β-casein. These observations will be reviewed before considering the potential application of targeting gene expression to the mammary gland.

β-Lactoglobulin

The Edinburgh group selected β-lactoglobulin for study as it is the whey protein present at highest concentration in ruminant milk. Four clones were isolated from a genomic library of a sheep, one of which (clone SS1) has been characterized in detail (Ali & Clark, 1988; Harris *et al.*, 1988). The sequence of SS1 revealed that the β-lactoglobulin gene has 7 exons. To discover whether the clone contained all of the *cis*-acting sequences necessary for efficient tissue specific expression of the gene, the clone was introduced into mice (Simons *et al.*, 1987). Nine transgenic mice were selected for study; 5 of the mice were female, milk was collected from them and analysed for the presence of β-lactoglobulin. Three of these animals produced β-lactoglobulin; the 2 that failed to do so carried less than one copy of the gene per cell. Transgenic lines were established from 2 of the expressing females and from 3 males. In 4 of the 5 lines, females secrete β-lactoglobulin into their milk. In the other line, it is inferred that the gene was incorporated into the Y chromosome since all transgenic mice of this line are male. When studied by SDS polyacrylamide electrophoresis, the β-lactoglobulin in mouse milk co-migrated with that in sheep milk and with purified β-lactoglobulin. In the most abundantly expressing line, the concentration of β-lactoglobulin was 23 mg/ml, 5 times that in the milk of control sheep. The pattern of expression of the transgene was analysed in 3 lines by Northern blotting of RNA prepared from mammary, salivary and lachrymal glands, liver, kidney and spleen. Abundant transcripts were found only in mammary gland, although after long exposures very low levels of RNA were detected in other tissues. Production of this novel protein had no apparent side-effects upon the mothers or the young.

The developmental regulation of the transgene has been studied in 2 lines of mice (Harris *et al.*, 1990). The changes in the level of expression of β-lactoglobulin paralleled that of the endogenous β-casein gene. This pattern is similar to that of β-lactoglobulin in sheep, but different from the endogenous mouse whey acidic protein. These observations show that clone SS1 contains all of the *cis*-acting sequences that are essential for high level tissue-specific expression of the gene. In addition, it seems that these sheep sequences can be interpreted correctly by the mammary gland of mice despite the fact that β-lactoglobulin is absent from mice and that there are differences between mice and sheep in the hormonal regulation of lactation. In subsequent experiments, the nature of the *cis*-acting sequences is being studied by deletion of 5′ sequences before gene transfer. Preliminary results suggest that all of the essential upstream sequences lie within 0·8 kb of the 5′ end (Harris *et al.*, 1990).

This gene and fragments of it have been used to direct production of human therapeutic proteins to the mammary gland (see section on production of pharmaceutical proteins, below). The proteins chosen for expression are clotting factor IX and α_1-antitrypsin (also known as α_1-protease inhibitor). In the first experiments, the entire β-lactoglobulin gene was used, contained in a fragment of 10·5 kb which functions well in transgenic mice (Simons *et al.*, 1987). A fragment from a human factor IX cDNA clone was inserted into the 5′ untranslated region of the β-lactoglobulin clone (Clark *et al.*, 1989). The human sequences encode the entire pre-pro-protein except for a rearrangement of the first AUG of the RNA, but have been successfully used for the production of biologically active factor IX from cells in tissue culture (Anson *et al.*, 1985). At the 5′ junction the sheep and human sequences were fused in their 5′ untranslated regions, while the human fragment includes a stop codon, but no polyadenylation site. The predicted fully processed transcript would contain the coding sequences for both proteins with the factor IX termination codon preceding the β-lactoglobulin initiation codon by 167 nucleotides. Four transgenic sheep were produced by microinjection of the β-lactoglobulin–factor IX hybrid gene into 1-cell eggs (Simons *et al.*, 1988).

Two of the transgenic sheep were female and their milk has been examined for the presence of factor IX (Clark *et al.*, 1989). Radioimmunoassay of freeze-dried whole milk samples showed the milk of both ewes to contain approximately 25 ng factor IX/ml. This is around 1/250th the concentration in human plasma and 1/100 000th the concentration of β-lactoglobulin in sheep milk. The factor IX was enriched from the milk by monoclonal antibody affinity chromatography and was shown to be active in a clotting assay. The steady-state level of the fusion gene RNA was approximately 1/1250th that of the endogenous β-lactoglobulin mRNA. A similar construct containing α_1-antitrypsin cDNA in place of the factor IX sequences has also been introduced into sheep and expression has been studied in 2 ewes. One ewe produced α_1-antitrypsin at a considerably greater concentration than had been observed with factor IX (approximately 5 μg/ml). While this compares favourably with the expression of factor IX, the level of β-lactoglobulin is about 1000 times greater than that of human α_1-antitrypsin in the milk of this animal.

The level of expression of these genes is too low to be of commercial value. In order to optimize expression, several new fusion genes are being studied in mice. Preliminary results suggest that one fusion gene directs considerably higher levels of α_1-antitrypsin production than the first construct. This gene contains only the 5′ sequences of β-lactoglobulin and a genomic fragment of α_1-antitrypsin. The higher levels of expression add to the evidence that introns may have significant effects upon gene expression in transgenic animals (Brinster *et al.*, 1988).

Whey acidic protein

Whey acidic protein is the whey protein present at greatest concentration in mouse and rat. The genes have been cloned (Campbell *et al.*, 1984), and mouse sequences used to direct gene expression to the mammary epithelium. The protein is encoded by a 2·8 kb gene having 4 exons. A 2·5 kb fragment containing the promoter and 5′ sequences has been shown to direct tissue-specific hormone dependent expression of 2 oncogenes (Andres *et al.*, 1987, 1988) and of human tissue plasminogen activator (Gordon *et al.*, 1987; Pittius *et al.*, 1988). The fusion genes contained 2·6 kb of 5′ sequences terminating 24 bp downstream of the transcription start site. This fragment contains binding sites for proteins that are specific to lactating mammary gland and which may be *cis*-acting regulatory elements (Lubon & Hennighausen, 1987). These sequences were capable of directing tissue-specific expression of human Ha-*ras* and murine c-*myc* gene expression to the mammary gland of transgenic mice (Andres *et al.*, 1987, 1988). The level of RNA of the 2 fusion genes was appreciably lower than that of the endogenous whey acidic protein gene (between 2 and 10%), but followed similar profiles during lactation. In both cases expression of the fusion gene was marked by the development of mammary tumours although the speed of development was more rapid and the incidence more frequent with the *myc* construct. Tumour formation in these mice was dependent on them having gone through lactation, but in the tumours with the *myc* oncogene, expression of the endogenous milk protein genes was independent of hormonal stimulation (Schonenberger *et al.*, 1988). In animals with the whey acidic protein–*myc* construct, endogenous whey acidic protein and the transgene were only expressed during lactation in normal mammary tissue. By contrast, in tumours from the same mice, whey acidic protein and β-casein genes were expressed after lactation and following transfer to nude mice.

The same 5′ whey acidic protein sequences have been used to direct expression of human tissue plasminogen activator in the mammary glands of transgenic mice. The whey acidic protein sequences were linked to the 5′ untranslated region of a cDNA coding for tissue plasminogen activator. The plasminogen activator cDNA encoded the entire protein, including its secretion signal sequence. At the 3′ end of the construct, a polyadenylation signal derived from SV40 was incorporated (Gordon *et al.*, 1987). Milk from 6 transgenic lines of mice was analysed and 4 were found to produce biologically active protein. As determined by ELISA the concentration of tissue plasminogen activator was around 50 μg/ml in the line with the greatest concentration (Pittius *et al.*, 1988). In that line the steady state level of mRNA from the fusion gene was approximately 1% that of the endogenous whey acidic protein RNA. RNase protection assays showed that the transcription start site was the same for the endogenous gene and the transgene. During pregnancy the level of expression both of fusion and endogenous genes was greatly increased in the mammary gland, by approximately 100- and 10 000-fold, respectively. The transgene was expressed predominantly in the mammary gland, but low levels of expression were detected in sublingual gland, tongue and kidney. The level of expression of the two genes did not change significantly in other tissues during pregnancy.

This fragment apparently contains some of the sequences necessary for tissue-specific hormonally induced expression, but perhaps not all (Pittius *et al.*, 1988). The relative concentration of whey acidic protein mRNA increased to a greater extent than did that of the fusion gene (10 000-fold compared with 100-fold) and at a later stage of gestation (Day 14–16 compared to Day 10). There may be sequences that govern response to hormone in a more distant 5′ site, in the coding region or the 3′ sequences.

β-Casein

β-Casein is the milk protein present at highest concentration. A rat genomic clone was isolated and transferred into mice (Lee *et al.*, 1988). The gene is 7·5 kb long and has 9 exons, the first and last of which are non-coding (Rosen, 1987). In transgenic mice the gene was expressed in a developmentally regulated and tissue specific manner, but at only 0·01–1% of the level of the endogenous mouse β-casein gene (Lee *et al.*, 1988). An RNase protection assay confirmed that transcription was initiated at the authentic start site. The low level of expression may reflect an effect of site of integration of the transgenes (only 3 lines were studied) or the fact that important *cis*-acting regulatory elements lie outside the region cloned.

Applications

Production of pharmaceutical proteins

Several human proteins are now produced by recombinant microorganisms (see Primrose, 1987; Dalboge *et al.*, 1987; Sarmeintos *et al.*, 1989), but there are other proteins that cannot be manufactured in this way because microorganisms are unable to complete the necessary post-translational modifications. At present proteins that require modification, such as clotting factors, are prepared from human or animal material. There are other potential means of producing these proteins including tissue culture of insect cells transfected by baculovirus (Miller, 1988) or mammalian cells (Cartwright, 1987) and transgenic farm animals.

Production of therapeutic proteins in the milk of farm animals has several potential advantages. Preparations of clotting factors from human blood have in the past contained the agents that cause AIDS and hepatitis, and place patients requiring such treatment at high risk of infection. By contrast, the preparations from milk would be free of contamination by such infectious agents. It would be essential, however, to ensure that no other infectious agents were present in material prepared from animals.

Transgenic animals may ultimately be a cheaper source of recombinant proteins than mammalian cells maintained in tissue culture. Lines of transgenic animals are expensive to create, but then cheap to multiply and use. By contrast, large scale tissue culture of mammalian cells will always be difficult and expensive. The relative advantages of these two approaches remain to be determined. There is every reason to expect that production of the proteins could be directed to the blood in transgenic livestock, as for example, factor IX has been produced in the liver of transgenic mice and secreted into blood (Choo *et al.*, 1987). Milk has the advantage that it is very easy to collect and the volume of milk that could be collected would be greater than the available volume of blood. In addition, milk is relatively isolated from the animal and so there are less likely to be side-effects upon the animal of producing proteins, such as clotting factors, which might otherwise have a deleterious biological effect in the producer animal.

The ease and expense of purification of the product must also be borne in mind. New processes will have to be developed for extraction of proteins from milk, but there is no reason to suspect that this will be more taxing than purification of proteins from other sources such as bacteria. The therapeutic use of recombinant DNA-derived protein will depend upon production of biologically active materials and purification to homogeneity. The experiments performed so far suggest that transgenic animals have produced biologically active clotting factor IX (Clark *et al.*, 1989) and tissue plasminogen activator (Gordon *et al.*, 1987) in their milk. In these experiments, the amounts of protein produced were insufficient to provide a suitable source of therapeutic proteins. Before proteins may be used therapeutically, considerable work will be required to verify their structure and activity and to demonstrate efficacy.

Two factors are likely to have major effects upon the development of this approach in farm animals. First, the concentration of the protein produced by the present sheep is low. This may be a reflection of the site of integration of the transgene as the number of animals studied is small. However, it is likely to result from the use of what will come to be seen as primitive gene construct design. Future experiments will provide a great deal of information needed for the design of effective fusion genes. This would include the identification of regulatory sequences for full tissue specific expression of a particular milk protein gene, clarification of the role of introns in gene expression, comparisons between different milk protein genes and different species of dairy animal. Finally, much remains to be learned about the ability of the mammary gland to perform post-translational modifications. Biologically active clotting factor IX has been produced at very low levels by a number of cell lines (Anson *et al.*, 1985; Brownlee 1987). In this case a major limitation seems to arise from the relative inability of the cells to carry out gamma-carboxylation (Kaufman *et al.*, 1986). When larger amounts of protein were produced only a small proportion was biologically active. The potential of the mammary gland in this respect remains to be determined.

The principle is established that the mammary gland is able to make some proteins of therapeutic value, but much more remains to be learned. The very high value of the products make this an avenue of development that will be followed up vigorously. Once established it may then be exploited for the production of other proteins.

Milk proteins

There is the opportunity to make changes to the milk proteins that are present, either by introducting extra copies of existing genes, by using regulatory sequences from a protein gene that is expressed at a higher level, by transferring milk protein genes from another species or by transferring modified genes.

The relative concentration of a particular protein might be increased by transferring additional copies of the gene. In circumstances in which production of a protein is limited by the quantity of mRNA, such transfer would be expected to increase the relative concentration of the particular protein. If the protein in question is expressed at a low level, the regulatory sequences of a different protein gene could be linked to the coding sequences of the protein in question to increase expression.

Such changes might be contemplated in order to change the response of milk to processing (see Kang *et al.*, 1986; Jimenez-Flores & Richardson, 1988). The stability of casein micelles is important in the processing of milk, and changes to milk protein composition have been suggested with a view to increasing micelle stability. Micelles are formed by interactions between calcium phosphate and the casein molecules, and stabilized by a coating of κ-casein. Increasing the amount of κ-casein in milk would be expected to reduce micelle size, enhance the thermal stability of the milk and so reduce the danger of coagulation in sterilized products. Addition of β-casein to milk reduces the time for rennet coagulation, increases the rate of whey expulsion and increases the firmness of the resulting curd, apparently because of the increased interaction between the casein and calcium (reviewed by Jimenez-Flores & Richardson, 1988).

Changes to caseins would also be expected to affect the properties of the cheese formed. Deletion of phosphate groups would be expected to yield a softer, moist cheese, while conversely, an increase in phosphorylation would be anticipated to produce a firmer cheese. Maturation of cheese results from breakdown of proteins and it may be possible to modify the proteins to increase the number of sites available for proteolysis. A change of one nucleotide in the α_{s1}-casein coding region would lead to the incorporation of phenylalanine instead of isoleucine at amino acid 71. This would create an additional site for cleavage by acid proteases (Jimenez-Flores & Richardson, 1988). More rapid maturation of texture would be a considerable economic advantage.

There may be benefit in directing the production of human milk proteins to the milk of farm animals in order to make the milk a better substitute for human milk. In particular, it has been suggested that lactoferrin may be important (Mercier, 1986).

Lactose and fat

While all babies are able to digest lactose, this ability is retained by only a minority of adult humans, the Caucasians. In a part of the population of all other races consumption of lactose can cause digestive upset and diarrhoea. In principle, genetic manipulation could be used to reduce the concentration, either by causing breakdown of the sugar after its synthesis or by preventing its formation. Secretion of lactase into milk would be expected to cause the breakdown of lactose into the two constituent monosaccharides, glucose and galactose. It may be possible to direct synthesis and secretion of lactase as described above for pharmaceutical proteins. In this case low level expression may be sufficient. It may be possible to prevent production of lactose by abolishing α-lactalbumin expression. The most likely route by which this may be achieved is by targeted mutation of the α-lactalbumin gene, techniques for which are not yet fully developed. While there is a potential application for these approaches, there is also the possibility of undesirable side effects. Lactose is the main osmotic constituent of milk and the effect on milk production and volume of changing lactose concentration is not known. It is possible that a reduction in volume will result, without a change in the yield of protein or fat, which could be beneficial in requiring less frequent milking.

A major component of milk is fat; it has been suggested (Gannon, 1986) that it may be possible to change the nature of milk fat by expression of Δ12 desaturase. Expression of Δ12 desaturase is predicted to result in the conversion of oleic acid to the polyunsaturated linoleic acid.

Antibacterial proteins to combat mastitis

Mastitis is a cause of distress to the animal and of financial loss to the producer. Production of antibacterial proteins in milk may reduce the incidence of the problem. Lysostaphin is an enzyme produced by *Staphylococcus simulans* in the stationary growth phase under certain culture conditions, that degrades the cell wall of many staphylococcal species. The gene has been cloned (Recsei *et al.*, 1987) and recombinant protein eliminated *Staphylococcus aureus* intramammary infections when administered into the gland (Sears *et al.*, 1988). Lysostaphin protein is being produced in the milk of transgenic mice to investigate this approach (J. A. Bramley & C. M. Williamson, personal communication).

Conclusions

Methods of gene transfer are very new and still particularly expensive in farm animals. As discussed above, improvements in efficiency and reductions in cost are to be expected. However, the great cost of the present procedures will limit their exploitation for some time and initially ensure that they are used for the particularly high value products, such as pharmaceutical proteins. Experience will be gained from attempting such applications and in time it seems probable that these techniques will be used to modify the composition of milk for liquid consumption or food processing. While in this paper some of the potential of gene transfer for the modification of milk has been outlined, it seems probable that many of the potential applications of this technology have still to be considered.

We thank Pharmaceutical Proteins Ltd for financial help during part of the project in Edinburgh.

References

Ali, S. & Clark, A.J. (1988) Characterisation of the gene encoding ovine β-lactoglobulin. *J. molec. Biol.* **199,** 415–426.

Andres, A.C., Schonenberger, C.A., Groner, B., Hennighausen, L., LeMeur, M. & Gerlinger, P. (1987) Ha-*ras* oncogene expression directed by a milk protein gene promoter: tissue specificity, hormonal regulation, and tumor induction in transgenic mice. *Proc. natn. Acad. Sci. USA* **84,** 1299–1303.

Andres, A.C., van der Valk, M.A., Schonenberger, C.A., Fluckiger, F., LeMeur, M., Gerlinger, P. & Groner, B. (1988) Ha-*ras* and c-*myc* oncogene expression interfere with morphological and functional differentiation of mammary epithelial cells in single and double transgenic mice. *Genes and Development* **2,** 1486–1495.

Anson, D.S., Austen, D.E.G. & Brownlee, G.G. (1985) Expression of active human clotting factor IX from recombinant DNA clones in mammalian cells. *Nature, Lond.* **315,** 683–685.

Biery, K.A., Bondioli, K.R. & DeMayo, F.J. (1988) Gene transfer by pro-nuclear injection in the bovine. *Theriogenology* **29,** 224, abstr.

Bissel, M.J. & Hall, H.G. (1987) Form and function in the mammary gland. The role of the extracellular matrix. In *The Mammary Gland, Development, Regulation and Function*, pp. 97–146. Eds M. C. Neville & C. W. Daniel. Plenum Press, New York.

Bradley, A., Evans, M., Kaufman, M.H. & Robertson, E. (1984) Formation of germ-line chimaeras from embryo-derived teratocarcinoma cell lines. *Nature, Lond.* **309,** 255–256.

Brinster, R.L., Allen, J.M., Behringer, R.R., Gelinas, R.E. & Palmiter, R.D. (1988) Introns increase transcriptional efficiency in transgenic mice. *Proc. natn. Acad. Sci. USA* **85,** 836–840.

Brownlee, G.G. (1987) The molecular pathology of haemophilia B. *Biochem. Soc. Trans.* **15,** 1–8.

Campbell, S.M., Rosen, J.M., Hennighausen, L., Strech-Jurk, U. & Sippel, A.E. (1984) Comparison of the whey acidic protein genes of the rat and mouse. *Nucleic Acids Res.* **12,** 8685–8697.

Cartwright, T. (1987) Isolation and purification of products from animal cells. *Trends in Biotechnol.* **5,** 25–30.

Choo, K.H, Raphael, K., McAdam, W. & Peterson, M.G. (1987) Expression of active human clotting factor IX in transgenic mice: use of cDNA with complete mRNA sequence. *Nucleic Acids Res.* **15,** 871–884.

Clark, A.J., Bessos, H., Bishop, J.O., Brown, P., Harris, S., Lathe, R., McClenaghan, M., Prowse, C., Simons, J.P., Whitelaw, C.B.A. & Wilmut, I. (1989) Expression of human anti-hemophilic factor IX in the milk of trans-genic sheep. *Bio/Technology* **7,** 487–492.

Dalboge, H., Dahl, H.-H.M., Pedersen, J., Hansen, J.W. & Christensen, T. (1987) A novel enzymatic method for production of authentic hGH from an *Escherichia coli* produced hGH precursor. *Bio/Technology* **5,** 161–166.

Doetschman, T., Gregg, R.G., Maeda, N., Hooper, M.L., Melton, D.W., Thompson, S. & Smithies, O. (1987) Targeted correction of a mutant HPRT gene in mouse embryonic stem cells. *Nature, Lond.* **330,** 576–578.

Evans, M.J. & Kaufman, M.H. (1981) Establishment in culture of pluripotential cells from mouse embryos. *Nature, Lond.* **292,** 154–156.

Gannon, F. (1986) Transgenics. In *Exploiting New Technologies in Animal Breeding*, pp. 54–58. Eds C. Smith, J. W. B. King & J. C. McKay. Oxford University Press, Oxford.

Gordon, K., Lee, E., Vitale, J.A., Smith, A., Westphal, H. & Hennighausen, L. (1987) Production of human tissue plasminogen activator in transgenic mouse milk. *Bio/Technology* **5,** 1183–1187.

Harris, S., Ali, S., Anderson, S., Archibald, A.L. & Clark, A.J. (1988) Complete nucleotide sequence of the genomic ovine β-lactoglobulin gene. *Nucleic Acids Res.* **16,** 10379–10380.

Harris, S., McClenaghan, M., Simons, J.P., Ali, S. & Clark, A.J. (1990) Gene expression in the mammary gland. *J. Reprod. Fert.* **89,** 707–715.

Hennighausen, L.G. & Sippel, A.E. (1982) Mouse whey acidic protein is a novel member of the family of "four-sulfide core" proteins. *Nucleic Acids Res.* **10,** 2677–2684.

Jaehner, D., Kirsten, H., Mulligan, R. & Jaenisch, R. (1985) Insertion of the bacterial gpt gene into the gene line of mice by retroviral infection. *Proc. natn. Acad. Sci. USA* **82,** 6927–6931.

Jenness, R. (1982) Inter-species comparison of milk proteins. In *Developments in Dairy Chemistry*, Vol. 1, pp. 83–114. Ed. P. F. Fox. Applied Sciences Publishers, London.

Jimenez-Flores, R. & Richardson, T. (1988) Genetic engineering of the caseins to modify the behaviour of milk during processing: a review. *J. Dairy Sci.* **71,** 2640–2654.

Joyner, A.L., Skarnes, W.C. & Rossant, J. (1989) Production of a mutation in mouse *En-2* gene by homologous recombination in embryonic stem cells. *Nature, Lond.* **338,** 153–156.

Kang, Y., Jimenez-Flores, R. & Richardson, T. (1986) Casein genes and genetic engineering of the caseins. In *Genetic Engineering of Animals*, pp. 95–111. Eds J. W. Evans & A. Hollaender. Plenum Press, New York.

Kaufman, R.J., Wasley, L.C., Furie, B.C., Furie, B. & Shoemaker, C.B. (1986) Expression, purification and characterisation of recombinant gamma-carboxylated factor IX synthesised in Chinese Hamster Ovary cells. *J. biol. Chem.* **261,** 9622–9628.

Kuhn, N.J. (1983) The biosynthesis of lactose. In *Biochemistry of Lactation*, pp. 159–176. Ed. T. B. Mepham. Elsevier, Amsterdam.

Lathe, R., Clark, A.J., Archibald, A.L., Bishop, J.O., Simons, P. & Wilmut, I. (1986) Novel products from livestock. In *Exploiting New Technologies in Animal Breeding*, pp. 99–102. Eds C. Smith, J. W. B. King & J. C. McKay. Oxford University Press, Oxford.

Lee, K.-F., DeMayo, J., Aitee, S.H. & Rosen, J.M. (1988) Tissue-specific expression of the rat β-casein gene in transgenic mice. *Nucleic Acids Res.* **16,** 1027–1041.

Lovell-Badge, R.H., Bygrave, A.E., Bradley, A., Robertson, E., Evans, M.J. & Cheah, K.S.E. (1985) Transformation of embryonic stem cells with the human type-II collagen gene and its expression in chimeric mice. *Cold Spring Harbor Symp. Quant. Biol.* **50,** 707–711.

Lubon, H. & Hennighausen, L. (1987) Nuclear proteins from lactating mammary glands bind to the promoter of a milk protein gene. *Nucleic Acids Res.* **15,** 2103–2121.

Marx, J.L. (1988) Cloning in sheep and cattle embryos. *Science, NY* **239,** 463–464.

Mather, I.H. (1987) Proteins of the Milk-fat-globule membrane as markers of mammary epithelial cells and apical plasma membrane. In *The Mammary Gland; Development, Regulation and Function*, pp. 217–267. Eds M. C. Neville & C. W. Daniel. Plenum Press, New York.

Mercier, J.-C. (1986) Genetic engineering applied to milk producing animals: some expectations. In *Exploiting New Technologies in Animal Breeding*, pp. 121–131. Eds C. Smith, J. W. B. King & J. C. McKay. Oxford University Press, Oxford.

Mercier, J.-C., Gaye, P., Soulier, S., Hue-Delahaie, D. & Vilotte, J.-L. (1985) Construction and identification of recombinant plasmids carrying cDNAs coding for α_{s1}, α_{s2}, κ-casein and β-lactoglobulin. *Biochimie* **67,** 959–971.

Miller, L.K. (1988) Baculoviruses for foreign gene expression in insect cells. In *Vectors: a Survey of Molecular Cloning Vectors and their Uses*, pp. 457–465. Eds R. L. Rodriguez & D. T. Denhardt. Butterworth, Boston.

Palmiter, R.D. & Brinster, R.L. (1986) Germline transformation of mice. *Ann. Rev. Genetics* **20,** 465–499.

Papiz, M.Z., Sawyer, L., Eliopolous, E.E., North, A.C.T., Findlay, J.B.C., Sivaprasadarao, R., Jones, T.A., Newcomer, M.E. & Kraulis, P.J. (1986) The structure of β-lactoglobulin and its similarity to retinol binding protein. *Nature, Lond.* **324,** 383–385.

Pittius, C.W., Hennighausen, L., Lee, E., Westphal, H., Nicols, E., Vitale, J. & Gordon, K. (1988) A milk protein gene promoter directs the expression of human tissue plasminogen activator cDNA to the mammary gland in transgenic mice. *Proc. natn. Acad. Sci. USA* **85,** 5874–5878.

Primrose, S.B. (1987) *Modern Biotechnology*. Blackwell Scientific Publications, Oxford.

Recsei, P.A., Gruss, A.D. & Novick, R.P. (1987) Cloning, sequence, and expression of the lysostaphin gene from *Staphylococcus simulans*. *Proc. natn. Acad. Sci. USA* **84,** 1127–1131.

Robertson, E., Bradley, A., Kuehn, M. & Evans, M. (1986) Germ-line transmission of genes introduced into cultured pluripotential cells by retroviral vector. *Nature, Lond.* **323,** 445–448.

Rosen, J.M. (1987) Milk protein gene structure and function. In *The Mammary Gland; Development, Regulation and Function*, pp. 301–322. Eds M. C. Neville & C. W. Daniel. Plenum Press, New York.

Sarmientos, P., Duchesne, M., Denefle, P. Boiziau, J., Fromage, N., Delporte, N., Parker, F., Lelievre, Y., Mayaux, J.-F. & Cartwright, T. (1989) Synthesis and purification of active human tissue plasminogen activator from *Escherichia coli*. *Bio/Technology* **7,** 495–501.

Schonenberger, C.-A., Andres, A.-C., Groner, B., van der Valk, M., LeMeur, M. & Gerlinger, P. (1988) Targeted *c-myc* gene expression in mammary glands of transgenic mice induces mammary tumours with constitutive milk protein gene transcription. *EMBO Jl* **7,** 169–175.

Sears, P.M., Smith, B.S., Polak, J., Gusik, S.N. & Blackburn, P. (1988) Lysostaphin efficacy for treatment of *Staphylococcus aureus* intramammary infection. *J. Dairy Sci.* **71,** Suppl. 1, p. 244, abstr.

Simons, J.P., McClenaghan, M. & Clark, A.J. (1987) Alteration of the quality of milk by expression of sheep β-lactoglobulin in transgenic mice. *Nature, Lond.* **328,** 530–532.

Simons, J.P., Wilmut, I., Clark, A.J., Archibald, A.L., Bishop, J.O. & Lathe, R. (1988) Gene transfer into sheep. *Bio/Technology* **6,** 179–183.

Smith, L.C. & Wilmut, I. (1989) Influence of nuclear and cytoplasmic activity on the development *in vivo* of sheep embryos after nuclear transfer. *Biol. Reprod.* **40,** in press.

Solari, R. & Kraehenbuhl, J.-P. (1987) Receptor-mediated transepithelial transport of polymeric immunoglobulins. In *The Mammary Gland; Development, Regulation and Function*, pp. 269–298. Eds

M. C. Neville & C. W. Daniel. Plenum Press, New York.

Soriano, P., Cone, R.D., Mulligan, R.C. & Jaenisch, R. (1986) Tissue-specific and ectopic expression of genes introduced into transgenic mice by retroviruses. *Science, NY* **234,** 1409–1413.

Stewart, C.L., Vanek, M. & Wagner, E.F. (1985) Expression of foreign genes from retroviral vectors in mouse teratocarcinoma chimeras. *EMBO Jl* **4,** 3701–3709.

van der Putten, H., Botteri, F.M., Miller, A.D., Rosenfeld, M.G., Fan, H., Evans, R.M. & Verma, I.M. (1985) Efficient insertion of genes into the mouse germ line via retroviral vectors. *Proc. natn. Acad. Sci. USA* **82,** 6148–6152.

Varmus, H. (1988) Retroviruses. *Science, NY* **240,** 1427–1435.

Wagner, E.F., Vanek, M. & Vennstrom, B. (1985) Transfer of genes into embryonal carcinoma cells by retrovirus infection: efficient expression from an internal promoter. *EMBO Jl* **4,** 663–666.

Walstra, P. & Jenness, R. (1984) *Dairy Chemistry and Physics.* John Wiley & Sons, New York.

Ward, K.A.,, Franklin, I.R., Murray, J.D., Nancarrow, C.D., Raphael, K.A., Rigby, N.W., Byrne, C.R., Wilson, B.W. & Hunt, C.L. (1986) The direct transfer of DNA by embryo microinjection. *Proc. 3rd Wld Congr. Genetics Applied to Livestock Production, Lincoln,* **12,** 6–21.

Zimmer, A. & Gruss, P. (1989) Production of chimaeric mice containing embryonic stem (ES) cells carrying a homeobox *Hox1-1* allele mutated by homologous recombination. *Nature, Lond.* **338,** 150–153.

TRANSGENIC POULTRY

Chairman
B. Calnek

J. Reprod. Fert., Suppl. **41** (1990), 149–152

Use and desired properties of poultry vaccines

P. M. Biggs

Willows, London Road, St Ives, Huntingdon, Cambs PE17 4ES, UK

Keywords: disease; immunity; poultry; vaccines; vaccine production

Introduction

Vaccines have been used for the prevention and control of infectious disease in poultry since Pasteur developed in 1880 a vaccine for fowl cholera (Pasteur, 1880). Admittedly, this vaccine was not very safe and consequently it was not widely adopted. It was not until the period from the 1920s to the 1950s, when the poultry industry expanded and intensive systems of management were adopted resulting in an increase in the number and incidence of infectious disease, that there was an essential need for the prevention and control of infectious disease.

The major diseases that could not be controlled by means other than vaccination were diseases caused by viruses. As with most scientific endeavours there was a lag period between the perceived need and the appearance of vaccines. The main period of development and implementation of vaccines to control poultry diseases was the 1940s and 1950s. It could be said that the rapid expansion of intensive methods of poultry production thoughout the world could not have occurred without the pioneering work of the poultry research scientists who developed vaccines over this period.

Early developments relied on the use of virulent field virus either alone using abnormal routes of infection or in combination with antiserum. However, in most cases these methods were not very satisfactory because they put the vaccinee at risk and they spread virulent virus in the population. This was a serious matter in a population which was kept in large numbers and in high concentration. This led to the development and use of inactivated vaccines or live vaccines of suitably low virulence, either found naturally in the field, or modified by experimental procedures. Although inactivated vaccines are used for specific purposes the need for vaccination in poultry to be inexpensive and easy to administer has favoured live vaccines.

Vaccine use today

Vaccines for bacterial diseases such as fowl typhoid, fowl cholera, erysipelas and infectious coryza are not universally used. They are used where the disease is endemic and their use is appropriate. The efficacy of bacterial vaccines has been variable and difficulties are most serious when a range of serotypes exist such as in fowl cholera and infectious coryza. The most widely used vaccines are for viral diseases. For the domestic fowl there are different requirements for breeders, layers and broilers. In some circumstances domestic fowl can be vaccinated against 8 viral diseases. Vaccines for Marek's disease, Newcastle disease, infectious bronchitis, infectious avian encephalomyelitis and infectious bursal disease are widely and commonly used for breeders and layers. Vaccines for infectious laryngotracheitis, egg drop syndrome-76 and reoviruses are used less commonly. For broilers, whose life is rarely longer than 7 weeks, vaccination against Newcastle disease and infectious bronchitis are widely and commonly used and, in many parts of the world, vaccination against Marek's disease is also common. Vaccines are also used for turkeys and ducks to prevent and control the most important diseases of these species (Macpherson, 1981; Pattison, 1981).

Inactivated, or non-living, vaccines at present have to be administered parenterally to each individual. Live vaccines may also have to be adminstered to each individual by parenteral inoculation, e.g. Marek's disease. However, in some cases live vaccines can be mass administered through the drinking water or by aerosol, e.g. Newcastle disease and infectious bronchitis.

For reasons of cost, administration through the drinking water or by aerosol is desired when practicable for birds of relatively low value such as broilers. Individual parenteral administration produces a more even vaccinal response in a group of birds and is considered worth the expense for birds of high value such as breeders. The better response is important for breeders and supply flocks in which high concentrations of maternally derived antibody in the progeny are important because exposure to field virus is commonly at a young age. Also for this reason vaccines have to be used early in life and frequently in the face of maternally derived antibody.

Desirable properties

The desirable properties of poultry vaccines can be considered from the point of view of the user and of the producer.

User

The desirable properties from the user's standpoint can be considered under the headings of safety, quality and efficacy.

Safety. Live vaccines should be non-pathogenic and have no adverse effect on the host such as reducing growth rate and productivity. They should not spread to members of its host or other species. Shedding of vaccinal organism and local spread can be tolerated if it is only for a short period after vaccination. However, persistent shedding and consequent spreading from the vaccinated flock to the population at large results in a risk of reversion to virulence of the vaccine organism and interferes with monitoring for field infection in the population. Live vaccines should not revert or change to a pathogenic form. This is particularly important for vaccines that are disseminated from vaccinated animals. All vaccines should be free of contaminating agents and inactivation must be complete for killed vaccines containing whole organisms.

Quality. Vaccines should have a stated potency and there should be a reliable method of establishing and measuring potency. They should be stable in the conditions they are to be stored and used and should be free of unnecessary impurities and additives such as antibiotics. The production premises should fulfil criteria designated by licensing authorities.

Efficacy. Vaccines should provide a significant degree of protection against mortality and morbidity caused by the disease in question. In addition they should protect against subclinical disease such as effects on productivity, feed conversion and growth rate. Vaccines should ideally prevent infection with the challenging field organism and, if the vaccine does not prevent infection with the field organism, it should protect against its shedding and spread from the vaccinated bird.

Other. There are other desirable properties for which the user has an interest. Vaccines and vaccinated birds should be readily distinguishable from the field organism and birds infected with the field organism respectively. They should be easy to store, distribute and administer. Finally, they should be inexpensive.

Vaccine producer

The vaccine producer wishes to provide vaccines with the properties desired by the consumer but, in addition, there are other properties of particular interest to the producer.

The producer requires strains of organism and systems of propagation which provide high yields of antigen (Churchill, 1981). The systems of propagation available for poultry viral vaccines are restricted. Embryonating eggs and cell culture are the major systems although the hatched chicken is used in some cases. Avian cell lines are generally not available and so well characterized cell lines and systems of growing cells in suspension cannot be used. Because of these constraints, strains of organism and systems of propagation providing high yields of antigen are particularly important. They are also important because techniques for concentrating antigen are too expensive for use in the manufacture of poultry vaccines. Whatever system of propagation is used the substrate should be free from adventitious agents. The seed lot system of production enables characterization of the vaccine and production over time of a product with uniform properties of safety and efficacy. However, this also requires strains and propagation systems which provide high yields to allow a small number of passages between seed lot and production batch. Lastly, the systems used for vaccine manufacture should be of low cost.

Advantages and disadvantages of currently available vaccines

Currently available poultry vaccines are either live organisms or inactivated whole organisms. How do these vaccines fulfil the needs of the user and the producer? Each has advantages and disadvantages but there are few, if any, of either category that have all the attributes I have outlined. Before hearing about the opportunities recombinant DNA technology provides in the next paper it would be useful to consider the advantages and disadvantages of the classical live and inactivated vaccines currently available to the poultry industry.

Inactivated vaccines

Inactivated vaccines are relatively safe because inactivating agents have been used in their production. Even so, there are dangers from incomplete inactivation and from contaminating agents surviving the inactivating process. Inactivated vaccines do not have to be modified and therefore they have the potential for rapid development in response to urgent need. However, inactivated vaccines require adjuvants and have to be administered parenterally to each individual. They therefore tend to be expensive to manufacture and to administer. Finally, they stimulate humoral immunity more effectively than cell mediated immunity.

Live vaccines

Live vaccines generally induce protective immunity rapidly and more quickly than inactivated vaccines. They are more likely to stimulate all arms of the immune system than inactivated vaccines. They are relatively inexpensive to produce and can be easy and inexpensive to administer where mass administration techniques can be employed. However, there are distinct disadvantages to the use of live vaccines. There is a risk of contamination with unknown infectious agents and of reversion to virulence of the vaccine organism. There are difficulties in the field of differentiating between vaccine and field virus and immune responses to them. Finally, it can be difficult to ensure adequate attenuation of virulence to satisfy all circumstances for which the vaccine will be used and to achieve appropriate attenuation yet maintain adequate immunogenicity.

The future

Clearly neither live nor inactivated vaccines developed and produced using classical techniques fulfil all of the desirable properties required by the user and the producer. For this reason there is a need to use the opportunities provided by modern biotechnology and immunology to search for and develop improved methods of disease control. This can be through developments in transgenic technology discussed at this symposium, whether by introducing genes of the major histocompatibility complex coding for disease resistance factors, genes coding for products which block viral receptors or genes providing antisense nucleic acids. Although not mutually exclusive it will also be through the development of novel vaccines which can satisfy the needs of the poultry industry better than currently available vaccines. However, it is important to remember that when tailor making vaccines using recombinant DNA techniques they will need to be specifically designed for each disease, the design being based on a knowledge of the pathogenesis and immunogenesis of the disease.

Factors to be taken into consideration are the site and form of immunity. Is the immunity required mucosal, humoral, cell-mediated or a combination of two or all three of these? Is protection required in the alimentary, respiratory, urogenital tracts or systemic? Knowledge of immune cell trafficking, epitope characterization and mapping on each immunogen and of the function of adjuvants will be important. With the opportunity to tailor make vaccines it is important to recognize not only the needs of the user and producer but also those required to produce the most effective protective immunity for each disease.

References

Churchill, A.E. (1981) Recent advances in vaccine production. In *Avian Immunology*, pp. 389–398. Eds M. E. Rose, L. N. Payne & B. M. Freeman. British Poultry Science Ltd., Longman Group Ltd, Harlow.

Macpherson, I. (1981) Vaccination: the user's requirements. In *Avian Immunology*, pp. 429–441. Eds M. E. Rose, L. N. Payne & B. M. Freeman. British Poultry Science Ltd., Longman Group Ltd, Harlow.

Pasteur, L. (1880) De l'attenuation du virus due cholera des poules. *C. r. hebd. Séanc. Acad. Sci., Paris.* **91,** 673–680.

Pattison, M. (1981) Vaccination programmes for poultry. In *Avian Immunology*, pp. 399–410. Eds M. E. Rose, L. N. Payne & B. M. Freeman. British Poultry Science Ltd. Longman Group Ltd, Harlow.

J. Reprod. Fert., Suppl. **41** (1990), 153–162

Printed in Great Britain

Recombinant viruses as poultry vaccines

R. F. Silva and A. Finkelstein

USDA-Agricultural Research Service, Regional Poultry Research Laboratory, East Lansing, Michigan 48823, USA

Keywords: vaccines; poultry; herpesvirus; poxvirus; vector; recombinant DNA

Current vaccines

On average, 100 000 000 broiler chicks are hatched and 95 000 000 chicken eggs are produced every week in the United States. This $17 000 000 000-a-year poultry industry is greatly dependent upon vaccination to control disease. In addition, the low worth of an individual bird compared to the value of a cow or pig necessitates that the cost of avian vaccines and their application must be kept to a minimum.

Current vaccines have substantially reduced poultry condemnations and losses. However, they are not totally satisfactory. Inactivated or killed vaccines may contain residual infectious pathogens, thereby necessitating stringent and sometimes costly controls to ensure the safety of the product. In addition, sub-unit or killed vaccines do not stimulate cell-mediated immunity and therefore are inefficient against diseases that require a cell-mediated immune response for protection. Although adjuvants can potentiate the immune response, the response is not equal to the immunity provided by infectious agents.

The use of attenuated pathogens as vaccines poses the risk of reversion to a pathogenic form. In addition, the use of related, but non-pathogenic viable organisms as vaccines is often dependent upon the fortuitous isolation of appropriate field strains. As a result, the prospects for new and more efficient vaccines are limited by the existing genetic diversity. Finally, the use of live avirulent or attenuated viruses as vaccines in young chickens can be severely compromised by the presence of homologous maternal antibodies which interfere with viral replication. Consequently, the induction of active immunity following vaccination will be delayed and not provide sufficient protection against early exposure to the pathogen.

Table 1 lists some of the more important avian diseases and their vaccines (Hofstad *et al.*, 1984). Several of the vaccines are administered only as needed to control outbreaks. The efficiency of most of the vaccines is quite good. However, because of the large numbers of birds involved, a new vaccine that is capable of reducing condemnations by a few percentage points can result in substantial savings for the producers. Consequently, research is on-going to develop even more efficacious vaccines.

Vaccines for some of the avian diseases are rated as marginal and are especially appropriate targets for efforts to develop new vaccines. Some problems with current vaccines include the following. (1) The rapid appearance of new serotypes of infectious bronchitis virus constantly requires the development of new vaccines. (2) An avian influenza virus national eradication programme is in effect. Consequently, the difficulty of distinguishing between vaccinated birds and birds exposed to wild-type virus has drastically curtailed a widespread vaccination programme. (3) The usual mode of controlling coccidiosis is by administering anti-coccidial drugs in the drinking water. Often, the result is the appearance of drug-resistant strains. (4) Vaccines for infectious bursal disease generally provide good protection. However, circulating maternal antibodies can prevent the normal development of protection. (5) At present, no licensed vaccines exist for

Table 1. Selected avian diseases and their vaccines

Disease	Type of vaccine*	Usage†	Efficacy‡
Marek's disease	L	+	++
Fowlpox	L	−	++
Fowl cholera	L,K	−§	++
Infectious laryngotracheitis	L	−	++
Newcastle disease	L,K	+	++
Infectious bronchitis	L,K	+	+
Avian influenza	K	−¶	+
Coccidiosis	L	+	+
Infectious bursal disease	L,K	+	+
Avian leukosis	NA	NA	NA

*NA = not available; K = killed; L = live.
†− = Administered to control an outbreak; + = routinely administered.
‡+ = Marginally effective; ++ = effective.
§Routinely used in turkeys, occasionally in chickens.
¶Vaccination interferes with eradication programme.

retrovirus-induced avian leukosis. Killed preparations are ineffective and no attenuated variants have been isolated.

Recombinant DNA vaccines

New and potentially more efficient vaccines can be created through genetic engineering. Three general methods of constructing recombinant vaccines have been utilized. The first approach involves identifying and deleting regions of a pathogen's genome that are necessary for pathogenicity but are not required for replication. The resulting apathogenic organism can serve as a vaccine. One possible problem with this approach is that the deleted gene products may also play an important role in the host protective immune response. In addition, the identification and deletion of the pertinent portions of the genome must be repeated for each individual avian pathogen. A second approach involves cloning and expressing selected genes from pathogens to produce a sub-unit vaccine. However, sub-unit vaccines are rarely as efficacious as 'live' vaccines. This review will concentrate on a third approach that utilizes genetic engineering to construct avian virus vectors to deliver and express foreign genes.

Figure 1 shows the construction of a generic viral expression vector using a large DNA virus as the vector. The first step is to identify and clone a region of the prospective viral vector that is known to be non-essential for virus replication. A foreign gene is then inserted into the first clone such that the foreign gene is now flanked by the non-essential sequences from the virus. Finally, the clone containing the foreign gene is introduced, together with wild-type virus, into a susceptible cell. In the cell, the flanking viral sequences in the clone pair with their homologous sequences in the wild-type viral genome. Following homologous recombination, the foreign gene is inserted into the replicating viral vector genome (Mackett *et al.*, 1984; Roizman & Jenkins, 1985).

By using appropriately constructed viral vectors as vaccines, it should be possible to abrogate many of the problems with current poultry vaccines. Table 2 lists several expected advantages of using viral vectors as vaccine. In addition to these improvements, herpesviruses can induce latent infections, thereby making them attractive vectors to deliver and express a variety of genes. By using appropriate insertion sites and promoters, non-pathogenic avian herpesviruses that induce latent infections in chickens could possibly be designed to express gene products such as avian growth hormones and enzymes. Such a herpesvirus vector could serve as a vehicle to express foreign gene products at a low but continuous level for the life of the bird, and could be an alternative to the germ-line insertion of foreign genes.

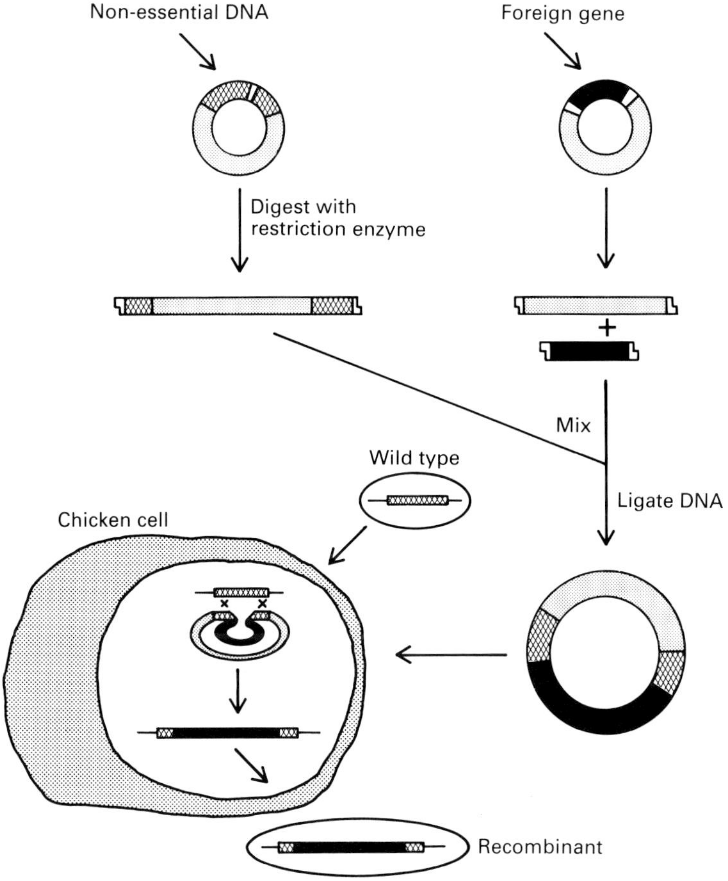

Fig. 1. Construction of a generic viral expression vector. DNA from a non-essential region of a large DNA virus is first cloned into a plasmid. The foreign gene, with appropriate regulatory sequences, is cloned into the plasmid such that the non-essential viral sequences flank the foreign gene. The final step is to introduce wild-type virus and the cloning plasmid into susceptible cells. *In vivo* recombination will occur between homologous sequences in the wild-type virus and the cloning plasmid. The result is an infectious recombinant virus that will express the foreign gene.

Table 2. Advantages of expression vectors as poultry vaccines

(1)	Expression vectors would not be affected by circulating maternal antibodies directed against the foreign gene product
(2)	Appropriate choices of promoters and foreign gene insertion sites would permit large quantities of antigen to be produced at appropriate times
(3)	Through the analysis of antibodies, it would be possible to distinguish between prior infection with a pathogen and infection with an expression vector based vaccine
(4)	Antigens expressed in viral vectors are properly processed and glycosylated
(5)	Expressed antigens can stimulate both humoral and cell-mediated immunity
(6)	In some cases, polyvalent expression vector based vaccines would be cheaper and easier to administer than conventional vaccines
(7)	Vectors expressing genes from only a portion of a pathogen's genome would be much safer than an attenuated vaccine

In mammals, various viruses including bovine papovavirus (Mitrani-Rosenbaum, 1988), simian virus 40 (Mulligan *et al.*, 1979), adenovirus (McDermott *et al.*, 1989; Prevec *et al.*, 1989), adeno-associated virus (Hermonat & Muzyczka, 1984), Sindbis virus (Xiong *et al.*, 1989), and members of the retrovirus family (Brown & Scott, 1987) have served as the basis for eucaryotic viral expression vectors. The small genome size of these viruses simplifies the construction of the recombinants. Unfortunately, their small size also restricts the amount of foreign DNA that can be inserted.

In contrast, the large size of the pox and herpesvirus genomes enables them to incorporate at least 24 000 base pairs of new foreign DNA (Smith & Moss, 1983; Roizman & Jenkins, 1985), which more than compensates for the increased difficulty in constructing the recombinants. Numerous recombinant vaccinia virus vectors expressing foreign immunogens have been created (Mackett *et al.*, 1985; Moss *et al.*, 1988). A growing list of vaccinia vectors includes recombinants expressing genes from such avian pathogens as Newcastle disease virus (Meulemans *et al.*, 1988), avian influenza virus (Chambers *et al.*, 1988), and infectious bronchitis virus (Tomley *et al.*, 1987).

Progress on constructing herpesvirus-based expression vectors has lagged slightly behind vaccinia virus vector research. Nevertheless, several different herpesvirus vectors have been constructed and include: (1) a pseudorabies virus vector expressing tissue plasminogen activator (Thomsen *et al.*, 1987); (2) a varicella-zoster vector expressing the EBV membrane glycoproteins gp359/220 (Lowe *et al.*, 1987); (3) a Herpesvirus saimiri vector expressing bovine growth hormone (Desrosiers *et al.*, 1985); and (4) herpes simplex virus (HSV) vectors. Some of the genes expressed in replication competent HSV vectors include the EBNA-1 gene of EBV (Hummel *et al.*, 1986), the hepatitis B virus S gene (Shih *et al.*, 1984) and the rabbit beta-globin gene (Smiley *et al.*, 1987). Small defective HSV vectors have been also used to express foreign genes (Kwong & Frenkel, 1985). It has been proposed that HSV vectors may be useful as vaccines (Roizman *et al.*, 1983).

Ideally, a recombinant poultry vaccine should have limited host range. Two large DNA viruses with host ranges restricted to avian species are fowlpox virus and the Marek's disease virus complex. Efforts to utilize these viruses as vectors to build recombinant poultry vaccines will be discussed in more detail.

Fowlpox virus vectors

Methodology developed to construct vaccinia vectors has been applied to the fowlpox virus (FPV) genome. As with vaccinia virus recombinants, the most commonly used insertion site for foreign sequences in the FPV genome is the thymidine kinase gene (Boyle & Coupar, 1986; Boyle *et al.*, 1987; Binns *et al.*, 1988; Schnitzlein *et al.*, 1988). Using the thymidine kinase gene of FPV as the insertion site, Boyle & Coupar (1988) were able to express the bacterial gene, xanthine guanine phosphoribosyl transferase (gpt), and the haemagglutinin gene of avian influenza. The gpt gene acted as a dominant selectable marker for recombinant FPV. Birds inoculated with the recombinant FPV produced anti-haemagglutinin.

The haemagglutinin gene from an avian influenza virus has also been inserted into an FPV vector. Chickens and turkeys inoculated with their recombinant FPV were protected against challenge by a virulent avian influenza virus (Taylor *et al.*, 1988a). When a rabies virus glycoprotein gene was inserted into the FPV genome, although FPV did not replicate in mammals, high doses of the recombinant FPV protected dogs and cats from a lethal rabies virus challenge (Taylor *et al.*, 1988b).

These early reports are encouraging and suggest that FPV vectors can be designed to function as beneficial poultry vaccines. More laboratories are expected to report on developing their own FPV vectors.

Marek's disease virus vectors

Marek's disease virus (MDV) is a herpesvirus with a host range restricted to chickens, quail and turkeys. Based upon agar-gel precipitation and virus neutralization, MDV has been classified into three serotypes. Serotype 1 viruses include oncogenic as well as attenuated variants. The serotype 2 viruses are the naturally occurring non-oncogenic chicken viruses. Serotype 3 viruses are the naturally occurring non-oncogenic turkey viruses. Serotype 3 viruses are also referred to as turkey herpesvirus or HVT (Kato & Hirai, 1985).

Attempts to utilize MDV as a vector are being made in several laboratories. The thymidine kinase gene of HVT has been identified and suggested as a likely insertion site for foreign DNA (Martin *et al.*, 1989). Although no manuscripts have been published to date, one group has reported constructing HVT vectors expressing the G protein of vesicular stomatitis virus, bacterial chloramphenicol acetyl transferase (CAT) and β-galactosidase (Bandyopadhyay *et al.*, 1988). Our efforts to construct MDV vectors are described below.

Generation of an MDV recombinant

A defective serotype 2 MDV that is closely related to a certain class of defective HSVs (Frenkel, 1980) has been identified. During cell culture passage at a high multiplicity of infection, a 4 kb fragment of the viral genome was excised. The 4 kb fragment replicates, in the presence of wild-type virus, as a high molecular weight concatemer. We cloned a 4 kb EcoRI fragment from this concatemer into pUC18. The clone was named pA5. Figure 2 shows the restriction endonuclease map of the 4 kb insert in pA5.

Wild-type MDV DNA and pA5 DNA were cotransfected into chicken embryo fibroblasts. The cells were immediately inoculated into day-one chickens. After 12 days, virus was isolated from the

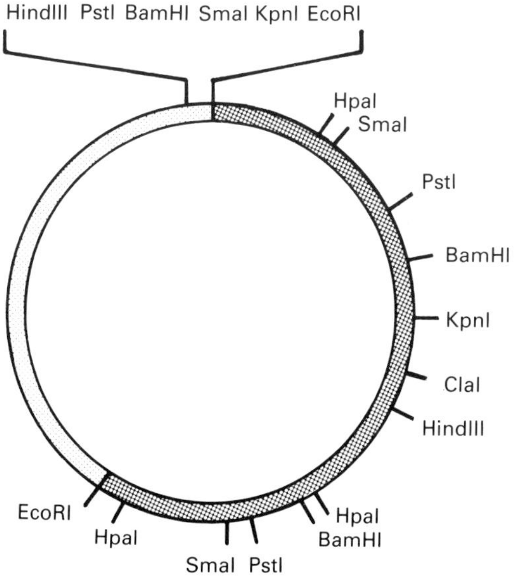

Fig. 2. Restriction endonuclease map of pA5. The amplified 4 kb EcoRI fragment from HP MDV was cloned into pUC18. The lightly shaded area represents pUC18 sequences, while the darker area represents the 4 kb viral sequences.

peripheral blood lymphocytes. Southern blotting analysis identified a recombinant MDV that contained two copies of pUC18 DNA. This suggests that the recombinant has occurred in a repeat region of the MDV genome.

The recombinant MDV replicates as well as the wild-type virus and is stable through multiple cell culture passages. Therefore, the pUC18 insertion site is non-essential for MDV replication. Various foreign genes are being inserted into this site. Since a particular foreign gene insertion site may also affect the activity of the promoters, new insertion sites are also being identified. In addition to locating insertion sites, the identification and ranking of various promoters will be crucial to optimize foreign gene expression.

Identification and ranking of various promoters

The ability of several herpesvirus promoters to direct gene expression in avian cells was compared by means of a transient assay which utilizes the activities of the *E. coli* enzymes, β-galactosidase and CAT. The genes encoding both enzymes are contained on one plasmid (*in cis*) (generous gift of E. K. Wagner) (Flanagan & Wagner, 1987). Several plasmids were constructed in which different herpesvirus promoters were inserted near the transcriptional start site of the LacZ gene. In all constructs, the SV40 promoter directed the transcription of the CAT gene (Fig. 3). Thus, when these plasmids are transfected into chicken embryo fibroblasts, the resultant β-galactosidase activity is a monitor of the relative strengths of the herpesvirus promotors. CAT activity was meant to serve as a control of transfection efficiency. Figure 4(a) shows the relative strengths of the cytomegalovirus immediate early (CMVIE) promoter, the herpes simplex virus alpha 4 (alpha 4) promoter, and the HSV thymidine kinase (βTK) promoter.

The two promoters do not appear to behave independently. The strength of the SV40 promoter seems to be reduced when it is adjacent to a strong herpesvirus promoter. Consequently, the highest CAT activity occurs in the control plasmid that lacks a promoter directing the LacZ gene (ø in Figs 4a & b). To test whether the putative competition was due to the close proximity of the two promoters, we linearized the plasmids by cutting between the two promoters. The competition effect was still observed even though approximately 10 kb separated the two promoters (Fig. 5). To examine this competition further, we co-transfected cells with our constructs and another plasmid, pON249 (Geballe *et al.*, 1986), that contains the CMV immediate early promoter driving LacZ. We found that the CMVIE promoter is able to reduce the activity of the SV40 promoter even when it is present on another plasmid (*in trans*) (Fig. 6). These results suggest that the weak SV40 promoter activity is a result of competition for transcription factors between the two promoters.

For these promoters to be useful, they must function in cells that have been infected with MDV. To investigate this, we co-transfected cells with the plasmids and MDV DNA. MDV increases the activity of all the promoters we have tested (data not shown). We are now integrating these plasmids into the MDV genome at the pUC18 insertion site so that we can evaluate whether the effects of MDV *in cis* differ from the effects of having MDV *in trans*.

Future prospects

FPV-based vectors expressing genes from poultry pathogens have already been constructed. Similar MDV based vectors are expected shortly. For any of these expression vectors to compete successfully with conventional poultry vaccines, the recombinants must be at least as efficacious. Available data suggest that avian viral vectors can be constructed to deliver, in a controlled fashion, sufficiently large amounts of foreign gene products. However, the key to developing truly efficacious recombinant vaccines will depend upon choosing the correct foreign gene to

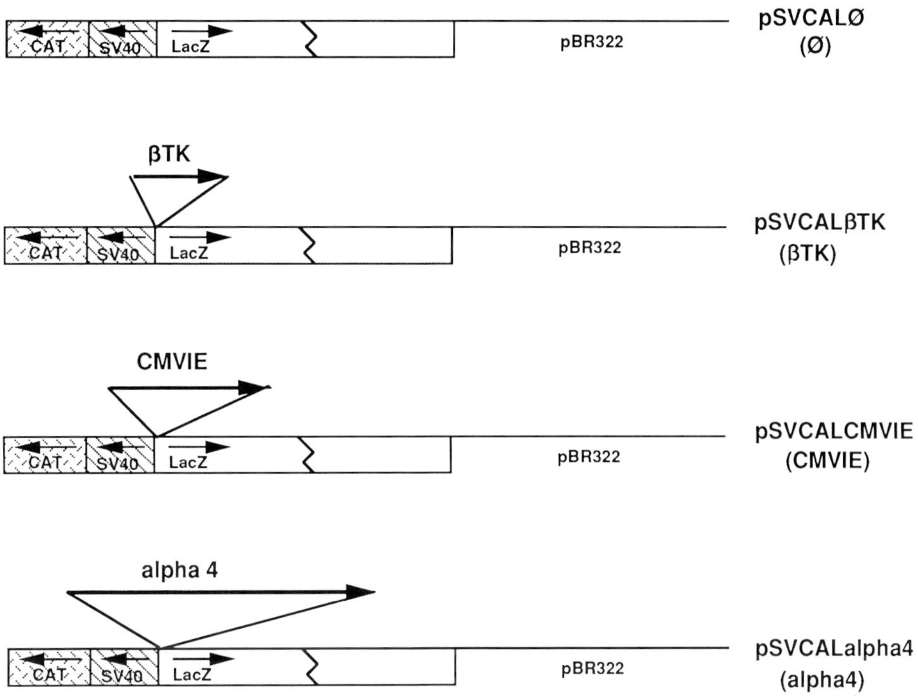

Fig. 3. Plasmid vectors used to rank promoters. In all plasmids, the SV40 early promoter was inserted in front of the CAT gene in pCAL5. We constructed plasmids pSVCALβTK, pSVCALCMVIE, and pSVCALalpha4 by inserting in front of LacZ the HSV thymidine kinase, cytomegalovirus immediate early, or HSV α-4 promoters, respectively.

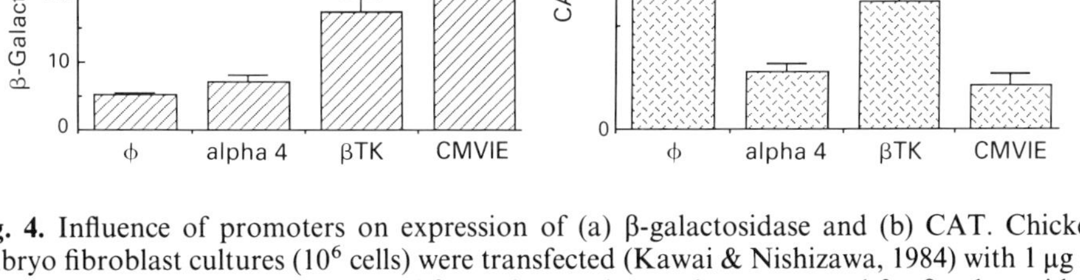

Fig. 4. Influence of promoters on expression of (a) β-galactosidase and (b) CAT. Chicken embryo fibroblast cultures (10^6 cells) were transfected (Kawai & Nishizawa, 1984) with 1 μg of the appropriate plasmid constructs. After 1 day, each sample was assayed for β-galactosidase activity (Spaete & Mocarski, 1985) and CAT activity (Neumann *et al.*, 1987). Protein concentrations were determined as described by Bradford (1976). The bars represent the mean ± s.e. of 5 samples. ø = pSVCALø; alpha 4 = pSVCALalpha4; βTK = pSVCALβTK; CMVIE = pSVCALCMVIE.

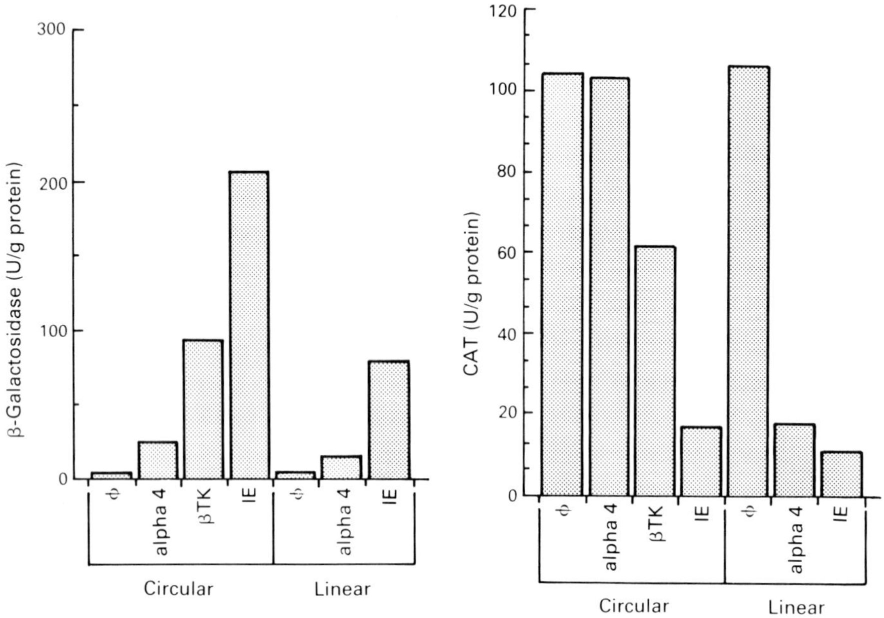

Fig. 5. Promoter competition in cis. Chicken embryo fibroblasts (10^6) were transfected, as described in Fig. 4, with 5 μg intact plasmid or with linear plasmid that had been cut once at a point immediately 5′ of the SV40 promoter. ø = pSVCALø; alpha 4 = pSVCALalpha4; βTK = pSVCALβTK; IE = pSVCALCMVIE.

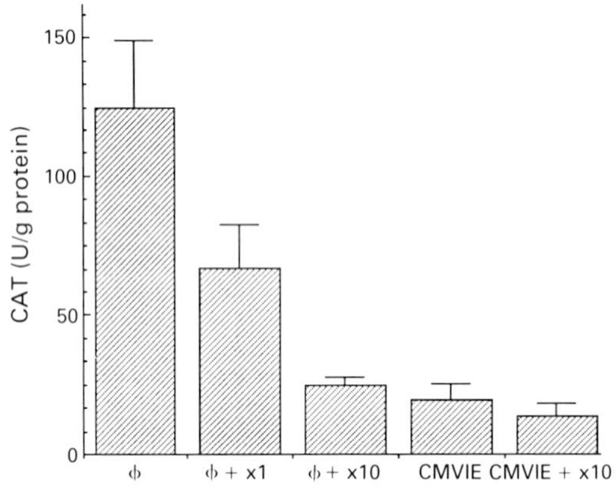

Fig. 6. Promoter competition in trans. Chicken embryo fibroblasts (10^6) were transfected, as described in Fig. 4, with 1 μg pSVCALø (ø) or pSVCALCMVIE (CMVIE) plasmid DNA. Some cultures were also transfected with 1 μg (× 1) or 10 μg (× 10) pON249. Salmon sperm DNA was added to the transfection mixture to give a total of 11 μg DNA/10^6 cells.

be expressed. Much more research remains to be done to identify the appropriate epitopes to be expressed.

This work is supported by the Agricultural Research Service and by a US Department of Agriculture competitive research grant (85-CRCR-1-1709). We thank Heidi Camp and Abby Schwartz for excellent technical assistance.

References

Bandyopadhyay, P., Martin, S., Aparisio, D., Doherty, D. & Florkiewicz, R.Z. (1988) Use of herpes virus of turkeys for the introduction of foreign genes into poultry. *Proc. 13th International Herpesvirus Workshop, Irvine* **323,** Abstr.

Binns, M.M., Tomley, F.M., Campbell, J. & Boursnell, M.E.G. (1988) Comparison of a conserved region in fowlpox virus and vaccinia virus genomes and the translocation of the fowlpox virus thymidine kinase gene. *J. gen. Virol.* **69,** 1275–1283.

Boyle, D.B. & Coupar, B.E.H. (1986) Identification and cloning of the fowlpox virus thymidine kinase gene using vaccinia virus. *J. gen. Virol.* **67,** 1591–1600.

Boyle, D.B. & Coupar, B.E.H. (1988) Construction of recombinant fowlpox viruses as vectors for poultry vaccines. *Virus Res.* **10,** 343–356.

Boyle, D.B., Coupar, B.E.H., Gibbs, A.J., Seigman, L.J. & Both, G.W. (1987) Fowlpox virus thymidine kinase: nucleotide sequence and relationships to other thymidine kinases. *Virology* **156,** 355–365.

Bradford, M.A. (1976) Rapid and sensitive method for the quantitation of microgram quantities of protein utilizing the principle of protein-dye binding. *Analyt. Biochem.* **72,** 248–254.

Brown, A.M.C. & Scott, M.R.D. (1987) Retroviral vectors. In *DNA Cloning,* Vol. III, pp. 189–212. Ed. D. M. Glover. IRL Press, Oxford.

Chambers, T.M., Kawaoka, Y. & Webster, R.G. (1988) Protection of chickens from lethal influenza infection by vaccinia-expressed hemagglutinin. *Virology* **167,** 414–421.

Desrosiers, R.C., Kamine, J., Bakker, A., Silva, D., Woychick, R.P., Sakai, D.D. & Rottman, F.M. (1985) Synthesis of bovine growth hormone in primates by using a herpesvirus vector. *Molec. cell. Biol.* **5,** 2796–2803.

Flanagan, W.M. & Wagner, E.K. (1987) A bi-functional reporter plasmid for the simultaneous transient expression assay of two herpes simplex virus promoters. *Virus Genes* **1,** 61–71.

Frenkel, N. (1980) Defective interfering herpesvirus. In *The Human Herpesvirus, an Interdisciplinary Perspective,* pp. 91–120. Eds A. J. Nahmias, W. R. Dowdle & R. F. Schinazi. Elsevier, New York.

Geballe, A.P., Spaete, R.R. & Mocarski, E.S. (1986) A cis-acting element within the 5′ leader of a cytomegalovirus B transcript determines kinetic class. *Cell* **46,** 865–872.

Hermonat, P.L. & Muzyczka, N. (1984) Use of adeno-associated virus as a mammalian DNA cloning vector: transduction of neomycin resistance into mammalian tissue culture cells. *Proc. natn. Acad. Sci. USA* **81,** 6466–6470.

Hofstad, M.S., Barnes, H.J., Calnek, B.W. & Yoder, H.W., Jr (1984) *Diseases of Poultry,* 8th edn. Iowa State University Press, Ames.

Hummel, M., Arsenakis, M., Marchini, A., Lee, L., Roizman, B. & Kieff, E. (1986) Herpes simplex virus expressing Epstein-Barr virus nuclear antigen 1. *Virology* **148,** 337–348.

Kato, S. & Hirai, K. (1985) Marek's disease virus. *Adv. Virus Res.* **30,** 255–277.

Kawai, S. & Nishizawa, M. (1984) New procedure for DNA transfection with polycation and dimethyl sulfoxide. *Molec. cell. Biol.* **4,** 1172–1174.

Kwong, A.D. & Frenkel, N. (1985) The herpes simplex virus amplicon. IV. Efficient expression of a chimeric chicken ovalbumin gene amplified within defective virus genomes. *Virology* **142,** 421–425.

Lowe, R.S., Keller, P.M., Keech, B.J., Davidson, A.J., Whang, Y., Morgan, A.J., Kieff, E. & Ellis R.W. (1987) Varicella-zoster virus as a live vector for the expression of foreign genes. *Proc. natn. Acad. Sci. USA* **84,** 3896–3900.

Mackett, M., Smith, G.L. & Moss, B. (1984) General method for production and selection of infectious vaccinia virus recombinants expressing foreign genes. *J. Virol.* **49,** 857–864.

Mackett, M., Smith, G.L. & Moss, B. (1985) The construction and characterization of vaccinia virus recombinants expressing foreign genes. In *DNA Cloning,* Vol. II, pp. 191–211. Ed. D. M. Glover. IRL Press, Oxford.

Martin, S.L., Aparisio, D.I. & Bandyopadhyay, P.K. (1989) Genetic and biochemical characterization of the thymidine kinase gene from herpesvirus of turkeys. *J. Virol.* **63,** 2847–2852.

McDermott, M.R., Graham, F.L., Hanke, T. & Johnson, D.C. (1989) Protection of mice against lethal challenge with herpes simplex virus by vaccination with an adenovirus vector expressing HSV glycoprotein B. *Virology* **169,** 244–247.

Meulemans, G., Letellier, C., Gonze, M., Carlier, M.C. & Burny, A. (1988) Newcastle disease virus F glycoprotein expressed from a recombinant vaccinia virus vector protects chickens against live-virus challenge. *Avian Pathol.* **17,** 821–827.

Mitrani-Rosenbaum, S. (1988) Use of a stable bovine papillomavirus vector to study inducible genes. *Intervirology* **29,** 108–114.

Moss, B., Fuerst, T.R., Flexner, C. & Hugin, A. (1988) Roles of vaccinia virus in the development of new vaccines. *Vaccine* **6,** 161–163.

Mulligan, R., Howard, B.H. & Berg, P. (1979) Synthesis of rabbit β-globin in cultured monkey kidney cells following infection with a SV40 β-globin recombinant genome. *Nature, Lond.* **277,** 108–114.

Neumann, J.R., Morency, C.A. & Russian, K.O. (1987) A novel rapid assay for chloramphenicol acetyltransferase gene expression. *Bio/Technology* **5,** 444–447.

Prevec, L., Schneider, M., Rosenthal, K.L., Belbeck, L.W., Derbyshire, J.B. & Graham, F.L. (1989) Use of adenovirus-based vectors for antigen expression in animals. *J. gen. Virol.* **70,** 429–434.

Roizman, B. & Jenkins, F.J. (1985) Genetic engineering of novel genomes of large DNA viruses. *Science, NY* **229,** 1208–1214.

Roizman, B., Meignier, B., Norrild, B. & Wagner, J.L. (1983) Bioengineering of herpes simplex virus variants for potential use as live vaccines. In *Modern Approaches to Vaccines,* pp. 275–281. Eds R. M. Chanock & R. A. Lerner. Cold Spring Harbor Laboratory, Cold Spring Harbor.

Schnitzlein, W.M., Ghildyal, N. & Tripathy, D.N. (1988) A rapid method for identifying the thymidine kinase genes of avipoxviruses. *J. virol. Methods* **20,** 341–352.

Shih, M., Arsenakis, M., Tiollais, P. & Roizman, B. (1984) Expression of hepatitis B virus S gene by herpes simplex virus type 1 vectors carrying α- and β-regulated gene chimeras. *Proc. natn. Acad. Sci. USA* **81,** 5867–5870.

Smiley, J.R., Smibert, C. & Everett, R.D. (1987) Expression of a cellular gene cloned in herpes simplex virus: rabbit beta-globin is regulated as an early viral gene in infected fibroblasts. *J. Virol.* **61,** 2368–2377.

Smith, G.L. & Moss, B. (1983) Infectious poxvirus vectors have the capacity for at least 25,000 base pairs of foreign DNA. *Gene* **25,** 21–28.

Spaete, R.R. & Mocarski, E.S. (1985) Regulation of cytomegalovirus gene expression: α and β promoters are trans activated by viral functions in permissive human fibroblasts. *J. Virol.* **56,** 135–143.

Taylor, J., Weinberg, R., Kawaoka, Y., Webster, R.G. & Paoletti, E. (1988a) Protective immunity against avian influenza induced by a fowlpox virus recombinant. *Vaccine* **6,** 504–508.

Taylor, J., Weinberg, R., Languet, B., Desmettre, P. & Paoletti, E. (1988b) Recombinant fowlpox virus inducing protective immunity in non-avian species. *Vaccine* **6,** 497–503.

Thomsen, D.R., Marotti, K.R., Palermo, D.P. & Post, L.E. (1987) Pseudorabies virus as a live virus vector for expression of foreign genes. *Gene* **57,** 261–265.

Tomley, F.M., Mockett, A.P.A., Boursnell, M.E.G., Binns, M.M., Cook, J.K.A., Brown, T.D.K. & Smith, G.L. (1987) Expression of the infectious bronchitis virus spike protein by recombinant vaccinia virus and induction of neutralizing antibodies in vaccinated mice. *J. gen. Virol.* **68,** 2291–2298.

Xiong, C., Levis, R., Shen, P., Schlesinger, S., Rice, C.M. & Huang, H.V. (1989) Sindbis virus: An efficient, broad host range vector for gene expression in animal cells. *Science, NY* **243,** 1188–1191.

J. Reprod. Fert., Suppl. **41** (1990), 163–171

Printed in Great Britain

Expression of retroviral genes in transgenic chickens

L. B. Crittenden* and D. W. Salter†

**US Department of Agriculture, Agricultural Research Service, Regional Poultry Research Laboratory, 3606 East Mount Hope Rd, East Lansing, MI 48823, USA; and †Department of Microbiology and Public Health, Michigan State University, East Lansing, MI 48824, USA*

Summary. The development of transgenic technology in poultry has lagged behind that in mammals because of the unique reproductive system of birds. Therefore, we chose to use wild-type and recombinant replication-competent avian leukosis viruses to determine whether these retroviruses could be artificially introduced into the germ line by injecting them near the blastoderm of fertile eggs just before incubation. We generated 23 proviral inserts that were stably inherited through two generations. Twenty-one inserts coded for complete avian leukosis virus. Two interesting inserts failed to produce complete virus. One coded for envelope glycoprotein only and the other coded for the group-specific antigen and envelope glycoprotein. Cell culture and animal studies showed that one of these inserts was very resistant to infection and oncogenesis by subgroup A field strains of avian leukosis virus. Therefore, this represents a model system for introducing genes from the pathogen into the host to induce host resistance to the pathogen. Future studies should be aimed at developing more efficient systems for introducing replication-defective retroviral vectors or cloned DNA into the germ line of poultry so that the regulation of gene expression can be studied and transgenic technology can be applied to the improvement of this highly reproductive group of farm animals.

Keywords: avian leukosis virus; transgenic; chickens; resistance; viral interference

Introduction

The development of transgenic technology for poultry has lagged behind technology for mammals because the newly fertilized ovum is located on the surface of the yolk within the oviduct and access and manipulation are difficult (Crittenden & Salter, 1985; Freeman & Messer, 1985). Recently a technique has been developed to grow chicks from fertilization to hatching *in vitro* (Perry, 1988). While this is a promising approach, the success rate is low and few chicks can be hatched at one time. The in-vitro transfection of stem-cell lines with cloned DNA, growth in culture and transplantation to the embryo has led to the development of germ-cell chimaeras in mice and thus transgenic animals. This appears to be an ideal approach because of the possibility of assay for expression in the stem cells and selection for cells with specific sites of integration (Thomas & Capecchi, 1987). While inter-breed germ-cell chimaeras have been produced in chickens by injecting foreign cells from the blastoderm of newly laid fertile eggs of one breed into the blastoderms of another breed, no method has yet been developed for the culture and manipulation of stem cells of the chicken so that specific cloned genes can be introduced (Petitte *et al.*, 1989).

Because of these limitations and the fact that avian retroviral genes have entered the germ line of the chicken during evolution (Smith, 1986), we chose to attempt to infect germ cells of chickens with replication-competent recombinant and wild-type avian leukosis viruses (ALV) (Salter *et al.*, 1986). Retroviruses are attractive for gene transfer because these RNA viruses replicate through chromosomally integrated DNA proviruses and act as vectors for the insertion of foreign genes in nature (Varmus & Swanstrom, 1982). Replication-competent and replication-defective vectors have been constructed for the avian system, and replication-defective vectors have been used to insert model genes into the germ line (Hughes *et al.*, 1990; Bosselman *et al.*, 1990). In this paper we discuss our successful insertion of ALV into the germ line, the spontaneous inactivation of some proviruses, and the expression viral genes by those proviruses.

Methods

Our approach has been described in detail elsewhere (Hughes *et al.*, 1986; Salter *et al.*, 1986, 1987; Salter & Crittenden, 1989; Crittenden *et al.*, 1989). We used high-titre ALV stocks of two recombinant viruses that had the long terminal repeats (LTR) of the endogenous ALV, RAV-0, and the envelope gene of a subgroup A virus of exogenous origin. Such viruses have been shown to have the antigenic and interference characteristics of typical field strains of ALV but are much less oncogenic because of the lower promoter and enhancer activity of the RAV-0 LTR. These viruses were called 882/-16,RAV-0 (Hughes *et al.*, 1986) and RAV-0-A(1) (Wright & Bennett, 1986). The third virus was RAV-1, an exogenous virus with an exogenous LTR and subgroup A envelope.

We injected 0·05 ml virus stock or infected cellular material near the blastoderm of fertile line 0 eggs just before incubation. Line 0 is free of endogenous ALV proviruses that are highly homologous to the ALV injected. Therefore, DNA dot-blot hybridizations with complete ALV cDNA probes could be used to screen for infection or germ-line transmission. The generation 0 (G-0) day-old chicks were screened by dot-blot hybridizations and the positive males were reared for mating and transmission studies. Males are not known to transmit ALV congenitally to progeny (Spencer *et al.*, 1980). Therefore, male transmission was putative evidence for germ-line transmission. Infected G-0 males were mated to uninfected line 0 females and the G-1 chicks were screened by dot-blot hybridization. Southern blots of *Sac*I-digested DNA from positive G-1 chicks hybridized with complete ALV cDNA probes were scanned for clonal proviral–host DNA junction fragments suggestive of germ-line integration. Positive G-1 males and females were mated with line 0 to determine whether DNA junction fragments of the same molecular weight as carried by the G-1 parent were transmitted to the G-2 progeny as expected with Mendelian inheritance.

Assays for infectious ALV, *gag* and *env* expression were conducted by published procedures (Crittenden *et al.*, 1979a, 1987). Virus challenges for resistance were conducted by using avian sarcoma viruses in cell culture and with ALV in chickens.

Results

Germ-line transmission of ALV proviruses

Table 1 summarizes proviral transmission through two generations. Of 37 G-0 males, 9 transmitted to their G-1 progeny and were possible germ-line chimaeras. The overall transmission rate from viraemic G-1 males was $<2\%$. Each G-1 chick that was dot-blot positive was shown to have clonal provirus–host junction fragments on Southern blotting. Twenty positive G-1 birds were reared to sexual maturity and mated to line 0 chickens to produce G-2 progeny. In each case clonal junction fragments, of the same molecular weight as found in G-1 DNA, were inherited in about a 1:1 ratio, clearly confirming germ-line transmission. The 20 parents transmitted 23 unique inserts that were numbered *alv*1 through *alv*23. One G-1 female carried 3 and another carried 2 inserts that were segregated in subsequent generations and represented different integration events.

Table 1 also shows that two types of non-Mendelian transmission took place. First, there were G-2 progeny that were dot-blot positive but lacked clonal junction fragments on Southern blotting. These were, almost without exception, progeny of female G-1 parents. We think that this represents somatic infection by congenital transmission from the dam, but not from the sire, confirming

Table 1. Transmission of ALV proviral DNA in two generations of progeny of male chicken hatched from eggs that had been infected with ALV just before incubation

ALV stock	G-0 No. of males Tested	G-0 No. of males Trans-mitting	G-1* Segregation of progeny† +	G-1* Segregation of progeny† −	G-1* No. of (sex) progeny tested‡	G-2* Segregation of inserts§ +	G-2* Segregation of inserts§ −	G-2* No. with new inserts¶	G-2* Congenital infection‖
RAV-0-A(1)	14	4	21	794	6 (M)	109	107	30	0
					9 (F)	218	233	2	35
882/-16, RAV-0	9	4	5	532	2 (M)	63	83	9	1
					2 (F)	36	39	0	32
RAV-1	14	1	2	551	1 (M)	13	20	0	0
Total	37	9	28	1877	20	439	482	41	68

*G-1 and G-2 represent the first and second backcross generations to line 0 respectively.
†Positive (+) or negative (−) for proviral DNA by dot-blot hybridization.
‡Number of G-1 birds that survived to maturity and were sucessfully progeny tested. All G-1 birds had clonal junction fragments after digestion of their DNA with *Sac*I.
§+, Progeny whose DNA had *Sac*I junction fragments of the same molecular weight as the parent; −, Progeny with no junction fragments of the same molecular weight.
¶Chickens whose DNA had *Sac*I junction fragments of a different molecular weight.
‖Number of chickens that were dot-blot positive but lacked clonal *Sac*I junction fragments.

earlier observations that congenital transmission is restricted to the female (Spencer *et al.*, 1980). Secondly, there were G-2 progeny, largely produced by two G-1 males, that had clonal provirus–host DNA junction fragments that were not carried by their parents. Many of these proviruses were shown to be inherited and we think that they arose through reinfection of the germ cells during the development of the bird carrying a fully infectious provirus. We have shown that families carrying these proviruses continue to generate new inherited proviruses for several generations (Crittenden & Salter, 1988).

These results clearly show that replication-competent ALV can be used to infect germ-line cells artificially to generate stably inherited proviruses by the relatively simple procedure of injecting live ALV blindly near the blastoderm of fertile eggs just before incubation.

Expression of retroviral genes

G-1, 2 and 3 chicks were assayed for infectious virus production: 21 out of 23 of the *alv* genes coded for complete virus. Two, *alv*6 and *alv*11, did not code for infectious virus. These interesting exceptions were studied in more detail for expression of ALV genes. Table 2 gives the results. Cell cultures and birds carrying *alv*6 lacked p27, a protein encoded by the retroviral *gag* gene, but did express ALV envelope glycoprotein encoded by the *env* gene. Reverse transcriptase assays for expression of the *pol* gene were not conducted. Electron microscopy of cell cultures carrying *alv*6 did not reveal typical ALV particles. Southern blotting with several enzymes failed to reveal gross internal proviral alterations. Therefore, *alv*6 must have a minor defect in the provirus or a cis-acting restriction due to the chromosomal location in the host that leads to a defect in viral protein processing or packaging. In contrast, *alv*11 codes for both *gag* and *env* expression and virus particle production as visualized by electron microscopy. Restriction enzyme mapping revealed a deletion of about 0·5 kb in *pol*. Therefore, we believe that a defect in virion reverse transcriptase accounts for the lack of infectivity of the ALV particles generated.

These results suggest that most germ-line inserts generated in chickens by replication-competent ALV using these methods will produce infectious virus. Therefore, the use of replication-competent vectors for the introduction of foreign genes would, in many cases, be expected to generate transgenics that also produce complete, possibly pathogenic, virus.

Resistance of envelope expressing inserts to ALV infection and oncogenesis

Previous studies of endogenous viral (*ev*) genes that express the envelope glycoproteins of subgroup E ALV showed that cell cultures and chickens were very resistant to infection with ALV belonging to subgroup E but not other subgroups (Robinson *et al.*, 1981). This resistance is apparently due to specific interference with infection by viruses belonging to the same ALV subgroup (Vogt & Ishizaki, 1966). Therefore, we thought that the expression of subgroup A envelope proteins by *alv*6 and 11 might interfere with subgroup A ALV infection and that these inserts could act as dominant genes for resistance to the common pathogenic subgroup A viruses found in the field (Crittenden & Salter, 1985).

We made chick-embryo fibroblasts from embryos carrying *alv*6 and 11 as well as line 0 embryos. Table 3 gives the results of titration of the subgroup A and B avian sarcoma viruses, RSV(RAV-1) and RSV(RAV-2) respectively, in these cultures. Cells carrying *alv*6 were very resistant to focus formation by subgroup A virus but showed little resistance to subgroup B virus in these representative data (Salter & Crittenden, 1989). Cells carrying *alv*11 were less resistant to subgroup A than were *alv*6-carrying cells and showed a more variable phenotype. This variability could be due to the fact that these cells produce defective virus particles that interfere with infection and perhaps particle production varies from culture to culture, producing a more variable phenotype than found for *alv*6 that codes for envelope glycoprotein expressed in the cell membrane.

We have conducted extensive in-vivo studies of the resistance of *alv*6-carrying chickens to infection and oncogenesis by ALV. Table 4 gives the results of an experiment on the segregating progeny generated by mating males hemizygous for *alv*6 with line 0 females known not to transmit ALV congenitally. The chicks were bled at 1 day of age to determine their *alv*6 status by dot-blot hybridization, and inoculated with 10^4 infectious units of the field strain of subgroup A ALV, RPL-42. No live ALV was detected in the *alv*6-carrying chickens and none died with bursal lymphomas or other neoplasms by 40 weeks of age. In contrast, all the *alv*6-negative chicks became infected and more than half died with neoplasms. These chicks were intermingled throughout the experiment so that there was ample opportunity for the *alv*6-carrying chickens to become infected by contact as well as by the original injection. In a parallel experiment chickens were inoculated with the subgroup B ALV, RAV-2. No difference in infection rate or oncogenesis was found, confirming the specificity of resistance to subgroup A (Crittenden *et al.*, 1989). The expression of subgroup A envelope protein in these chicks did not appear to reduce the immune response to subgroup B ALV as might be expected due to immune tolerance to envelope antigens shared by these subgroups (Crittenden *et al.*, 1987).

Congenital transmission of ALV from the dam to progeny through the egg is a major mode of transmission, and is responsible for maintaining infection in breeding populations of chickens because congenitally infected chickens usually become immunologically tolerant to the virus, remain viraemic, and congenitally transmit to the next generation (Spencer *et al.*, 1980). To test the possibility that introduction of the dominant gene for resistance, *alv*6, by the male could block congenital transmission, we mated males hemizygous for *alv*6 to line 0 females that had been inoculated as embryos with the field strain of subgroup A ALV, RPL-41. The females used were all viraemic and immunologically tolerant to subgroup A ALV. Table 5 gives the frequency of p27 detection at 1 day of age, virus isolation and mortality from bursal lymphomas to 35 weeks of age. The rate of congenital infection was clearly reduced by the presence of *alv*6 in the progeny but was not eliminated, and viraemia in some of the *alv*6-positive progeny persisted to 30 weeks of age. We think that virus infection may have been established in some of the *alv*6-positive embryos very early in development before the full expression of the subgroup A envelope antigen coded for by *alv*6, accounting for the reduced resistance to congenital infection compared to infection at 1 day of age.

Table 2. ALV gene expression and proviral structure of two germ-line inserts (*alv*6 and 11), that are defective for complete virus production

Insert	ALV gene expression		Virus particles‡	Proviral structure§
	*gag**	*env*†		
*alv*6	−	+	−	Complete
*alv*11	+	+	+	Deletion of 0·5 kb in *pol*

*Determined by direct enzyme-linked immunosorbant assay (ELISA) for the *gag* protein p27 (Smith *et al.*, 1979).
†Determined by a phenotypic mixing assay (Crittenden *et al.*, 1979a).
‡Visualized by electron microscopy of cell culture pellets (Crittenden *et al.*, 1989).
§Determined by restriction enzyme mapping of internal proviral fragments generated by *Bam*HI, *Eco*RI, and *Hin*dIII and the use of a complete ALV cDNA probe (Crittenden *et al.*, 1989).

Table 3. Relative resistance of chick embryo fibroblasts (CEFs) carrying *alv*6 and 11 that fail to produce complete ALV but express envelope protein to the subgroup A and B avian sarcoma viruses, BH-RSV(RAV-1) and BH-RSV(RAV-2)

Sarcoma virus		*alv*				Mean*
		6	11	11	11	Line 0
BH-RSV (RAV-1)	Titre†	$4{\cdot}0 \times 10^1$	$1{\cdot}7 \times 10^5$	$3{\cdot}4 \times 10^3$	$1{\cdot}6 \times 10^3$	$3{\cdot}6 \times 10^6$
	Relative titre‡	0·000011	0·047222	0·000944	0·000444	1·000000
BH-RSV (RAV-2)	Titre†	$4{\cdot}9 \times 10^5$	$3{\cdot}1 \times 10^6$	$6{\cdot}6 \times 10^5$	$8{\cdot}1 \times 10^5$	$1{\cdot}6 \times 10^6$
	Relative titre‡	0·306250	1·900000	0·412500	0·506250	1·000000

*Mean titre for 3 separate CEFs.
†Titres per ml of virus stock were estimated by averaging the focus counts of the plates with the 2 highest dilutions that had foci.
‡Ratio of titre on the test embryo to the titre on line 0 cell cultures.

Table 4. Interference with subgroup A ALV infection and oncogenicity in transgenic chickens carrying the defective proviral insert, *alv*6, after infection with RPL-42 at 1-day of age

Progeny	Viraemia*				Bursal lymphomas*†
	2 weeks	7 weeks	16 weeks	40 weeks	
*alv*6+	0/36 (0)	0/24 (0)	0/27 (0)	0/27 (0)	0/36 (0)
*alv*6−	39/39 (100)	20/23 (87)	23/25 (92)	0/1 (0)	22/39 (56)

*Ratio of positive/total observed (%).
†Mortality to 40 weeks of age.

Table 5. Interference with subgroup A ALV infection and oncogenicity in transgenic chickens carrying the defective proviral insert, *alv*6, hatched from dams that were viraemic with RPL-41

Progeny	p27*† (1 day)	Viraemia* 6 weeks	16 weeks	30 weeks	Bursal lymphomas*‡
*alv*6+	17/94 (18)	26/71 (37)	18/67 (27)	16/60 (27)	5/68 (7)
*alv*6−	71/85 (84)	55/57 (96)	44/48 (92)	25/26 (90)	25/53 (47)

*Ratio of positive/total observed (%).
†Direct assay for p27 (Smith *et al.*, 1979).
‡Mortality to 35 weeks of age.

Further studies on *alv*6

The *alv*6 provirus has been maintained in the hemizygous state for 4 generations and as a homozygous line for 2 generations, and we have not detected any gross alteration in the provirus nor the production of complete virus or p27. Apparently *alv*6 has not disrupted the expression of a gene crucial for development because the homozygotes and heterozygotes appear to survive and reproduce equally well. Surprisingly, about 15% of the older breeding birds have died of bursal lymphomas histologically similar to those that are induced by ALV. So far, we have not detected the presence of ALV of any subgroup in these birds. Some developed neoplasms even though they were held in filtered air-positive pressure isolators. The explanation for these observations must await further molecular and virological studies.

All of our experiments with *alv*6-carrying birds have been conducted in a line 0 background free of *ev* genes with close homology to ALV. All commercial chickens carry *ev* genes and recombination among ALVs, ALVs and *ev* gene transcripts and among *ev* gene transcripts have been observed in the laboratory (Coffin, 1982). For this reason we are studying the consequences of crossing birds carrying *alv*6 with those that carry *ev*21, a gene that codes for complete subgroup E virus that is common in egg-laying chickens (Smith & Crittenden, 1988). We have isolated subgroup A ALV at a high frequency from birds carrying the two genes. Such viruses are certainly generated by phenotypic mixing. Studies in progress will probably yield recombinant viruses, with subgroup A envelopes and the LTRs of RAV-0, very similar to those used for generating our transgenic chickens.

These studies, taken together, indicate that the insertion of a replication-competent subgroup A ALV and generation, by chance, of a transgenic with expression of the subgroup A envelope but not infecting ALV is not a practical approach for introducing ALV resistance into poultry populations. It is, however, an interesting model that suggests that animals transgenic and expressing viral genes, particularly those of the virus coat, may be resistant to superinfection by the same virus. It remains to be determined whether inserting and expressing the subgroup A envelope alone in chickens by using a replication defective vector will generate commercially useful transgenics for ALV control.

Discussion

The development of transgenic chickens that express the envelope gene of subgroup A ALV and are resistant to infection by homologous viruses is a special example of the principle developed by Sanford & Johnston (1985). They proposed the general strategy of producing pathogen-resistant hosts by transforming them with modified genes derived from the parasite. This strategy is being used in plants (Nelson *et al.*, 1988) and *alv*6 illustrates its use in animals. This principle should hold at least for the envelope genes of other retroviruses, and has been extended, in cell culture studies, to the reticuloendotheliosis virus of the fowl (Delwart & Panbanigan, 1989; Federspiel *et al.*,

1989). The general applicability of the envelope system for other classes of viruses remains to be demonstrated, but other modified viral genes may also block viral infection or replication as pointed out by Sanford & Johnston (1985).

The avian leukosis virus model illustrates one potential problem for the use of this general approach in animals which have a well developed immune system. The gene from the pathogen to be inserted may code for a protein that not only interferes with infection or replication but also induces a protective immune response. If that gene is expressed in embryos the animal may develop immune tolerance to homologous pathogens and not be capable of mounting a protective immune response upon infection or vaccination. This is one example of a situation in which the inserted gene could have both beneficial and detrimental effects (Crittenden *et al.*, 1987; Salter & Crittenden, 1989).

Another important question concerns the expression of retroviral genes inserted in the germ line and, more importantly, the expression of genes inserted into the germ line by retroviral vectors. There are many examples of the down-regulation of retroviral genes in very early embryos. One example is the methylation and delayed expression of murine leukaemia viruses inserted into the germ line of mice (Jaenisch *et al.*, 1981). In the chicken *ev* genes that code for complete virus, *ev*2, 10, 11, 12, and 21, are all down-regulated when they occur in a genetic background that is not permissive for infection with subgroup E virus, but the viruses grow to a high titre in cells or animals permissive for infection (Crittenden *et al.*, 1974, 1979b; Smith & Crittenden, 1988). Apparently the transcription of viral RNA is restricted by methylation in the germ-line genomic location but escapes methylation when the provirus is inserted in new locations upon somatic infection of permissive host cells (Conklin & Groudine, 1986). It is not clear whether this restriction holds for all inserts that code for complete virus or whether only restricted germ-line proviruses survive natural selection. Artificial insertion of murine leukaemia viruses suggests that they would be restricted to different extents depending on chromosomal location (Jaenisch *et al.*, 1981). All of our newly inserted *alv* proviruses are in a genetic background permissive for subgroup A. It would be of interest to determine whether they are all restricted in a subgroup A resistant background. In contrast to *ev* genes that code for complete virus, at least some *ev* genes that code for viral antigens only are highly expressed and at least some of the viral RNAs are transcribed at high levels (Hayward *et al.*, 1980; Baker *et al.*, 1981; Conklin & Groudine, 1986). The *alv*6 and 11 genes also appear to be well expressed. These observations are consistent with the idea that genes inserted into the germ line colinearly with a completely expressed retroviral provirus may be down regulated, but that if the provirus is defective for replication by design or by chance the foreign gene may be expressed. Further studies of the efficiency of transcription and translation of genes inserted into the germ line by retroviral vectors are needed to answer these important questions.

In these studies we have shown that retroviral genes can be inserted and expressed in the germ line of chickens using replication-competent retroviral vectors. Future studies should centre around the development of replication-defective vectors that are truly defective and can be produced in a high titre, and on the development of methods for direct insertion of cloned DNA in the chicken system. Once such systems are developed for the efficient production of transgenic fowl, the development of optimum systems for regulation of gene expression can be addressed.

We thank Leonard Provencher and Kent Helmer for the pedigree mating work, DNA dot-blotting, and assays for virus expression; Marilyn Newton and Carrie Cantwell who extracted large numbers of DNA samples and produced high quality Southern blots; and Stuart Pankratz, Department of Microbiology and Public Health, Michigan State University, for the electron microscopy. This research was supported in part by grant 84-CRCR-1-1725 from the US Department of Agriculture, grant US-811-84 from BARD, The United States–Israel Binational Agricultural Research and Development Fund, and a grant in aid from the Campbell Soup Company.

References

Baker, B., Robinson, H., Varmus, H.E. & Bishop, J.M. (1981) Analysis of endogenous avian retrovirus DNA and RNA: viral and cellular determinants of retrovirus gene expression. *Virology* **114,** 8–22.

Bosselman, R.A., Hsu, R.-Y., Briskin, M.J., Boggs, T., Hu, S., Nicolson, M., Souza, L.M., Schultz, J.A., Rishell, W. & Stewart, R.G. (1990) Transmission of exogenous genes into the chicken. *J. Reprod. Fert., Suppl.* **41,** 183–195.

Coffin, J. (1982) Endogenous viruses. In *RNA Tumor Viruses: Molecular Biology of Tumor Viruses*, pp. 1109–1203. Eds R. Weiss, N. Teich, H. Varmus & J. Coffin. Cold Spring Harbor Laboratory.

Conklin, K.F. & Groudine, M. (1986) Varied interactions between proviruses and adjacent host chromatin. *Molec. cell. Biol.* **6,** 3999–4007.

Crittenden, L.B. & Salter, D.W. (1985) Genetic engineering to improve resistance to viral diseases of poultry: a model for application to livestock improvement. *Can. J. Anim. Sci.* **65,** 553–562.

Crittenden, L.B. & Salter, D.W. (1988) Insertion of retroviral vectors into the avian germ line. *Proc. 2nd Int. Conf. Quantitative Genetics, Raleigh,* pp. 207–214. Eds B. S. Weir, E. J. Eisen, M. M. Goodman & G. Namkoong. Sinauer, Sunderland, Massachusetts.

Crittenden, L.B., Smith, E.J., Weiss, R.A. & Sarma, P.S. (1974) Host gene control of endogenous avian leukosis virus production. *Virology* **57,** 128–138.

Crittenden, L.B., Eagen, D.A. & Gulvas, F.A. (1979a) Assays for endogenous and exogenous lymphoid leukosis viruses and chick helper factor with RSV(-) cell lines. *Infect. Immun.* **24,** 379–386.

Crittenden, L.B., Smith, E.J., Gulvas, F.A. & Robinson, H.L. (1979b) Endogenous virus expression in chicken lines maintained at the Regional Poultry Research Laboratory. *Virology* **95,** 434–444.

Crittenden, L.B., McMahon, S., Halpern, M.S. & Fadly, A.M. (1987) Embryonic infection with the endogenous avian leukosis virus Rous-associated virus-0 alters responses to exogenous avian leukosis virus infection. *J. Virol.* **61,** 722–725.

Crittenden, L.B., Salter, D.W. & Federspiel, M.J. (1989) Segregation viral phenotype, and proviral structure of 23 avian leukosis virus inserts in the germ line of chickens. *Theor. appl. Genetics* **77,** 505–515.

Delwart, E.L. & Panbanigan, A.T. (1989) Role of reticuloendotheliosis virus envelope glycoprotein in superinfection interference. *J. Virol.* **63,** 273–280.

Federspiel, M.J., Crittenden, L.B. & Hughes, S.H. (1989) Expression of avian reticuloendotheliosis virus envelope confers host resistance. *Virology* **173,** 167–177.

Freeman, B.M. & Messer, L.I. (1985) Genetic manipulation of the domestic fowl—a review. *World's Poultry Sci. J.* **41,** 124–132.

Hayward, W.S., Braverman, S.B. & Astrin, S.M. (1980) Transcriptional products and DNA structure of endogenous avian proviruses. *Cold Spring Harb. Symp. quant. Biol.* **44,** 1111–1121.

Hughes, S.H., Kosik, E., Fadly, A.M., Salter, D.W. & Crittenden, L.B. (1986) Design of retroviral vectors for the insertion of foreign DNA into the avian germ line. *Poultry Sci.* **65,** 1459–1462.

Hughes, S.H., Petropoulos, C.J., Federspiel, M.J., Sutrave, P., Forry-Schaudies, S. & Bradac, J.A. (1990) Vectors and genes for improvement of animal strains. *J. Reprod. Fert., Suppl.* **41,** 39–49.

Jaenisch, R., Jahner, D., Nobis, P., Simon, I., Lohler, J., Harbers, K. & Grotkopp, D. (1981) Chromosomal position and activation of retroviral genomes inserted into the germ line of mice. *Cell* **24,** 519–529.

Nelson, R.S., McCormick, S.M., Delanny, X., Dube, P., Layton, J., Anderson, E.G., Kaniewaka, M., Proksh, R.K., Horsch, R.B., Rogers, S.G., Fraley, R.T. & Beachy, R.N. (1988) Virus tolerance, plant growth, and field performance of transgenic tomato plants expressing coat protein from tobacco mosaic virus. *Bio/Technology* **6,** 403–409.

Perry, M.M. (1988) A complete culture system for the chick embryo. *Nature, Lond.* **331,** 70–72.

Petitte, J.N., Etches, R.J. & Clark, M.E. (1989) Evidence of germ-line chimerism after transfer of embryonic stem cells in the chick. *Poultry Sci.* **68,** Suppl. 1, 112, abstr.

Robinson, H.L., Astrin, S.M., Senior, A.M. & Salazar, F.H. (1981) Host susceptibility to endogenous viruses: defective, glycoprotein-expressing proviruses interface with infections. *J. Virol.* **40,** 745–751.

Salter, D.W. & Crittenden, L.B. (1989) Artificial insertion of a dominant gene for resistance to avian leukosis virus into the germ line of the chicken. *Theor. appl. Genetics* **77,** 457–461.

Salter, D.W., Smith, E.J., Hughes, S.H., Wright, S.E., Fadly, A.M., Witter, R.L. & Crittenden, L.B. (1986) Gene insertion into the chicken germ line by retroviruses. *Poultry Sci.* **65,** 1445–1458.

Salter, D.W., Smith, E.J., Hughes, S.H., Wright, S.E. & Crittenden, L.B. (1987) Transgenic chickens: insertion of retroviral genes into the chicken germ line. *Virology* **157,** 236–240.

Sanford, J.C. & Johnston, S.A. (1985) The concept of parasite-derived resistance—deriving resistance genes from the parasites own genome. *J. theor. Biol.* **113,** 395–405.

Smith, E.J. (1986) Endogenous avian leukemia viruses. In *Avian Leukosis*, pp. 101–120. Ed. G. F. deBoer. Martinus Nijhoff, Boston.

Smith, E.J. & Crittenden, L.B. (1988) Genetic cellular resistance to subgroup E avian leukosis virus (ALV) in slow-feathering dams reduces congenital transmission of an endogenous retrovirus encoded at locus *ev*21. *Poultry Sci.* **67,** 1668–1673.

Smith, E.J., Fadly, A. & Okazaki, W. (1979) An enzyme-linked immunoabsorbant assay for detecting avian leukosis-sarcoma viruses. *Avian Dis.* **23,** 698–707.

Spencer, J.L., Gavora, J.S. & Gowe, R.S. (1980) Lymphoid leukosis virus: natural transmission and non-neoplastic effects. *Proc. Cold Spring Harbor Conf. on Cell Proliferation*, Vol. **7,** pp. 553–564.

Thomas, K.R. & Capecchi, M.R. (1987) Site-directed mutagenesis by gene targeting in mouse embryo-derived stem cells. *Cell* **51,** 503–512.

Varmus, H. & Swanstrom, R. (1982) Replication of retroviruses. In *RNA Tumor Viruses: Molecular Biology of Tumor Viruses*, 2nd edn, pp. 369–512. Eds R. Weiss,

N. Teich, H. Varmus & J. Coffin. Cold Spring Harbor Laboratory.

Vogt, P.K. & Ishizaki, R. (1966) Patterns of viral interference in the avian leukosis and sarcoma complex. *Virology* **30,** 368–374.

Wright, S.E. & Bennett, D.D. (1986) Region coding for subgroup specificity of envelope of avian retroviruses does not determine lymphogenicity. *Virus Res.* **6,** 173–180.

J. Reprod. Fert., Suppl. **41** (1990), 173–182

Vectors, promoters, and expression of genes in chick embryos

H. Y. Chen, E. A. Garber, E. Mills*, J. Smith*, J. J. Kopchick, A. G. DiLella and R. G. Smith

*The Department of Growth Biochemistry & Physiology, Merck Sharp & Dohme Research Laboratories, Rahway, New Jersey 07065, USA; and *Hubbard Farms, Walpole, New Hampshire 03608, USA*

Summary. Transgenic chickens were produced by injecting the Day-1 egg with 10^5 infectious particles of a replication-competent virus based on the Schmidt–Ruppin A strain of Rous sarcoma virus. The chickens were resistant to transforming subgroup A virus containing the *src* gene but not the corresponding subgroup B virus. Transgenic chickens producing bovine growth hormone (bGH) were generated using a modified virus containing the Bryan high titre polymerase gene. The virus was constructed with the bGH gene and the mouse metallothionein promoter in the reverse orientation relative to the viral structural genes. Two male chickens produced serum concentrations of approximately 100 ng bGH/ml; the birds were larger than controls and matured more rapidly. Transgenic mice required for the analysis of skeletal muscle-specific expression *in vivo* were produced using 5′-flanking regions of the chicken α-skeletal actin promoter linked to a luciferase reporter gene to determine the region essential for tissue-specific expression. The defined promoter sequences are to be used in experiments designed to direct expression of growth-promoting genes in skeletal muscle of chickens.

Keywords: genes; chick; embryo; expression

Introduction

Our object is to express transgenes in poultry to improve growth performance and disease resistance. Although generation of transgenic mice by microinjection of DNA into the zygote has become relatively routine, the production of transgenic poultry has been frustrated by the inaccessibility of the single-celled fertilized egg. The Day 1 chick embryo consists of 10 000–50 000 cells. We have used a retrovirus containing a transgene to infect embryos for the generation of transgenic chickens. The feasibility of this approach was established using a replication-competent virus based on the subgroup A Rous sarcoma virus.

Materials and Methods

Construction of plasmids. Standard molecular biological techniques were used. Brief details of the cloning are provided in 'Results and Discussion'.

Cell culture and virus propagation. Primary chicken embryo fibroblasts (CEF) were derived from SPAFAS C/E Day 10 embryos. The CEF cultures were established by trypsinization followed by plating of cells in Ham's F10-medium 199 (50/50) containing 5% tryptose phosphate broth and 4% newborn calf serum. Avian leukosis virus was propagated for 5–14 days after transfection of CEF with proviral DNA. Medium was removed from CEF and cellular debris separated by centrifugation; aliquants of the medium were frozen at −70°C. Viral titres were determined by endpoint dilution assay (Lannett & Schmidt, 1980).

Preparation of chicken genomic DNA. Chicken genomic DNA was prepared by homogenizing 50 mg tissue or 50 μl packed red blood cells in a 3 ml solution containing proteinase K (0·5 mg/ml), 1% SDS, 5 mM-EDTA, and 10 mM Tris, pH 7·5. The solution was incubated overnight at 42°C. After two extractions with equal volumes of phenol and chloroform, DNA was precipitated with ethanol at −20°C. Precipitated DNA was collected by centrifugation, washed with 70% ethanol and resuspended in 2·0 ml sterile H_2O.

Restriction endonuclease digestion and DNA hybridization. Genomic DNA (10 μg) was digested with *Hind* III or *Pvu* II (50 units) overnight at 37°C and subjected to 1% agarose gel electrophoresis. Following electro-blotting to gene screen plus membranes (DuPont/NEN) at 42 V in 0·33 × TAE (0·04 M-Tris–acetate, 0·001 M-EDTA, pH 7·5), membranes were exposed to a ^{32}P-labelled DNA hybridization probe (6 × 10^5 c.p.m./ml of hybridization solution according to the manufacturer's suggestions).

Hybridization probes. For detection of the bGH gene in chicken genomic DNA, a *Pvu* II DNA fragment from pBGH-10 was used. Exogenous or endogenous proviral DNA sequences in chicken genomic DNA were monitored by using a 250 base pair fragment derived from the U_3 region of an exogenous viral clone, SR-RSV-A, or a 146 base pair fragment derived from the U_3 region of an endogenous viral clone, RAV-O. The exogenous and endogenous cloned viral LTR DNAs were kindly provided by L. Crittenden (USDA, East Lansing, Michigan, USA). Probes with specific activities of at least 1 × 10^9 c.p.m./μg were used in the analyses.

Virus assay. Virus stock was rapidly thawed at room temperature and 10-fold dilutions prepared in cold cell culture media. For each dilution 1 ml was inoculated into each of 6 primary CEF cultures (60 mm dish). Cultures were incubated at 37°C in an atmosphere of 5% CO_2 for 3–5 days at which time cells were passaged, grown for an additional 3–5 days and the supernatant from lysed cells assayed for p27 antigen by ELISA. Rabbit anti-p27 and rabbit anti-p27 conjugated to horseradish peroxidase used in the assay were obtained from Spafas (Storrs, CT, USA). Virus titres were calculated as described by Lannett & Schmidt (1980).

Virus assays on blood, meconium and cloacal swabs. Virus was isolated from 1-day-old chicks by expressing meconium from the chick into a tube containing 1·5 ml cell culture medium with 5 times the normal amount of antibiotics. Samples were frozen at −70°C. Samples were thawed and centrifuged (10 000 ***g***) and 0·4 ml of the supernatant was inoculated into primary CEF cultures made from the SPAFAS C/E embryos. Cultures were incubated at 37°C in an atmosphere of 5% CO_2 for 5 days, passaged and grown for an additional 5 days. The supernatants from lysed cells were assayed for p27 antigen by ELISA.

Older birds were screened for virus by using blood samples or cloacal swabs. Cloacal swabs were placed in 1·5 ml cold cell culture medium containing 5 × antibiotics and frozen at −70°C. Blood and cloacal swab samples were inoculated into primary CEF cultures and the cultures were treated as described above.

Blastoderm injection with retrovirus. Eggs to be injected were collected and stored at 15°C (large end up) for 7–8 days to allow the air cell to form and for the blastoderm to rotate to the large end of the eggs. The air cell was outlined in pencil and a small section (approximately 1 cm^2) of shell was cut and removed. Eggs were examined and those in which the blastoderm could be seen through the membrane were saved for injection.

Approximately 10^5–10^6 infectious units of virus were injected adjacent to and slightly below the blastoderm. Volumes of 10–50 μl were injected using glass pipettes or repeating micropipettes with a 27-gauge needle. All injection procedures were carried out in a biological safety cabinet. After being injected, eggs were sealed with cellophane tape and the large end dipped in melted paraffin wax. Eggs were placed in a commercial incubator and, at hatching, chicks were screened for virus using meconium samples. All virus-positive chicks were saved for DNA hybridization assays.

Results and Discussion

Generation of transgenic chickens by using permuted RSV-based vectors

A non-transforming permuted proviral vector based on the Schmidt–Ruppin A strain of Rous sarcoma virus (RSV) was constructed with the bGH inserted downstream of the virus structural genes. Virus was generated by transfection of this vector into CEF cells, and approximately 10^5 were injected into chicken eggs adjacent to the blastocyst to produce viraemic chicks. After maturation, viraemic hens were selected to produce progeny so that after fertilization the resulting embryo was exposed to virus as it passed through the oviduct. Using this approach, viraemic females (G_0) were crossed with non-viraemic males and G_1 offspring screened for the presence of exogenous virus. Male offspring produced from viraemic hens were then tested for their ability to transmit viraemia to their progeny. Males do not generally transmit viruses congenitally to their offspring. Production of viraemic offspring therefore implies that proviral DNA sequences are present in the germ line of the parent. The transmission frequency, following mating of such G_1 transgenic males with non-viraemic females to produce G_2 progeny, varied from 0 to 40%. These data, together with the results of Southern blot analysis (Fig. 1), showed that not all of the germ cells in the G_1 males were infected by the virus and that integration occurred at different sites. Thus, in the G_1 transgenic males infection had not occurred at the zygote stage. Southern blot analysis (Fig. 2) of subsequent generations showed that each line faithfully transmitted its unique proviral integration site. Transmission frequency was approximately 50%, confirming stable germ-line integration of exogenous retroviral DNA sequences into the chicken genome. Southern blot analysis showed that the bGH gene was not retained and subsequent studies showed that bGH was deleted before integration of the provirus (data not shown).

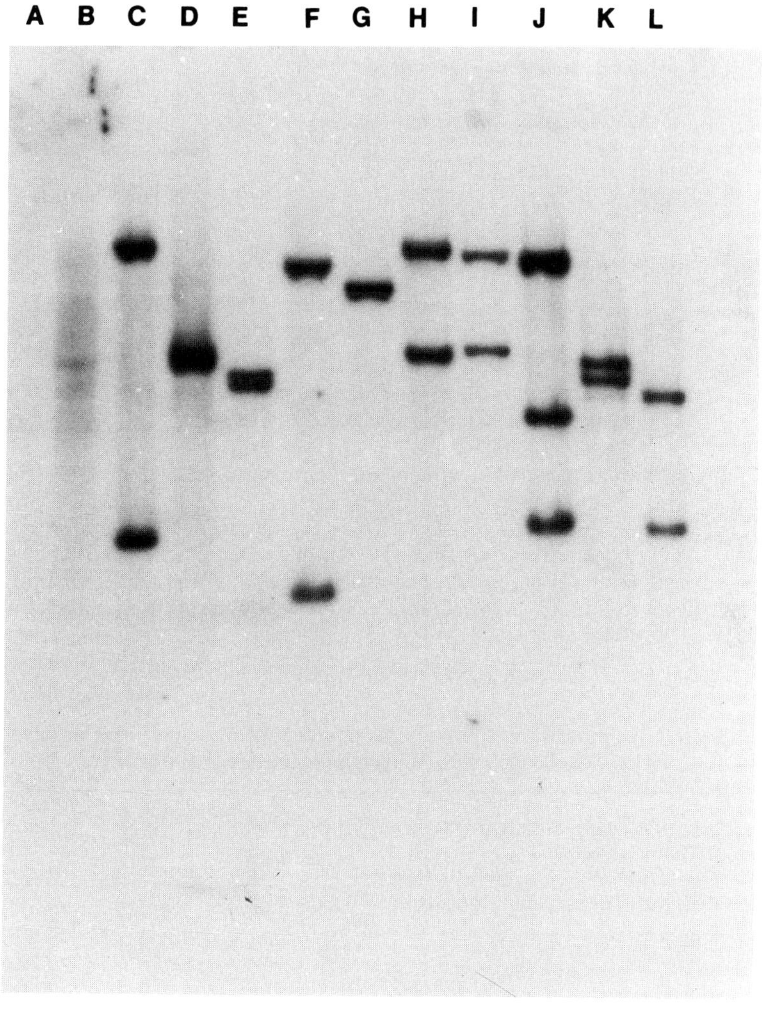

Fig. 1. Southern blot analysis of family members from G_o, G_1 and G_2 generations. Chicken genomic DNA was prepared as described in 'Materials and Methods'. After digestions with a 10-fold excess of Hind III, 10 µg DNA were resolved by 1% agarose gel electrophoresis. After transfer to a membrane filter, hybridization was carried out as described in 'Materials and Methods' using an exogenous virus-specific probe. Lanes A–L represent the following individuals: A = G_o female; B = G_1 male: C–L = G_2 chickens.

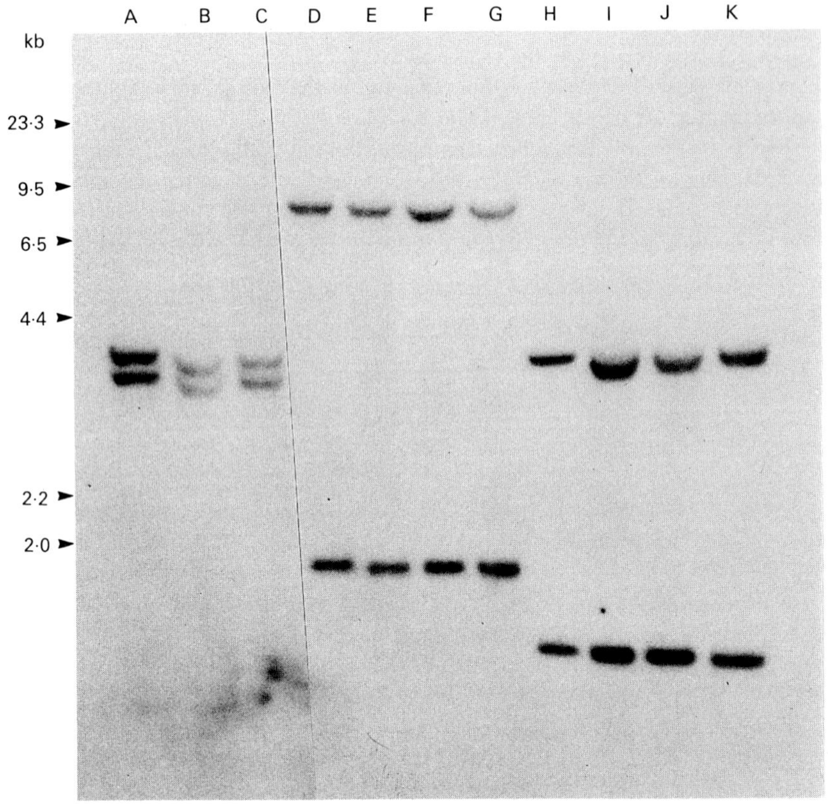

Fig. 2. Southern blot analysis of G_2 parents and G_3 offspring. Genomic DNA preparation, Hind III digestion, 1% agarose gel electrophoresis, transfer and hydridization protocols are described in 'Materials and Methods'. DNA from G_2 parents 1, 2 and 3 are represented in Lanes A, D, and H respectively. DNA from G_3 offspring of G_2 parent 1 are shown in Lanes B and C; G_3 offspring from G_2 parent 2 are shown in Lanes E, F and G; G_3 offspring from G_2 parent 3 are shown in Lanes I, J and K; molecular weight markers in kilobase pairs (kb) are indicated.

Table 1. Comparison of tumour development in transgenic and control chickens challenged with Subgroup A and Subgroup B viruses

			Tumour incidence			
			Subgroup A		Subgroup B	
	No. of birds	Viraemia	2 weeks	3 weeks	2 weeks	3 weeks
Transgenic	24	+	0/24	1/23	19/24	18/23
Control	11	—	9/11	9/11	9/11	8/11

In the above experiments a non-transforming permuted proviral vector based on the Schmidt–Ruppin A strain of RSV was used. Since we had demonstrated that the viral sequences were introduced into the germ line, we were curious to learn whether endogenous expression of the subgroup A *env* gene product would prove protective against an exogenously administered transforming subgroup A virus. Hence transgenic and control chickens were challenged with subgroup

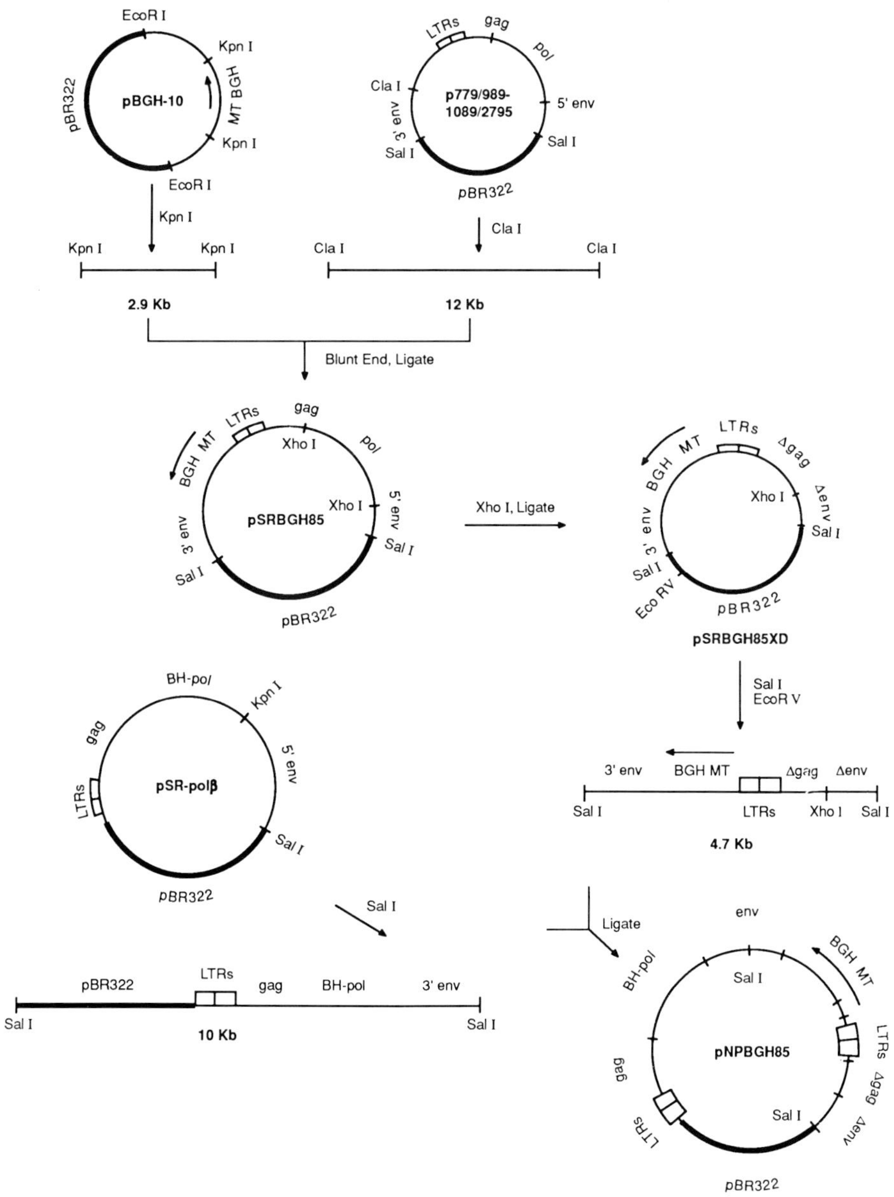

Fig. 3. Construction of pNPBGH85. pNPBGH85 was constructed in three steps from intermediate plasmids, pSRBGH85 and pSRBGH85XD. pSRBGH85 was constructed from a 2·9 kb KpnI restriction fragment from bGH expression vector pBGH-10 and a 12 kb Cla I fragment from the Schmidt-Ruppin RSV-A-derived vector p779/789-1089/2795. The detailed procedure is described in 'Results and Discussion'.

A and subgroup B Rous sarcoma viruses carrying the *src* gene. Table 1 indicates that, after injection of subgroup A and subgroup B viruses into alternate wing webs, only subgroup B virus produced tumours in the transgenic birds, whereas both subgroup A and B viruses produced tumours in the control birds. These results suggested that protection occurs by interference caused by expression of the *env* A gene product. Moreover, they illustrate the potential utility of such an

Table 2. Stable expression of bGH in pNPBGH5-transfected cells

DNA transfected	bGH conc.* (ng/ml) Days after transfection 5	10	15	20	25
None	3·8	4·9	5·7	2·1	4·5
pNPRAV	3·6	5·7	4·0	2·2	4·4
pNPBGH85	149·6	227·0	179·0	182·8	183·5

*Average of duplicate samples.

approach to confer disease resistance traits on livestock. Similar results have also been reported by Salter & Crittenden (1989).

Generation of transgenic chickens by using non-permuted RSV-based vectors

Having demonstrated the feasibility of production of transgenic birds expressing exogenous viral genes, it was necessary to modify the proviral vector in an attempt to integrate and express non-viral genes. The modified vector pNPBGH85 (Fig. 3) contained a non-permuted proviral copy of a modified subgroup A Schmidt–Ruppin strain RSV in which the reverse transcriptase gene from the Bryan high-titre strain of RSV (BH pol) was substituted for the Schmidt–Ruppin RSV-A *pol* gene. The *src* gene of RSV was replaced by the bGH gene linked to the murine metallothionein promoter in an antisense orientation relative to the viral structural genes.

Construction of pNPBGH85. pNPBGH85 was constructed in three steps from intermediate plasmids, pSRBGH85 and pSRBGH85XD. pSRBGH85 (Fig. 3) was constructed from a 2·9 kb Kpn I restriction fragment from the bGH expression vector pBGH-10 (Kelder *et al.*, 1989) and a 12 kb Cla I fragment from the Schmidt–Ruppin RSV-A-derived vector p779/789-1089/2795 (Hughes & Kosik, 1984). The two fragments were ligated after blunting their overhanging ends. Ligated DNA was used to transform DH5 to ampicillin resistance. The pSRBGH85 clone was identified by restriction mapping. pSRBGH85 was digested to completion with Xho I, then religated and used to transform DH5. A resultant clone, pSRBGH85XD, in which the 4·5 kb Xho I fragment had been deleted from pSRBGH85 was selected.

To construct pNPBGH85, pSRBGH85XD was digested with Sal I and the purified 4·7 kb fragment ligated to the 10 kb Sal I linearized fragment from pSR-pol-beta (Sudol *et al.*, 1986). After transformation of DH5 a clone was selected corresponding to pNPBGH85 (Fig. 3).

The efficiency with which pNPBGH85 produced infectious virus was tested by transfection of chick embryo fibroblasts using a calcium phosphate procedure (Cross & Hanafusa, 1983). Virus was harvested from fully infected cultures and used to infect fresh cultures of CEF cells. Virus was produced by transfected cells as early as 1 day after transfection. Virus spread, assayed by p27 *gag* ELISA (using a Spafas p27 ELISA for avian lymphoid leukosis virus), was maximal 5–7 days after transfection. Culture fluids were assayed for secreted bGH by double antibody radioimmunoassay (Kopchick *et al.*, 1985). The attainment of maximal GH levels was consistent with virus spread.

Stable expression of bGH in pNPBGH85-transfected cells. Secondary CEF cells were transfected by calcium phosphate precipitation with either no DNA, pNPRAV (a construct with a non-permuted proviral structure like pNPBGH85, but with no foreign gene insert) or pNPBGH85. Transfected cultures were passaged every 3–5 days for a total of 25 days. Every 5 days, cell-free supernatants were harvested from confluent cultures and assayed by radioimmunoassay for bGH. Genomic DNA was prepared from parallel transfected cultures. Table 2 shows the average bGH

Table 3. Expression of bovine growth hormone in transgenic chickens

Bird no.	Sex	Virus assay	DNA (copies/cell) Viral	bGH	Serum bGH (ng/ml)	Age (weeks)
9667	M	+	2	1	97	4
					76	18
9818	M	+	2	1	123	4
					86	35
146	F	+	2	0·2	23	4
					38	32
195	M	+	1	1	20	4
					8	31
309	M	+	2	1	61	4
					25	28
332	F	+	0·5	0·5	55	4
					55	27
450	M	+	1	1	46	4
					38	26
612	M	+	1	0·5	49	4
					80	21
646	M	+	2	0·1	48	4
					40	21
710	M	+	1	0·1	36	4
					43	21
878	F	+	1	0·1	16	4
					11	19
911	M	+	0·5	0·5	24	4
					19	19
987	M	+	1	0·5	18	4
					10	18

levels in duplicate cultures. Expression of bGH was specific to pNPBGH85-transfected cells and was stable during long-term culture of fully infected cells. DNA dot-blot analysis of genomic DNA verified the retention of bGH sequences and integrated provirus (data not shown).

The non-permuted vector pNPBGH85 is superior to the permuted vectors such as pSRBGH85 used in our initial studies. In particular, high titre virus is produced very quickly following transfection, presumably because the former carries the Bryan high titre polymerase gene. Since less time is required for full infection of transfected cells, the likelihood of contamination of the virus stock by viruses that have lost the bGH gene insert by recombination is reduced.

Production of transgenic chickens expressing the bGH gene. Viral stocks (10^6–10^7 particles/ml) were generated by transfection of proviral vector pNPBGH85 into CEF cells. The cell supernatant (50 µl) was injected beneath the blastoderm of Day-1 eggs to infect the chicken embryo. After hatching the chickens, meconium samples were taken for virus assay using p27 *gag* ELISA. Blood samples were also taken for DNA isolation and subsequent dot-blot hybridization and serum bGH assays. There was an excellent correlation between the presence of virus in meconium and viral DNA sequences in red blood cells. However, serum bGH concentrations did not correlate with bGH DNA copy number in red blood cells, suggesting that bGH was being produced by other cells. In most chickens bGH expression was maintained as the bird aged (Table 3), consistent with stability of integration. Two male chickens who expressed high levels of bGH (approximately 100 ng/ml) had longer legs (Table 4) and produced spermatozoa at 13–14 weeks instead of the normal 16–18 weeks of age. These effects were not observed in chickens producing lower amounts of bGH. Clearly, these observations need to be confirmed using a larger population of bGH transgenic birds.

Table 4. Comparison of leg length between bGH transgenic chickens and normal control chickens

	Bird no.	Leg length (cm)
bGH transgenic	9667M	20·2
	9818M	20·1
Controls	50–069M	18·4
	50–071M	17·5
	39–031M	17·3
	46–037M	17·0

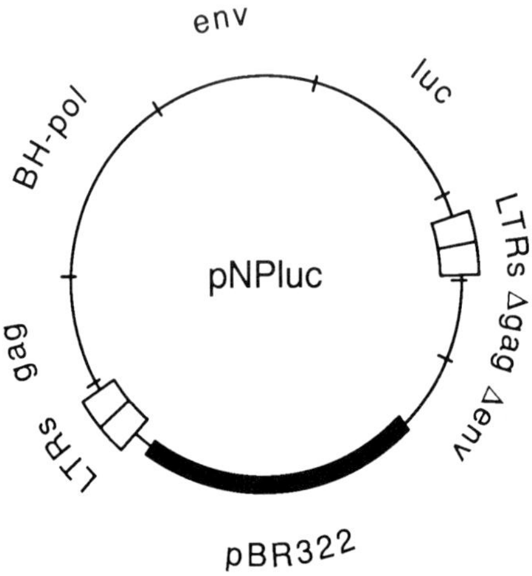

Fig. 4. Replication-competent luciferase vector. The strategy for construction of pNPluc is similar to that for pNPBGH85 shown in Fig. 3. The firefly luciferase gene was described by deWet *et al.* (1987).

Improving the efficiency of infection of chicken embryos

The prevalent method of gene transfer into chicken embryos involves injection of recombinant avian retrovirus into fresh fertile eggs near the blastoderm. In our initial attempts to infect Hubbard Leghorn line 139 eggs efficiency was relatively low (approx. 1–5%). Furthermore, the Day-1 blastoderm consists of approximately 50 000 cells and if only a small portion of the cells were infected, a high proportion of mosaic animals would be generated. Consequently, germ-line transmission of the integrated foreign gene would be infrequent, making it difficult to identify germ-line progeny.

To improve the efficiency of infection of embryos we developed a procedure to visualize the blastoderm so that virus could be directly injected onto the embryo. Fresh fertile eggs were placed large-end up on a rack. A small area (1 cm in diameter) of the egg shell was cut out from the top. A small piece of inner membrane (0·5 cm in diameter) was removed allowing visualization of the blastoderm. To allow us to monitor the efficiency of infection and expression of a transgene with high sensitivity, a proviral construct was prepared containing the firefly luciferase gene driven by the RSV LTR (Fig. 4). Volumes (20 μl) containing approximately 10^5 virus particles harvested from CEF cells transfected with the luciferase-carrying construct were injected underneath the blastoderm using a 26-gauge needle connected to a micropipette. The eggs were resealed with surgical tape and placed in an incubator to allow the embryo to develop. Using this procedure, the

Table 5. Efficiency of infecting Day-1 chick embryos using NP-luciferase viral vector*

Treatment	Mixed commercial Leghorns	Hubbard Line 139 Leghorns
Virus	5/31 = 16%	4/12 = 33%
Virus + cells	9/34 = 26%	10/19 = 53%

*Day-1 chicken embryos were injected with cleared virus stock or virus stock supplemented with $2{\cdot}5 \times 10^5$ virus-producing cells. Day-10 embryos were removed for measurement of luciferase activity.

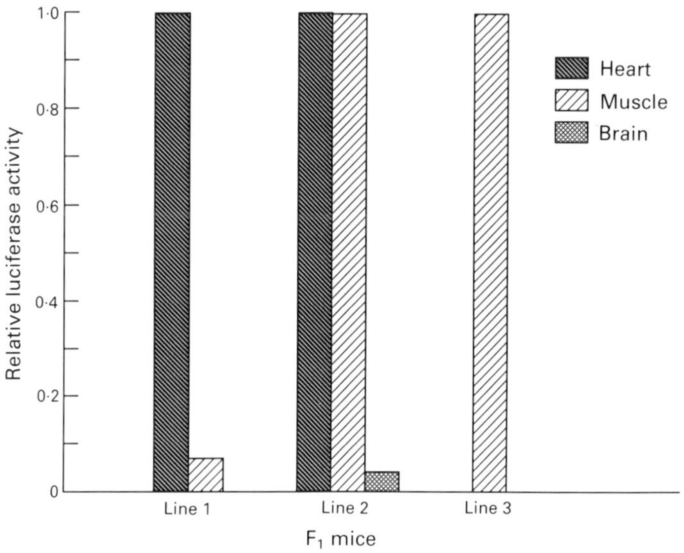

Fig. 5. Luciferase activity in tissues isolated from transgenic mice. Line 1 and line 2 represent F_1 mice in which the 0·2 kb α-skeletal actin promoter is linked to the luciferase reporter gene. Line 3 represents F_1 mice in which the 1·5 kb promoter is linked to the reporter gene. Expression in liver and kidney was not detectable in all mice.

efficiency of infecting embryos was increased to 33%. When embryos were injected with a virus stock supplemented with $2{\cdot}5 \times 10^5$ of virus-producing cells the efficiency was improved to 50% (Table 5). The use of luciferase as a reporter gene provided a very sensitive and rapid method to monitor integration and expression of a transgene in organs from Day-17 embryos. We were able to produce embryos in which relatively high levels of expression were detected in all tissues examined (brain, liver, spleen, heart, skeletal muscle), indicating widespread infection of the embryos by the virus, and suggesting a high probability of germ-line integration of the transgene.

Targetting expression of transgenes to skeletal muscle

One of our goals is to improve muscle growth in chickens by transgenic means. It was therefore necessary to select a muscle-specific promoter and define the minimum sequences of the promoter region required *in vivo* for skeletal muscle-specific expression. Because routine screening of various DNA constructs in transgenic chickens is not yet feasible, the promoters were evaluated in transgenic mice. To maximize efficiency of screening for expression the luciferase reporter gene was linked to each promoter. The promoters used were 1·5 and 0·2 kb of the 5′-flanking regions of the chicken α-skeletal actin genes (Bergsma *et al.*, 1986). The constructs were initially tested by transfection into cultures of chick myoblasts using RSV-CAT as a control for relative transfection efficiency. Following transfection, cultures were assayed for luciferase and CAT activity both before and after serum deprivation. After serum deprivation the cells fused and formed multinucleated myotube structures. While CAT activity was constant both before and after-differentiation, luciferase activity was expressed at very low levels in myoblasts but at extremely high activity in myotubes (data not shown). These results confirmed that both promoters were active specifically in mature muscle cells.

Each construct was then injected into the mouse zygote to generate transgenic mice for testing specificity of each promoter *in vivo*. Mice expressing luciferase activity were identified by assaying tail specimens (DiLella *et al.*, 1988). Figure 5 shows a comparison of expression of luciferase activity in heart, skeletal muscle and brain of 3 different lines of transgenic mice. In two different F_1 lines in which the 0·2 kb promoter was fused to the reporter gene, high level expression is observed in the heart as well as in skeletal muscle. In contrast, a line in which the 1·5 kb promoter was used expressed luciferase exclusively in skeletal muscle. These results suggest that, to produce transgenic animals in which expression is confined mainly to skeletal muscle, the 1·5 kb promoter is preferred.

References

Bergsma, D., Grichnik, J., Gossett, L. & Schwartz, R. (1986) Delimitation and characterization of cis-acting DNA sequences required for regulated expression and transcriptional control of the chicken skeletal α-actin gene. *Molec. cell. Biol.* **6,** 2462–2475.

Cross, F. & Hanafusa, H. (1983) Local mutagenesis of Rous sarcoma virus: the major sites of tyrosine and serine phosphorylation of $p60^{src}$ are dispensable for transformation. *Cell* **34,** 597–607.

deWet, J., Wood, K., DeLuca, M., Helinski, D. & Subramani, S. (1987) Firefly luciferase gene: structure and expression in mammalian cells. *Molec. cell. Biol.* **7,** 725–737.

DiLella, A., Hope, D., Chen, H., Trumbauer, M., Schwartz, R. & Smith, R. (1988) Utility of firefly luciferase as a reporter gene for promoter activity in transgenic mice. *Nucleic Acids Res.* **16,** 4159.

Hughes, S. & Kosik, E. (1984) Mutagenesis of the region between *env* and *src* of the SR-A strain of Rous sarcoma virus for the purpose of constructing helper-independent vectors. *Virology,* **136,** 89–99.

Kelder, B., Chen, H. & Kopchick, J. (1989) Activation of the mouse metallothionein-I promoter in transiently transfected avian cells. *Gene* **76,** 75–80.

Kopchick, J., Malavarca, R., Livelli, T. & Leung, F. (1985) Use of avian retroviral-bovine growth hormone DNA recombinants to direct expression of biologically active growth hormone by cultured fibroblast. *DNA* **4,** 23–31.

Lannett, E.H. & Schmidt, N.J. (1980) In *Diagnostic Procedures for Viral and Rickettsial Infections*, 4th edn, pp. 2–65. Aneum Public Health Association, New York.

Salter, D. & Crittenden, L. (1989) Artificial insertion of a dominant gene for resistance to avian leucosis virus into the germline of the chicken. *Theor. appl. Genet.* **77,** 457–461.

Sudol, M., Lerner, T. & Hanafusa, H. (1986) Polymerase-defective mutant of the Bryan high-titer strain of Rous sarcoma virus. *Nucleic Acids Res.* **14,** 2391–2405.

J. Reprod. Fert., Suppl. **41** (1990), 183–195

Printed in Great Britain

Transmission of exogenous genes into the chicken

R. A. Bosselman, R.-Y. Hsu, M. J. Briskin, Tina Boggs, Sylvia Hu, Margery Nicolson, L. M. Souza, J. A. Schultz*, W. Rishell* and R. G. Stewart*†

*Amgen Inc., 1900 Oak Terrace Lane, Thousand Oaks, CA 91320, USA; and *Arbor Acres Farm, Inc., Marlborough Road, Glastonbury, CT 06033, USA*

Summary. Injection of infectious non-replicating REV vector directly beneath the chicken blastoderm leads to infection of embryonic stem cells. Vector sequences are present in a variety of specialized tissues of embryos and mature birds derived from infected blastoderms. Breeding studies show that replication-defective REV vectors can transfer heritable, non-viral genetic information into the chicken germ line.

Keywords: gene transfer; retrovirus vectors; chicken

Introduction

Gene transfer into chickens has been limited to the use of retrovirus vectors (Souza *et al.*, 1984; Shuman & Shoffner, 1986; Salter & Crittenden, 1987, 1989; Salter *et al.*, 1986, 1987; Bosselman *et al.*, 1989a, b). Replication competent virus injected through the shell into the vicinity of the embryo can lead to a productive infection even if only a single cell of the embryo is initially infected (Salter *et al.*, 1986). Infected embryos yield birds mosaic with respect to provirus insertion sites. Outbreeding of a small percentage of these birds gives rise to individuals with a single proviral insertion in their germ line. A number of chickens transgenic for derivatives of replication-competent Rous sarcoma virus (RSV) have been generated in this way (Salter *et al.*, 1986, 1987; Salter & Crittenden, 1989). In some instances, the use of these replicating viruses has resulted in fortuitous transfer of non-replicating derivatives present as single proviral insertions in birds bred from originally infected embryos. One such insertion conferred genetic resistance to infection by the original virus, presumably by expression of the viral *env* gene product. Expression of this protein would bind to viral receptor sites needed for productive infection to occur (Salter & Crittenden, 1987, 1989). However, this approach normally leads to viraemia and associated health problems in infected birds and their progeny (Hayward *et al.*, 1981; Noori-Daloii *et al.*, 1981; Swift *et al.*, 1987). Since there are disadvantages associated with the use of replication-competent retroviruses, we developed an alternative method of gene transfer into the chicken (Bosselman *et al.*, 1989a). This method is based upon microinjection of replication-defective reticuloendotheliosis virus vectors directly beneath the blastoderm of the unincubated chicken embryo. The simple layered morphology of the embryo at this stage of development and the use of vector preparations with titres of $\geqslant 10^4$ eliminates the need for replication-competent virus, and circumvents the technical difficulties associated with manipulating the embryo at earlier stages. Analysis of infected embryos and their progeny demonstrates gene transfer into both somatic and germ-line stem cells. This paper describes the replication-defective REV vector system, the method of gene transfer, and transgenic chickens generated with this approach.

REV vector system

The development of packaging-defective helper proviruses has increased the utility of retrovirus vectors (Mann *et al.*, 1983; Watanabe & Temin, 1983). These specialized helper proviruses are

---

†Present address: Fieldale Farms Co., P.O. Box 558, Baldwin, GA 30511, USA.

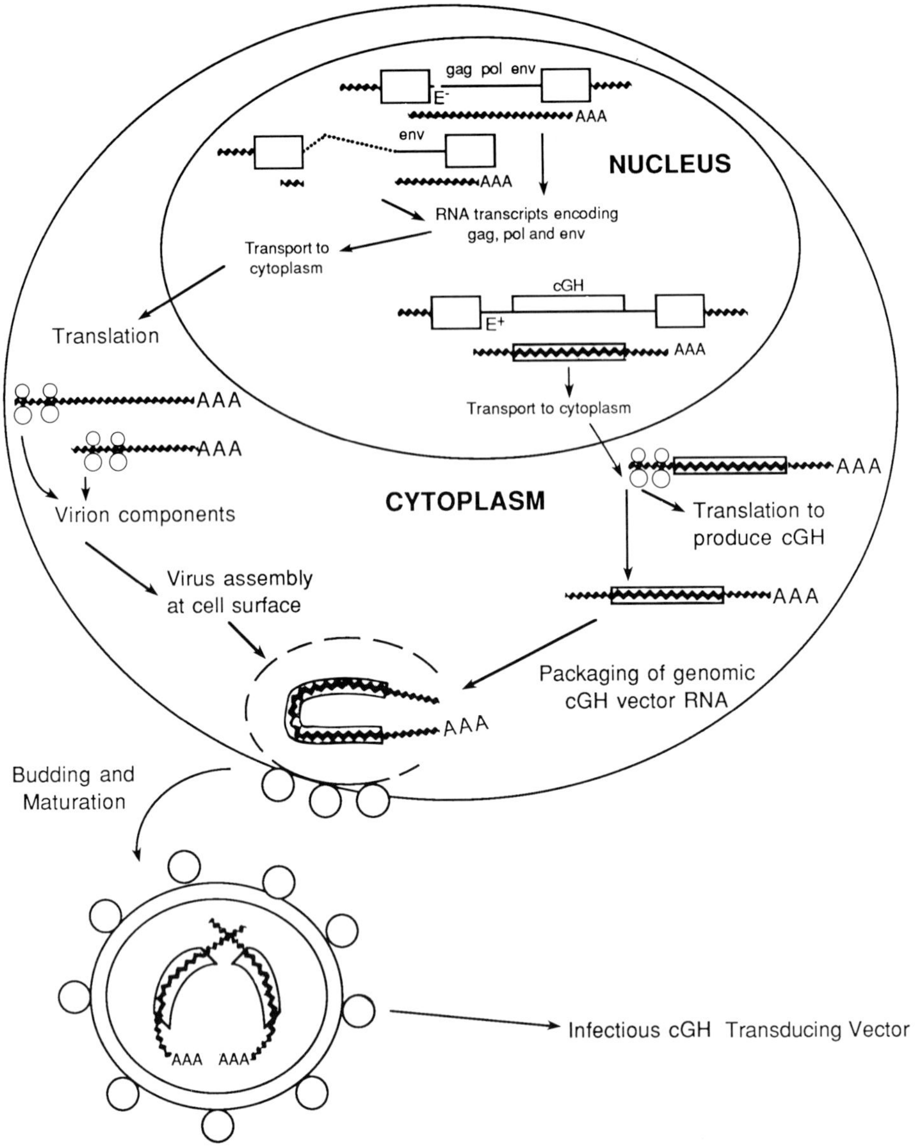

Fig. 1. Schematic representation of the REV vector system based upon two packaging-defective proviruses expressing *gag, pol* and *env*. The vector shown lacks viral structural genes and carries a chicken growth hormone sequence (cGH). E, encapsidation signal needed for efficient packaging of genomic RNA into virions. (From Bosselman *et al.*, 1988.)

derived by removal of *cis*-acting sequences needed for efficient incorporation of viral RNA transcripts into maturing virions. Viral structural genes remain intact, providing appropriate messenger RNA transcripts. In contrast, replication defective vectors derived from the same parental provirus have been made by substituting non-viral genes for the structural genes of the provirus while leaving the *cis*-acting packaging signal intact. When used together these two defective proviruses complement each other and result in formation of infectious, non-replicating vector which can

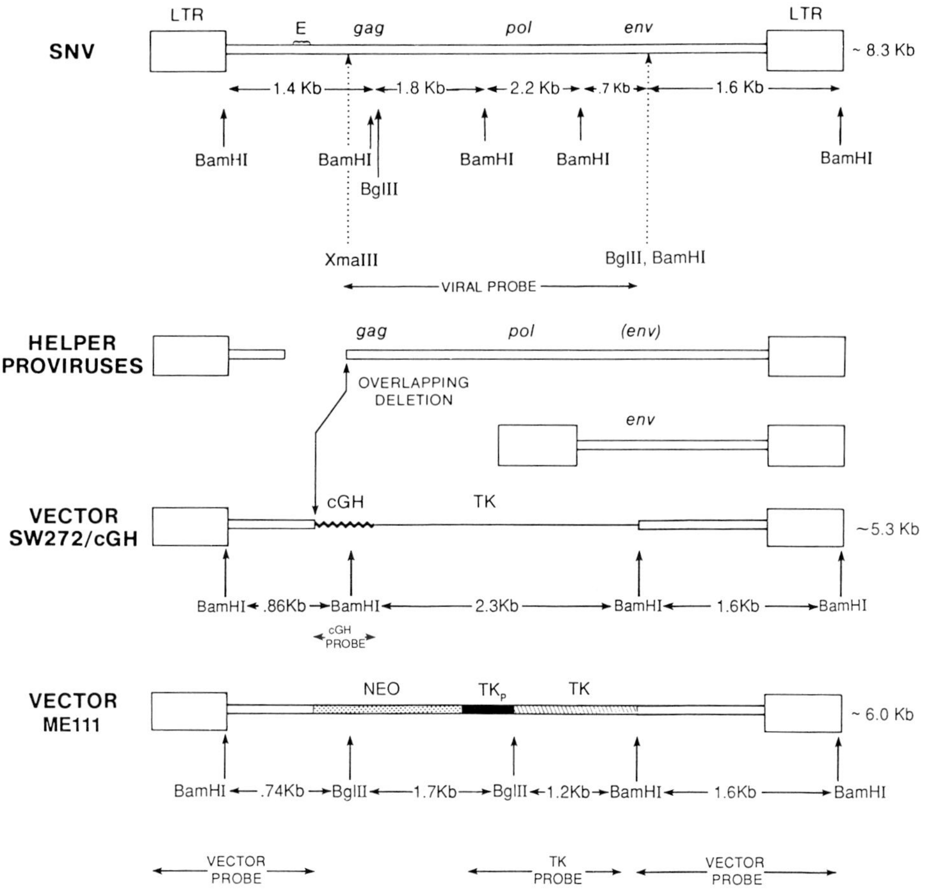

Fig. 2. Sequence relationships among the parental SNV provirus, the modified packaging-defective helper proviruses, and the vectors ME111 and SW272/cGH are shown. Relevant features of these proviruses include the LTRs, the structural genes of the virus (*gag, pol, env*), the approximate position of the packaging sequence (E), the cGH sequences, the HSV-1 tk gene promoter (TKp), the tk coding sequence (TK), and the neomycin phosphotransferase coding sequence (NEO). The *env* sequence in the larger of the two helper proviruses is presumably not expressed because of the removal of the 5′ splice donor. Overlapping deletions indicated between helper and vector sequences should reduce recombination between these genomes. Descriptions of the REV helper proviruses and the original TK transducing vector pSW272 and ME111 have been given (Watanabe & Temin, 1983; Emerman & Temin 1984). The 5′ LTRs of both helper proviruses are derived from SNV. Their coding sequences are derived from REV-A. The env helper provirus lacks viral splice donor and acceptor sequences. The first ATG is that of the *env* gene. The cGH vector is derived from SNV. REV-A and SNV share high sequence homology. Relative sizes (in kilobases) of *Bam*HI, *BGl*II restriction endonuclease fragments are indicated. Also given are the locations of viral, vector, TK, and cGH DNA probes. (From Bosselman *et al.*, 1989b.)

transduce non-viral genes into susceptible target cells. Figure 1 illustrates the interaction between complementing reticuloendotheliosis and spleen necrosis proviruses resulting in vector production (Watanabe & Temin, 1983). This system produces little or no replication-competent helper virus (Hu *et al.*, 1987).

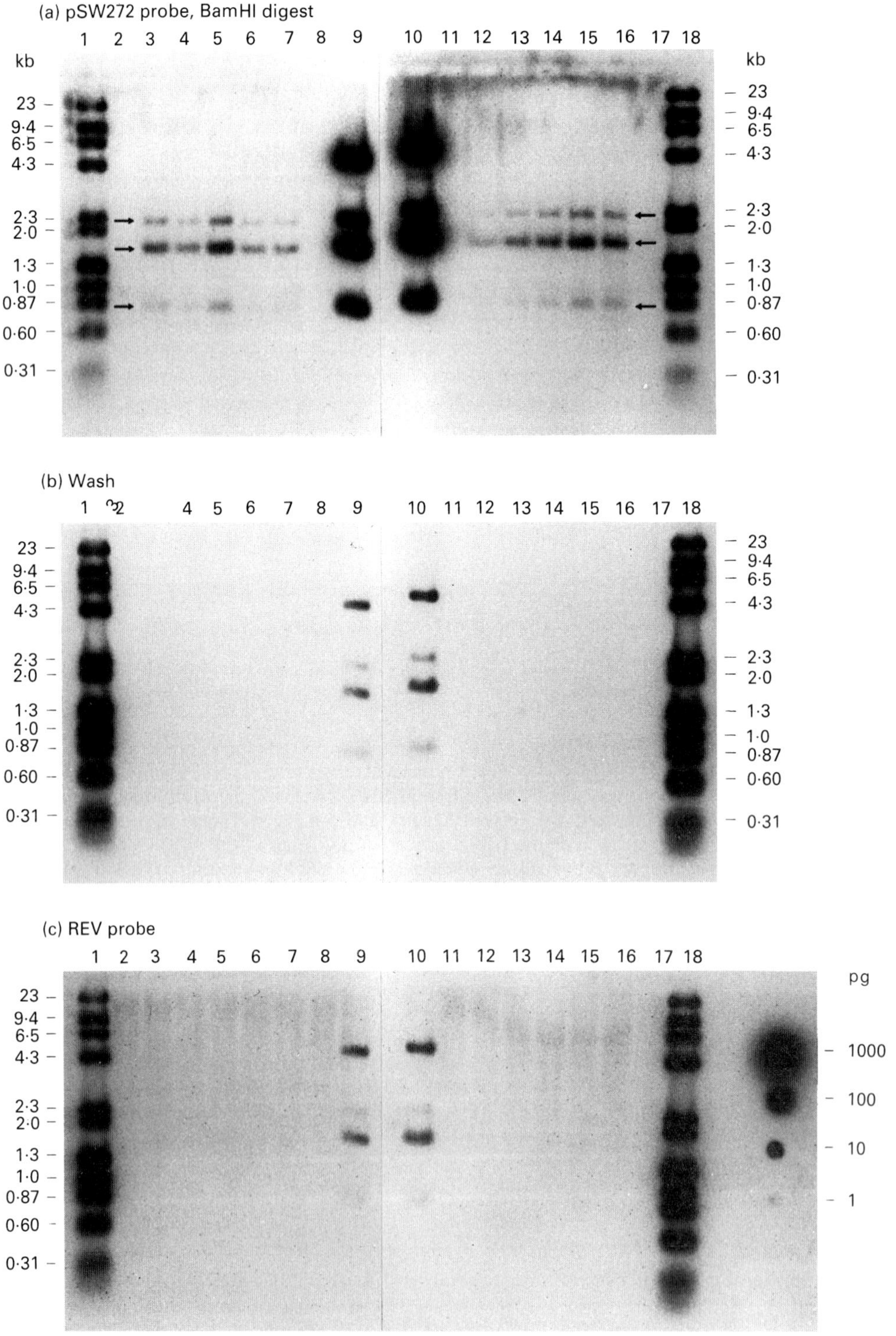
(a) pSW272 probe, BamHI digest
1 2 3 4 5 6 7 8 9 10 11 12 13 14 15 16 17 18
kb
23
9·4
6·5
4·3
2·3
2·0
1·3
1·0
0·87
0·60
0·31
(b) Wash
1 2 4 5 6 7 8 9 10 11 12 13 14 15 16 17 18
(c) REV probe
1 2 3 4 5 6 7 8 9 10 11 12 13 14 15 16 17 18
pg
1000
100
10
1

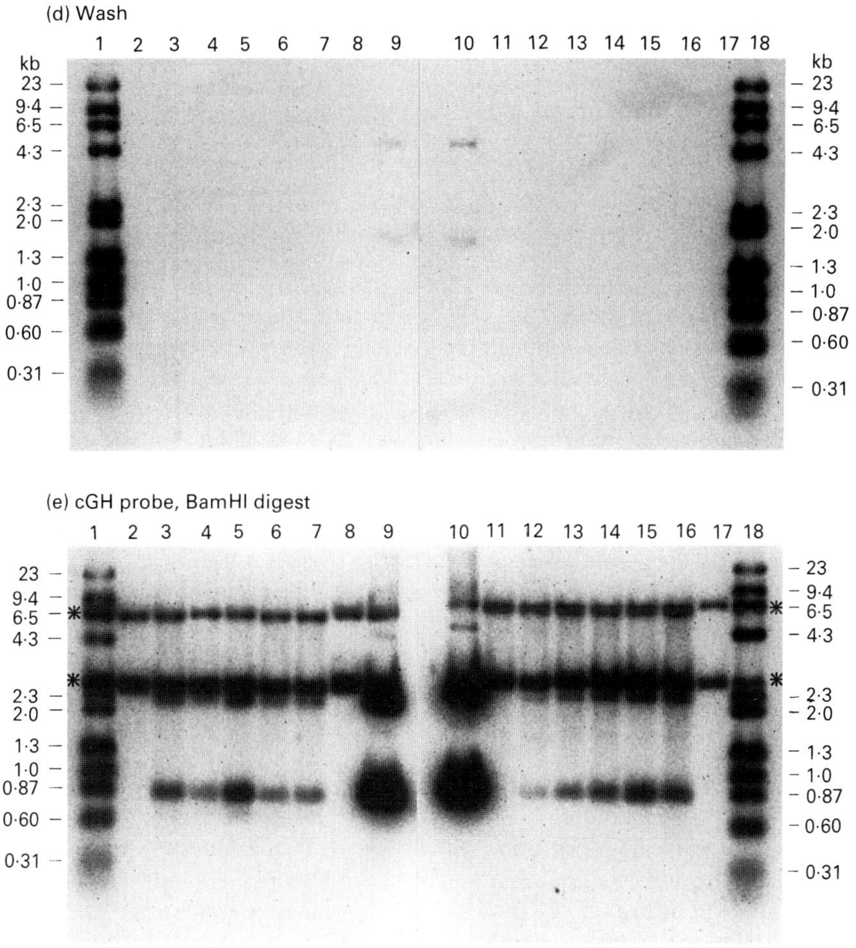

Fig. 3. Southern blot analysis of DNA from 7-day chicken embryos injected with 20 μl SW272/cGH vector before incubation. High-molecular-size DNA (15 μg) was digested with *Bam*HI before analysis. The same filter was hybridized to three different probes: pSW272 probe (a), probe removed from panel a (b), virus-specific probe (c), probe removed from (c) (d), and cGH-specific probe (e). Probe hybridized to vector DNA in lanes 9 and 10 of (a) could not be completely removed. Sequences recognized by these probes are illustrated in Fig. 2. Lanes 1 and 18: *Hind*III-digested lambda phage DNA. *Hae*III-digested ΦX174 DNA and *Bam*HI-digested uninjected chicken blood DNA. Lanes 2 and 11: DNA from uninjected embryos; 8 and 17, DNA from blood of uninjected chickens. Lanes 3–7 and 12–16; DNA from vector-injected embryos. Lanes 9 and 10: *Bam*HI-digested DNA of pSW272/cGH (1 ng) plus uninjected chicken blood DNA. *Bam*HI fragments internal to the proviral vector are marked with arrows in (a). *Bam*HI fragments containing the endogenous cGH sequence are marked by asterisks in (e). Dot blot on the right of (c) contains the indicated amounts of pSW253 containing the REV-A provirus (Chen *et al.*, 1981). Sizes are shown in kilobase pairs (kb). (From Bosselman *et al.*, 1989b.)

The vectors ME111 (Emerman & Temin, 1984) and SW272/cGH (Hu *et al.*, 1987; Bosselman *et al.*, 1989b) were used in this study. Figure 2 illustrates the structure and sequence relationships of these vectors with the parental spleen necrosis virus and the packaging-defective helper provirus. The NEO^r and cGH sequences present in these vectors are transcribed via the promoter of the viral LTR. The tk gene in each vector has its own promoter derived from HSV-1. These vectors have been described in detail elsewhere (Emerman & Temin, 1984; Hu *et al.*, 1987; Bosselman *et al.*, 1989b).

Embryo injection

The microinjection procedure described here allowed delivery of 5–20 µl of vector-containing conditioned medium directly beneath the embryonic blastoderm (Bosselman *et al.*, 1989a). Eggs were collected soon after being laid and were held horizontally with respect to their long axis for 5 h or more at 18–20°C. During this period the blastoderm becomes positioned beneath the topmost area of the shell. Egg shells were wiped with 70% ethanol before and after a 5–8-mm hole was made in the shell. A Dremel moto-tool fitted with an aluminium oxide grinding stone was used to remove the shell. The shell membrane was then removed with a scalpel just before injection. A Narishige micromanipulator and a Drummond 100 µl digital microdispenser fitted with a glass needle of 50–60 µm outer diameter were used to perform the injection. A Kopf Model 720 vertical pipette puller was used to make the needles. A 10 µl sample of an overnight harvest of cell culture medium containing the vector ME111 or SW272/cGH was injected beneath the surface of the blastoderm. Vector titres were about 10^4 TKTU/ml. A Wild M5A dissecting microscope was used to monitor the injection. Holes in the egg were resealed with a patch of shell membrane. The patch was air dried and covered with Devcon Duco cement. Eggs were incubated at 37·8°C and allowed to hatch (Bosselman *et al.*, 1989a).

Somatic cell gene transfer

Total DNA was isolated from injected and control embryos after 7 days of incubation: 10 embryo DNAs were positive for vector sequences as judged by dot-blot analysis and were further analysed by Southern blotting (Southern, 1975). *Bam*HI restriction endonuclease fragments predicted from the SW272/cGH sequence are represented in Fig. 2. *Bam*HI restriction fragments internal to the provirus are 0·86, 2·3 and 1·6 kilobase pairs (kb) in length. A 5′ fragment including the junction between the provirus and cellular DNA sequences would also be detected using the radiolabelled probes shown in Fig. 2. As shown in Fig. 3(a), lanes 3–7 and 12–16, DNAs from vector-infected embryos contain the expected *Bam*HI DNA fragments. The probe used in this instance was the complete SW272 plasmid DNA, which lacks the cGH sequence present in SW272-cGH. No *Bam*HI fragment containing the junction of cellular DNA and integrated vector DNA was observed, indicating multiple sites of vector provirus integration. No DNA fragments indicative of unintegrated vector or replication-competent helper virus were observed. DNA from uninjected embryos or from blood of uninjected chickens did not hybridize to the vector probe (Fig. 3a, lanes 2, 8, 11 and 17, respectively).

The SW272 probe was removed from the filter in Fig. 3(a) (see Fig. 3b), followed by rehybridization with viral-specific probe shown in Fig. 2. No virus-specific *Bam*HI fragments were observed (Fig. 3c).

Virus-specific probe was removed (Fig. 3d) and the filter rehybridized to cGH specific probe (Fig. 3e). The fragments of 0·86 and 2·3 kb in lanes 3–7 and 12–16 are the predicted cGH-containing vector sequences represented in Fig. 2. The asterisked bands of about 6·4 and 2·7 kb common to all lanes are *Bam*HI fragments derived from the endogenous cGH gene. The 1·6 kb fragment present in Fig. 3(a) is not present in Fig. 3(e) since this fragment does not contain cGH sequences (after Bosselman *et al.*, 1989b).

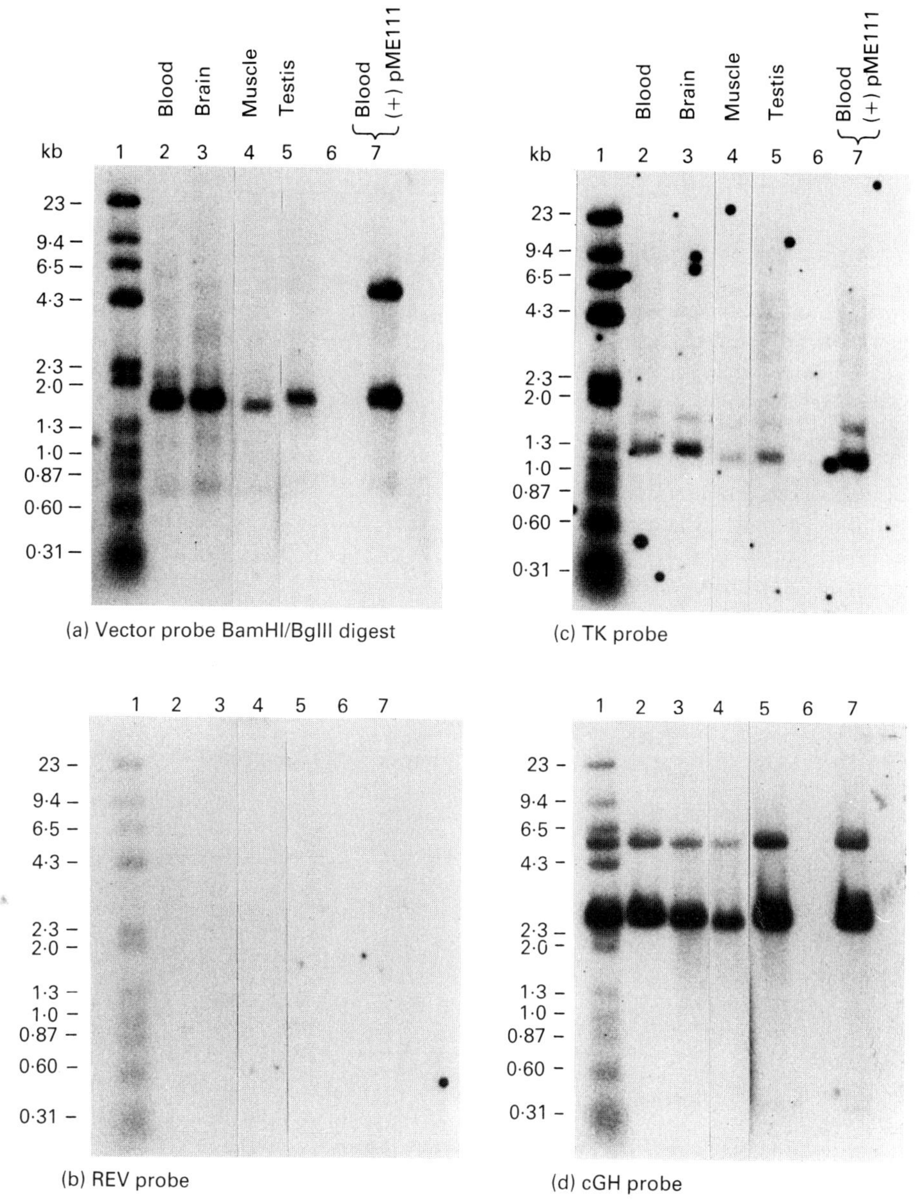

Fig. 4. Southern blot analysis of ~20 µg of *Bam*HI- and *Bgl*II-digested DNA from tissues of ME111-positive adult Male 87 725. Replicate blots hybridized with radiolabelled vector probe (a), virus probe (b), tk probe (c), and cGH probe (d). Lane 1: *Hind*III-digested lambda phage DNA, *Hae*III-digested ΦX174 DNA and *Bam*HI- and *Bgl*II-digested negative control chicken blood DNA. Lane 2: blood DNA. Lane 3: brain DNA. Lane 4: muscle DNA. Lane 5: testis DNA. Lane 6: blank. Lane 7: *Bam*HI- and *Bgl*II-digested pME111 (50 pg) and negative control chicken blood DNA. Sizes (in kilobases) are shown on the left of each panel. (From Bosselman *et al.*, 1989b.)

Analyses of tissues from adult birds derived from ME111 vector-injected embryos were also positive for vector sequences. Figure 4 shows Southern blot analysis of DNA from blood, brain, muscle and testis (of Bird 87 725). Four radiolabelled hybridization probes (see Fig. 2) were used

Table 1. GH concentrations in the serum of chicken embryos

SW272/cGH-injected embryos		Uninjected embryos	
Bird	GH (ng/ml)	Bird	GH (ng/ml)
1	51	31	<0·80
2	180	32	<0·80
3	100	33	<0·80
4	0·9	34	<0·80
5	41	35	<0·80
6	200	36	0·85
7	2·6	37	<0·80
8	80	38	<0·80
9	44	39	<0·80
10	106	40	<0·80
11	4·5	41	<0·80
12	18	42	1·2
13	8·6	43	1·0
14	1·1	44	<0·8
15	0·92	45	<0·8
16	2·2	46	<0·8
17	0·8	47	<0·8
18	254	48	<0·8
19	10·8	49	1·1
20	240	50	1·2
21	168	51	1·2
22	32	52	0·9
23	56	53	1·2
24	12	54	0·9
25	42	55	<0·8
26	0·86	56	<0·8
27	0·70	57	<1·1
28	3·4	58	<0·8
29	1·4	59	<0·8
30	0·76	60	<0·8
		61	<0·8
		62	<0·8
		63	<0·8
		64	<0·8
		65	<0·8

*Embryos of unincubated eggs were injected with 10 μl medium from cultures of C3-44 cells producing ~ 10^4 TKTU of the vector SW272/cGH. After 15 days of incubation serum from each embryo was assayed by RIA for cGH. (From Bosselman *et al.*, 1989b.)

which were derived from the 5′ and 3′ ends of the vector provirus, the HSV-1 tk sequences of the vector, REV structural genes, and endogenous cGH sequences. DNAs from tissues of Bird 87 725 contained *Bam*HI/*Bgl*II-generated 0·74 and 1·6 kb DNA fragments which hybridize to the vector probe, and 1·2 and 1·7 kb DNA fragments which hybridize to the tk probe (see Figs 4a and 4c, respectively). Blood and brain DNAs contained additional hybridizing fragments which may include junctions of vector and cellular DNA derived from sites of proviral integration (see Fig. 4a, lanes 2 and 3). No DNA fragments specific for replicating REV were observed in these tissue DNAs (Fig. 4b). Endogenous DNA fragments of ~2·7 kb and ~6·4 kb were revealed by hybridization with a chicken GH probe (Fig. 4d). Competent virus could not be detected even after blood from Bird 87 725 was co-cultured with D17 cells which were then passaged in culture for 4 weeks. Of 14

Table 2. Analysis of REV vector-mediated gene transfer in chickens (from Bosselman *et al.*, 1989a)

Embryos injected	Chicks hatched (%)	Chicks analysed	Blood-positive chicks (%)	Blood-positive males with vector in semen (%)
2599	995 (38)	760	173 (23)	33/82 (40)

G0 birds, 2 were virus-positive when assayed in this way (Hu *et al.*, 1987). These results are consistent with infection of embryonic stem cells by non-replicating REV vectors.

Expression of vector-encoded cGH was monitored by radioimmunoassay of chicken GH in the blood of 15-day-old embryos infected with vector (Souza *et al.*, 1984). Growth hormone concentrations were elevated in serum from 16 of 30 vector-injected embryos. Concentrations of GH were at least 10 times higher than in uninjected control embryos (Table 1). Since detectable plasma GH does not appear until Day 17 of embryonic development (Harvey *et al.*, 1979; Jozsa *et al.*, 1979; Scanes & Lauterio, 1984), these results are consistent with infection of stem cells in the blastoderm at the time of injection and subsequent expression of vector-encoded chicken GH (after Bosselman *et al.*, 1989b).

Germ-line gene transfer

We used the ME111 vector to demonstrate germ-line gene transfer into the chicken, since embryos which exhibit significant ecotropic expression of chicken GH did not hatch well. This vector and its relationship to competent SNV, the packaging-defective helper provirus and the SW272/cGH vector are represented in Fig. 2. Titres of the ME111 vector were similar to those of SW272/cGH ($\sim 10^4$ TKTU/ml). Of 2599 embryos, 987 (38%) were hatched after being injected with the ME111 vector before incubation. Blood DNA from 760 chicks was analysed by liquid hybridization with the vector-specific probe shown in Fig. 2: 173 of these chicks contained vector sequences. Of 82 males, 33 contained vector sequences in semen DNA. Analysis of REV vector-mediated gene transfer into G0 chickens injected as embryos is summarized in Table 2 (after Bosselman *et al.*, 1989a).

Vector-positive birds were tested for competent virus before breeding (Hu *et al.*, 1987): 2 of 14 G0 birds with vector-positive blood contained competent REV. Virus-negative birds were used to test for insertion of vector sequences into the chicken germ-line. G0 birds with vector-positive semen were used to inseminate control females artificially. Progeny chicks were analysed for the presence of vector sequences by liquid hybridization and Southern blotting.

Southern blot analysis of semen DNA from 3 G0 males and G1 blood DNA from their offspring are shown in Fig. 5. Semen DNAs from 3 vector-positive birds contain the predicted internal proviral *Bam*HI fragments of 3·7 and 1·65 kb (Fig. 5a, lanes 3–5). Other bands present probably represent the 5′ junctions between the integrated provirus and cellular DNA. DNA from semen of control birds did not hybridize to the vector probe. Semen DNAs from Males 28 428 and 87 620 are shown in lanes 3 and 4 of Fig. 5(a). Analysis of *Bam*HI- and *Bgl*II-digested blood DNA from G1 progeny of these G0 males is shown in Fig. 5(b). Lanes 4 to 6 of Fig. 5(b) contain blood DNA of progeny from Male 28 428. Lane 7 contains blood DNA from progeny of G0 Male 87 620. The probe used in this analysis derived from the 5′ and 3′ ends of the vector (see Fig. 2). The DNAs in Fig. 5(b), lanes 4, 5 and 7, show the expected internal vector DNA fragments of 1·65 and 0·74 kb. These fragments are also present in the positive control, lane 9. DNAs present in lanes 5, 6 and 7 each contain a single additional fragment representing the 5′ junction between proviral and cellular DNA. DNA in lane 4 contains two additional fragments, indicating that this bird carries two

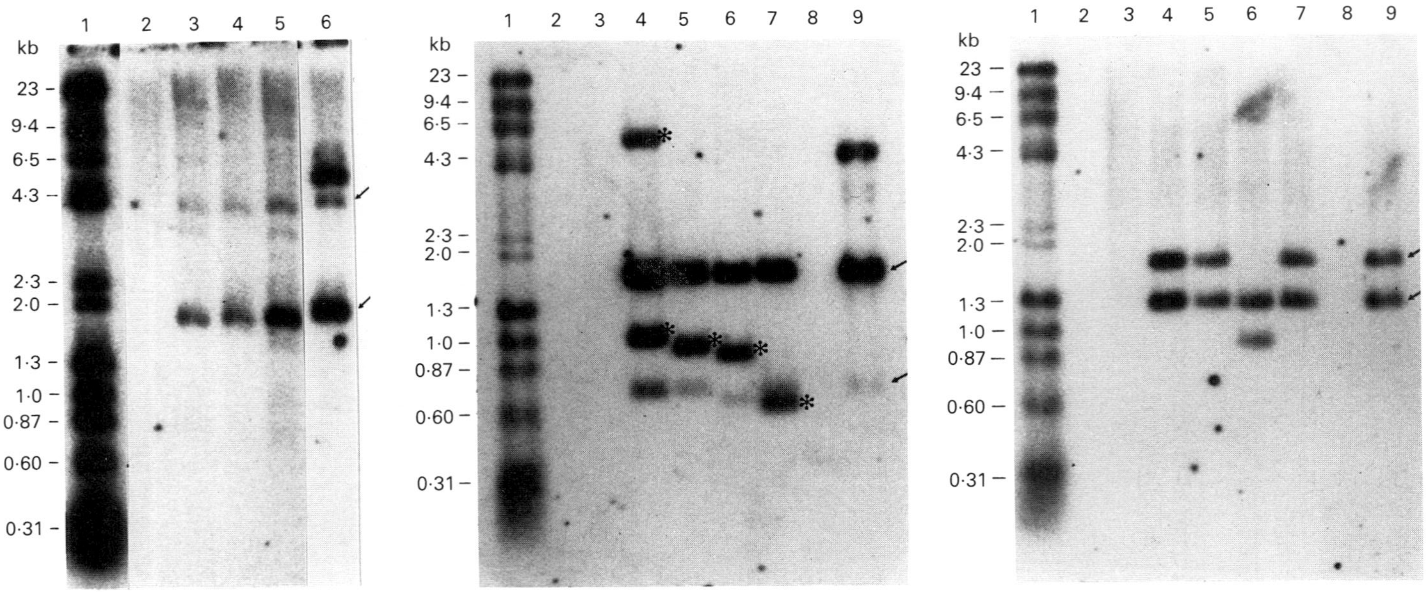

(a) Vector probe BamHI digest (b) Vector probe BamHI/BglII digest (c) TK probe BamHI/BglII digest

Fig. 5. DNA blot analysis of G_0 vector-positive semen and vector-positive G_1 progeny (Southern, 1975). (a) Hybridization with vector-specific probe. Lane 2: *Bam*HI-digested negative control semen DNA. Lanes 3–5: *Bam*HI-digested vector-positive semen DNA from three different G_0 chickens. Lane 6: *Bam*HI-digested pME111 DNA. (b) Hybridization with vector-specific probe. Lane 3: *Bam*HI/*Bgl*I-digested negative control blood DNA. Lanes 4–6: *Bam*HI/*Bgl*II-digested blood DNAs of vector-positive G_1 progeny of male G_0 bird No. 28 428. Lane 7: *Bam*HI/*Bgl*II-digested blood DNA of vector-positive G_1 progeny of male G_0 bird No. 87 620. Lane 9: *Bam*HI/*Bgl*II-digested negative control blood DNA and plasmid pME111. Lanes 2 and 8 are blank. (c) Hybridization with the TK-specific probe: Lanes as described for (b). Lane 1 of in (a), (b) and (c) contains *Hind*III-digested λ phage DNA and *Hae*III-digested ΦX174 DNA. DNA fragments that derive from internal regions of the vector are marked by arrows. Junction fragments containing the 5′ end of the vector and cellular DNA are marked by asterisks. The largest hybridizing fragments of DNA in lane 6 of (a) and lane 9 of (b) contain vector sequences that remain associated with plasmid DNA after restriction endonuclease digestion. (From Bosselman *et al.*, 1989a).

Table 3. Analysis of progeny from sires with vector in semen (from Bosselman *et al.*, 1989a)

	Sire			
	28 428	87 620	87 725	87 658
Fraction (%) of G1 progeny with vector	11/289 (3·8)	9/205 (4·4)	12/143 (8·4)	2/83 (2·4)

proviruses inserted at different sites within the chicken genome. These junction fragments marked by asterisks in Fig. 4(b) show the clonal nature of the inserted proviruses and demonstrate germ-line passage of vector sequences to G1 progeny. Unique junction fragments present in lanes 4, 5 and 6 also show that the germ-line of G0 sire 28 428 is mosaic with respect to vector insertions.

While lanes 4, 5 and 7 of Fig. 5(b) show the predicted pattern of restriction endonuclease fragments, lane 6 shows a fragment of ~0·65 kb rather than the 0·74 kb fragment present in the other lanes. Figure 5(c) shows the same DNAs hybridized with the TK probe. Again, lane 6 shows a fragment of ~0·9 kb, smaller than the expected fragment of 1·75 kb present in lanes 4, 5 and 7. These results indicate a deletion of ~0·8 kb sequences between the two *Bgl*II sites of the vector provirus in lane 6 (after Bosselman *et al.*, 1989a).

Analysis of progeny from 4 G0 sires is summarized in Table 3. A total of 720 offspring were screened. The percentage of vector positive offspring from each sire varied from 2·4% to 8·4%: 31 of these vector-positive offspring (each representing a unique insertion site) have been maintained and crossed with control chickens to generate G2 offspring. Detailed studies are currently underway which are aimed at understanding such issues as stability (rearrangements or deletions), modification (methylation) and expression of the ME111 transgenes. Initial results indicate that vector expression appears to vary in different organs tested and in different G2 offspring (data not shown).

Discussion

The generation of transgenic chickens is particularly suited to the use of retrovirus vectors as mediators of gene transfer. The reasons for this are numerous. The first 24 h of embryonic development take place *in utero* during egg formation. Manipulation of the large and fragile embryo during this stage is difficult. Gene transfer at this time would require surgical intervention and possibly transfer to female recipients capable of completing egg formation. As an alternative in-vitro culture is possible, but the hatching efficiency of these embryos is low (Rowlett & Simkiss, 1987; Perry, 1987). At about the time the egg is laid the embryo is disc-like in morphology and composed of many thousands of cells arranged in 1 or 2 layers (Eyal-Giladi & Kochav, 1976; Kochav *et al.*, 1980). The upper layer of cells or epiblast appears to be pluripotential with respect to subsequent development of the embryo (Eyal-Giladi & Spratt, 1965; Eyal-Giladi & Kochav, 1976; Eyal-Giladi *et al.*, 1976; Mitrani & Eyal-Giladi, 1982; Eyal-Giladi, 1984). Although the cellular complexity of the embryo at this stage precludes nuclear microinjection of DNA as a mode of gene transfer, easy access to the embryo and its layered morphology make gene transfer by retroviral vector infection a good alternative.

Retroviruses represent a well characterized example of natural gene transfer among the vertebrates, since they recombine with their host genomes to capture and transduce non-viral gene sequences (Hoelzer *et al.*, 1974; Hu *et al.*, 1987; Chen *et al.*, 1981; Schwartz *et al.*, 1983; Stephens *et al.*, 1983; Czernilofsky *et al.*, 1983; Wilhemsen *et al.*, 1984). The chicken genome already contains numerous retroviral insertions (Smith, 1986) which account for a variety of physiological effects

(Coffin, 1984). Retrovirus vectors derived by recombinant DNA technology can be divided into those that are replication competent and those that are replication defective. Examples of replication-competent vectors are derived from Rous sarcoma virus (Hughes *et al.*, 1986). These vectors carry genetic information in addition to the normal complement of viral genes, and have been used for both somatic and germ-line gene transfer in the chicken (Souza *et al.*, 1984; Salter *et al.*, 1986, 1987; Shuman & Shoffner, 1986; Salter & Crittenden, 1987, 1989). In contrast, replication-defective vectors used in the work described here lack viral structural genes, and rely on packaging-defective helper virus for gene products needed for their assembly. Demonstration that non-replicating vectors can effectively transduce new genetic information into the chicken provides a useful strategy in instances when the presence of competent virus is undesirable.

The ability to transfer new genetic information into the chicken germ-line represents an important step toward effective genetic manipulation of this agriculturally important species. The utility of the method we have described depends on development of vectors which do not result in productive infection (Yu *et al.*, 1986; Dougherty & Temin, 1987) and which allow stable germ-line insertion and controlled expression of genetic information. Currently, our understanding of the ability to express foreign genes in an animal is based largely on experiments with transgenic mice (Palmiter & Brinster, 1986). Both microinjection of DNA and infection with recombinant retroviruses have been used to generate transgenic mice, although the former has been more efficient in successfully expressing foreign genes. Germ-line insertions of the Moloney murine leukaemia provirus are not expressed or are expressed at very low levels. Inefficient levels of transcription may be due to methylation, sequences in the LTR, and/or negative transacting regulatory factors (Linney *et al.*, 1984; Jahner & Jaenish, 1985). Our preliminary results suggest that retroviruses introduced into the germ-line of chickens are more efficiently expressed than their counterparts in the mouse. Studies are underway to define further the stability and expression of REV transgenes in the chicken. Ultimately, practical avian gene transfer technology may derive from a combination of approaches, including other methods of gene transfer, in-vitro embryo culture (Perry, 1987), and the use of pluripotent embryonic stem cells (Robertson *et al.*, 1986; Petitte *et al.*, 1989).

References

Bosselman, R.A., Hu, S., Souza, L.M. & Nicolson, M. (1988) Retroviruses and avian gene transfer: a basic review. *Proc. 37th Western Poultry Disease Conference, Davis*, pp. 193–204.

Bosselman, R.A., Hsu, R.-Y., Boggs, T., Hu, S., Bruszewski, J., Ou, S., Kozar, L., Martin, F., Green, C., Jacobson, F., Nicolson, M., Schultz, J., Semon, K., Rishell, W. & Stewart, R.G. (1989a) Germline transmission of exogenous genes in the chicken. *Science, NY* **243,** 533–535.

Bosselman, R.A., Hsu, R.-Y., Boggs, T., Hu, S., Bruszewski, J., Ou, S., Souza, L., Kozar, L., Martin, F., Nicolson, M., Rishell, W., Schultz, J.A., Semon, K.M. & Stewart, R.G. (1989b) Replication-defective vectors of reticuloendotheliosis virus transduce exogenous genes into somatic stem cells of the unincubated chicken embryo. *J. Virol.* **63,** 2680–2689.

Chen, I.S., Mak, T.W., O'Rear, J.J. & Temin, H.M. (1981) Characterization of reticuloendotheliosis virus strain T by molecular cloning. *J. Virol.* **40,** 800–811.

Coffin, J. (1984) Endogenous viruses. In *RNA Tumor Viruses*, pp. 1109–1203. Eds R. Weiss, N. Teich, H. Varmus & J. Coffin. Cold Spring Harbor Laboratory, Cold Spring Harbor.

Czernilofsky, P.A., Levinson, A.D., Varmus, H.E., Bishop, J.M., Tischer, E. & Goodman, H. (1983) Corrections to the nucleotide sequence of the src gene of rous sarcoma virus. *Nature, Lond.* **301,** 736–739.

Dougherty, J.P. & Temin, H.M. (1987) A promoterless retroviral vector indicates that there are sequences in U3 required for 3′ RNA processing. *Proc. natn. Acad. Sci. USA* **84,** 1197–1201.

Emerman, M. & Temin, H.M. (1984) Genes with promoters in retrovirus vectors can be independently suppressed by an epigenetic mechanism. *Cell* **39,** 459–467.

Eyal-Giladi, H. (1984) The gradual establishment of cell commitments during the early stages of chick development. *Cell Differ.* **14,** 245–255.

Eyal-Giladi, H. & Kochav, S. (1976) From cleavage to primitive streak formation: a complementary normal table and a new look at the first stages of the development of the chick. *Devl Biol.* **49,** 321–337.

Eyal-Giladi, H. & Spratt, N.T., Jr (1965) The embryo-forming potencies of the young chick blastoderm. *J. Embryol. exp. Morph.* **13,** 267–273.

Eyal-Giladi, H., Kochav, S. & Menashi, M.K. (1976) On the origin of primordial germ cells in the chick embryo. *Differentiation* **6,** 13–16.

Harvey, S., Davison, T.F. & Chadwick, A. (1979) Ontogeny of growth hormone and prolactin secretion in the domestic fowl. *Gen. comp. Endocrinol.* **39,** 270–273.

Hayward, W.S., Neel, B.G. & Astrin, S.M. (1981) ALV-induced lymphoid leukosis: activation of a cellular *onc* gene by promoter insertion. *Nature, Lond.* **90,** 475–480.

Hoelzer, J.D., Franklin, R.B. & Bose, H.R., Jr (1974) Transformation by reticuloendotheliosis viruses: development of a focus assay and isolation of a non-transforming virus. *Virology* **93,** 20–30.

Hu, S., Bruszewski, J., Nicolson, M., Tseng, J., Hsu, R.-Y. & Bosselman, R. (1987) Generation of competent virus in the REV helper cell line C3. *Virology* **159,** 446–449.

Hu, S.S.F.M., Lai, M.C., Wong, T.C., Cohen, R.S. & Sevoian, M. (1981) Avian reticuloendotheliosis virus: characterization of genome structure by heteroduplex mapping. *J. Virol.* **40,** 800–811.

Hughes, S., Kosik, E., Fadly, A.M., Salter, D.W. & Crittenden, L.B. (1986) Design of retroviral vectors for the insertion of new information into the avian germ line. *Poult. Sci.* **65,** 1459–1467.

Jahner, D. & Jaenisch, R.L. (1985) Chromosomal position and specific demethylation in enhancer sequences of germline transmitted retroviral genomes during mouse development. *Molec. cell. Biol.* **5,** 2212–2220.

Jozsa, R., Scanes, C.G., Vigh, S. & Mess, B. (1979) Functional differentiation of the embryonic chicken pituitary gland studied by immunohistological approach. *Gen. comp. Endocrinol.* **39,** 158–163.

Kochav, S., Ginsburg, M. & Eyal-Giladi, H. (1980) From cleavage to primitive streak formation: a complementary normal table and a new look at the first stages of the development of the chick. *Devl Biol.* **79,** 296–308.

Linney, E., Davis, B., Overhauser, J., Chao, E. & Fan, H. (1984) Nonfunction of a Moloney murine leukaemia virus regulatory sequence in F9 teratocarcinoma cells. *Nature, Lond.* **308,** 470–472.

Mann, R., Mulligan, R.C. & Baltimore, D. (1983) Construction of a retrovirus packaging mutant and its use to produce helper-free defective retrovirus. *Cell* **33,** 153–159.

Mitrani, E. & Eyal-Giladi, H. (1982) Cells from early chick embryos in culture. *Differentiation* **21,** 56–61.

Noori-Daloii, J.R., Swift, R.A., Kung, H.-J., Crittenden, L.B. & Winter, R.L. (1981) Specific integration of REV proviruses in avian bursal lymphomas. *Nature, Lond.* **294,** 574–576.

Palmiter, R.D. & Brinster, R.L. (1986) Germline transformation of mice. *Ann. Rev. Genet.* **20,** 465–499.

Perry, M.M. (1987) A complete culture system for the chick embryo. *Nature, Lond.* **331,** 70–72.

Petitte, J.N., Guodong, L., Verrinder-Gibbons, A.M. & Etches, R.J. (1989) Germ-line chimeras can be produced by embryonic stem cell transfer in the chicken. *Cell Diff. & Devel.* **27,** S89.

Robertson, E., Bradley, A., Kuehn, M. & Evans, M. (1986) Germ-line transmission of genes introduced into cultured pluri-potential cells by retroviral vector. *Nature, Lond.* **323,** 445–448.

Rowlett, K. & Simkiss, K. (1987) Explanted embryo culture—*in vitro* and *in ovo* techniques for domestic fowl. *Br. Poult. Sci.* **28,** 91–101.

Salter, D.W. & Crittenden, L.B. (1987) Chickens transgenic for a defective recombinant avian leukosis proviral insert express subgroup A envelope glycoprotein. *Poult. Sci.* **66,** 170, abstr.

Salter, D.W. & Crittenden, L.B. (1989) Artificial insertion of a dominant gene for resistance to avian leukemia virus into the germline of the chicken. *Theor. appl. Genet.* **77,** 457–461.

Salter, D.W., Smith, E.J., Hughes, S.H., Wright, S.E., Fadly, A.M., Witter, R.L. & Crittenden, L.B. (1986) Gene insertion into the chicken germ line by retroviruses. *Poult. Sci.* **65,** 1445–1458.

Salter, D.W., Smith, E.J., Hughes, S.H., Wright, S.E. & Crittenden, L.B. (1987) Transgenic chickens: insertion of retroviral genes into the chicken germ line. *Virology* **157,** 236–240.

Scanes, C.G. & Lauterio, T.J. (1984) Growth hormone: its physiology and control. *J. exp. Zool.* **232,** 443–452.

Schwartz, D.E., Tizard, R. & Gilbert, W. (1983) Nucleotide sequence of Rous sarcoma virus. *Cell* **32,** 853–869.

Shuman, R.M. & Shoffner, R.N. (1986) Gene transfer by avian retroviruses. *Poult. Sci.* **65,** 1437–1444.

Smith, E.J. (1986) Endogenous avian leukemia virus. In *Developments in Veterinary Virology*, pp. 101–113. Eds Y. Becker & G. F. De Boer. Martinus Nijhoff, Boston.

Southern, E.M. (1975) Detection of specific sequences among DNA fragments separated by gel electrophoresis. *J. molec. Biol.* **98,** 503–517.

Souza, L.M., Boone, T.C., Murdock, D., Langley, K., Wypych, J., Fenton, D., Johnson, S., Lai, P.H., Everett, R., Hsu, R.-Y. & Bosselman, R. (1984) Application of recombinant DNA technologies to studies on chicken growth hormone. *J. exp. Zool.* **232,** 465–473.

Stephens, R.M., Rice, N.R., Hiebsch, R.R., Bose, H.R., Jr, & Gilden, R.V. (1983) Nucleotide sequence of v-rel: the oncogene of reticuloendotheliosis virus. *Proc. natn. Acad. Sci. USA* **80,** 6229–6233.

Swift, R.A., Boerkoel, C., Ridgway, A., Fujita, D.J., Dodgson, J.B. & Kung, H.-J. (1987) B lymphoma induction by reticuloendotheliosis virus: characterization of a mutated chicken syncytial virus provirus involved in *c-myc* activation. *J. Virol.* **61,** 2084–2090.

Watanabe, S. & Temin, H.M. (1983) Construction of a helper cell line for avian reticuloendotheliosis virus cloning vectors. *Molec. cell. Biol.* **3,** 2241–2249.

Wilhemsen, K.C., Eggleton, K. & Temin, H.M. (1984) Nucleic acid sequences of the oncogene v-rel in reticuloendotheliosis virus strain T and its cellular homologue, the Proto-oncogene c-rel. *J. Virol.* **52,** 172–182.

Yu, S.F., Ruden, T.U., Kantoff, P.W., Garber, C., Seiberg, M., Ristler, U., Anderson, W.R., Wagner, E.F. & Gilboa, E. (1986) Self-inactivating retroviral vectors designed for transfer of whole genes into mammalian cells. *Proc. natn. Acad. Sci. USA* **83,** 3194–3198.

FUTURE AND POTENTIAL

Chairman
R. McCarty

J. Reprod. Fert., Suppl. **41** (1990), 199–208
Printed in Great Britain

Animal production industry in the year 2000 A.D.

J. M. Massey

Granada Genetics, Inc., 100 Research Parkway, Suite 100, Texas A & M University Research Park, College Station, Texas 77840, USA

Summary. One can easily envision that, in the very near future, all bulls being progeny tested will be screened for genomic markers linked to economic traits and females may also be screened if enrolled as donors in a nuclear transfer programme. The concept of producing large numbers of genetically identical embryos, frozen, sexed, screened for economic traits and produced inexpensively from slaughterhouse by-products, is within our grasp. While large scale commercialization of these concepts is a function of time, knowledge and cost-effective biotechnologies, all of these concepts have already been demonstrated. The production of transgenic animals and embryos will be accelerated as gene mapping links genes to economic traits.

What will happen to protein production when commercial cow herds can be made up of one or more female clone lines mated to bulls of the same clone? The obvious answer is predictability of performance to a magnitude never before achieved in agriculture.

Keywords: transgenic; Y-specific probes; nuclear transplantation; restriction fragment length polymorphism

Introduction

Historically, animal scientists have been quick to take advantage of biotechnologies for the propagation of economically superior genetics. Artificial insemination has been a standard practice for the past 40 years in the dairy industry and has resulted in rapid improvement in milk production and milk components. In the past 5 years, the dairy industry has focussed on the improvement of milk solids such as protein and fats. These improvements through artificial insemination have occurred because of the selection pressure applied to the selection of superior bulls used by the artificial insemination organization. In the past 10 years embryo transfer technology has emerged as a practical method of accelerating genetic improvement by taking advantage of the number of offspring which can be produced from the mating of a superior female and progeny-proven sire. New embryo manipulation techniques are on the horizon in application to livestock improvements. Rapid development in molecular biology has resulted in the opportunity to develop superior animal lines which have been enhanced through insertion of a desirable gene into the genome. These developments have been reported for mice (Gordon *et al.*, 1980), sheep (Hammer *et al.*, 1985), pigs (Hammer *et al.*, 1986) and cattle (Biery *et al.*, 1988). These transgenic animals have been produced by the injection of a unique gene construct such as the human and rat growth hormone (Palmiter *et al.*, 1982, 1983) into early stage developing embryos. The development of molecular genetic markers detecting variations at the DNA level has opened the door to the mapping of not only the human genome but also the mapping of farm animals such as cattle (Womack, 1988). The intent of this paper will be to review the current and developing biotechnologies which will have an impact on the production of animals in agriculture into and beyond the year 2000 A.D.

Embryo transfer

The ability to manipulate embryos at early stages of development opens up a number of possibilities for increasing desirable genotypes in economically important food animals. Research into the techniques of embryo manipulation can also lead to basic knowledge about developmental biology.

The non-surgical recovery of embryos from superovulated donor cows (Elsden *et al.*, 1978) was a major step in the commercialization of embryo transfer technology in the cattle industry. Embryo transfer in both dairy and beef cattle is now an accepted commercial practice, producing a large number of embryo transfer offspring. The techniques have been applied to other food animals such as pigs, sheep, and goats, but acceptance of the procedure has been much slower because recovery of embryos from these species is still by surgical means, thus potentially limiting their reproductive life. The advancement of molecular biology will have an impact on the embryo transfer industry in the future with production of recombinant gonadotrophins for superovulation (Wilson, 1988). When recombinant bovine follicle-stimulating hormone (bFSH) was used to superovulate 205 cows the mean response was 11·1 total ova, of which 85% (9·4) were viable embryos (Table 1). When the best stimulation programme was reported, total ova and viable embryos increased slightly to 12·4 ova and 11·0 (89%) (Table 2) respectively. The means were not different ($P > 0{\cdot}05$). When a pituitary extract-FSH (FSP-P®) was used for superovulation 10 783 cows produced a mean of 10·32 total ova, of which 57% (5·85) were viable embryos (Table 3). When the best three FSH-P stimulation programmes were reported, total ova and viable embryos increased to 12·16 and 6·72 (55%) (Table 4) respectively. The means were not different ($P > 0{\cdot}05$). It was also reported that the percentage of Grade 1 and 2 embryos collected was greater for cows stimulated with bFSH than for cows stimulated with FSH-P. The percentages of Grade 1, 2 and 3 embryos were 53%, 39% and 8% respectively for bFSH, with Grade 1 being embryos which were at the proper stage of development and morphology, and embryos of Grades 2 and 3 being of lesser developmental quality. The percentages of Grade 1, 2 and 3 embryos were 39%, 40·3% and 20% respectively for FSH-P. While the total ova of 12·4 for bFSH and 12·16 for FSH-P are essentially the same, the viable embryos of 11·0 and 6·72 for bFSH and FSH-P, respectively, are different ($P < 0{\cdot}05$) (Duncan multiple range test). There could be several reasons for this difference beside the obvious one that the recombinant bovine FSH should be identical to the endogenous FSH of the cow. Another reason is that bFSH should be free of any other gonadotrophic hormone such as luteinizing hormone (LH). It can be postulated that additional exogenous LH is not needed to superovulate cattle or for ovum maturation.

Table 1. Embryo production from 205 donor cows stimulated with bFSH (May 1987–May 1988)* (data from Wilson, 1988)

	Total ova	Viable embryos	Unfertilized ova†	Degenerate embryos
	2267	1921	162	185
Mean/cow	11·1	9·4	0·8	0·9
%		85·0	7·2	8·1

*Doses ranging from 0·5 mg twice/day to 8·0 mg for 3–5 days.
†Four cows accounted for 60% of the total unfertilized ova.

The sex ratio

Controlling the sex ratio of offspring is of significant commercial value in agriculture and would best be controlled by sexed semen. There have been extensive efforts to produce and occasional

Table 2. Embryo production in 99 cows stimulated with FSH regimens which appear to be most efficacious and cost effective (data from Wilson, 1988)

	Total ova	Viable embryos	Unfertilized ova	Degenerate embryos
	1231	1089	67	75
Mean/cow	12·4	11·0	0·7	0·7
%		89·0	5·6	5·6

Table 3. Embryo production from 10 783 donor cows stimulated with FSH-P (1983–July 1988) (data from Wilson, 1988)

	Total ova	Unfertilized ova	Degenerate embryos	Viable embryos
	111 280	29 653	18 546	63 080
Mean/cow	10·32	2·75	1·72	5·85
%		26	17	57

Table 4. Embryo production from 5979 cows treated with 3 stimulation programmes after utilization of two 25-mg PGF injections (data from Wilson, 1988)

	Total ova	Unfertilized ova	Degenerate embryos	Viable embryos
	72 704	20 089	12 436	40 178
Mean/cow	12·16	3·36	2·08	6·72
%		28	17	55

claims of success in producing a sexed semen product, but today a reproducible technique for separating X- and Y-bearing spermatozoa has not been accomplished for farm animals (but see Johnson *et al.*, 1989). The economic potential of sexed semen is so extensive that it continues to warrant significant effort.

Sex ratio will first be controlled through embryo manipulation before embryo transfer. Cytological approaches to embryo sexing have been reviewed (King, 1984). Embryo splitting allows half the embryo to be karyotyped and the other half used for transfer. The pregnancy rate from split embryos is only slightly less than that for intact embryos. A major limitation is that half embryos are not easily frozen for later transfer, thus reducing the practical value of the approach. Karyotyping of embryos is also limited by the inability to generate mitotic structures sufficient to identify the sex chromsomes consistently due to the small number of cells available; sex determination is only reported 60% of the time.

It has also been reported that male-specific antigens are expressed as early as the 8-cell stage in embryos of mice, cattle, pigs and sheep (Wachtel, 1984) and that they can be identified by indirect immunofluorescence with antisera raised in female mice against spleen cells from males of the same inbred strain. A high degree of success has only been demonstrated in one laboratory (Anderson, 1987). This technique is highly variable and needs further verification.

The use of male-specific chromosomal DNA fragments to determine the sex of cattle embryos has been reported (Bondioli *et al.*, 1989). The identification of repetitive, male-specific chromo-

NA fragments allows use of DNA-probe technology to determine the sex of cattle embryos small embryonic sample. These Y chromosome-specific probes can be used in conjunction DNA-replicating techniques such as a polymerase chain reaction, thus decreasing the number cells required to sex an embryo, potentially to 1 or 2 cells. The removal of 1 or 2 cells essentially leaves the embryo intact for further manipulation such as embryo freezing. The accuracy of these techniques is essentially 100% with the limitation being a function of sufficient DNA sampling and recombinant DNA techniques.

In-vitro maturation and fertilization of oocytes

At the present time a large demand exists for cattle oocytes and early stage embryos for use in commercial operations and research. Basic research with cattle and other farm animals has been hampered by an inadequate source of competent oocytes and early stage embryos at an economically feasible cost. Research into the current biotechnologies for manipulating the genome of cattle or other farm animals also requires a large supply of oocytes and/or early stage embryos. A system to take advantage of oocytes from ovaries of commercially slaughtered cattle would be the best economic supply. In-vitro produced oocytes or embryos also have the advantage of known timing of development compared to those recovered *in vivo*. In-vitro culture of bovine primary oocytes taken from slaughterhouse ovaries and fertilized *in vitro* have been reported (Critser *et al.*, 1986) to produce live young. Currently variable efficiency exists in producing developmentally competent zygotes from the in-vitro maturation–in-vitro fertilization system (Critser *et al.*, 1986; Lu *et al.*, 1987; Xu *et al.*, 1987; Fukui & Ono, 1988).

A system for large scale oocyte production of slaughterhouse ovaries has been implemented by Granada Genetics, Inc. The laboratory currently receives 400 ovaries per day yielding about 1000 usable oocytes or 2·5 oocytes per ovary. The oocytes are utilized in an in-vitro fertilization programme and a nuclear transplantation programme. Approximately 20% of all oocytes matured and fertilized *in vitro* yield viable embryos, and approximately 50% of the in-vitro matured oocytes can be used as recipient oocytes for nuclear transfer, yielding about 20% viable embryos when fused and cultured for 5 days in a sheep oviduct. These results are generally comparable to in-vivo results but are much more cost effective.

Embryo splitting and nuclear transplantation

Microsurgical techniques developed for basic studies in mammalian embryology using mouse embryos have been applied to large domestic animals to increase the number of offspring with identical genotypes. Separation of blastomeres of early cleavage-stage embryos of cattle, horses, pigs and sheep have resulted in normal development from as little as a quarter of the normal complement of cells (Willadsen, 1979). Normal offspring can be produced by splitting embryos of late morula or early blastocyst stages into halves.

Another approach to multiplying the number of offspring with identical genotypes is by nuclear transplantation. Nuclear transfer involves the transplantation of living nuclei from typically embryonic cells to unfertilized eggs. The early research on vertebrates was performed in amphibians; nuclear material was taken from separated embryonic frog blastomere cells and introduced into enucleated frog oocytes (Briggs & King, 1952). Transplantation of nuclear material in mammals has proved very difficult to achieve, in part due to the microsurgical techniques used to manipulate embryos and eggs. The microsurgical techniques can be destructive to delicate cell structures necessary for later development. An alternative procedure is to deliver the nuclear material to a recipient egg by fusion of an intact cell or karyoplast consisting of a nucleus surrounded by a piece of plasma membrane.

Successful nuclear transplantation and cell fusion was achieved for sheep embryos when individual blastomeres from 8- and 16-cell embryos were used as the nuclear donor into enucleated or nucleated halves of unfertilized eggs (Willadsen,1986). The production of a live nuclear transplanted lamb was a clear demonstration that the nuclear material from at least the 8–16-cell sheep embryo was totipotent. The potential for cloning livestock had not been expected since in mouse experiments transplanted nuclei from 4-cell and older embryos did not support development to the blastocyst stage (McGrath & Solter, 1984). This may point to significant differences in mammalian embryos in the timing of activation of the embryonic genome and the interaction between nuclei and cytoplasm.

Bovine blastomeres from 2–32-cell-stage embryos recovered surgically were demonstrated to be totipotent when fused to a bovine enucleated oocyte (Prather *et al.*, 1987). Blastomeres from 32-cells to at least a compacted morula stage embryo (approximately 64 cells) (Table 5) collected from cattle non-surgically were found to be totipotent and produced live young when transferred to an enucleated oocyte and later a recipient (unpublished data). It has also been demonstrated that deep-frozen donor embryos produce similar results when transplanted into enucleated oocytes (Fig. 1). Embryos that were produced through nuclear transplantation procedures have in turn been used as nuclear donor cells for subsequent fusion to recipient eggs (Fig. 2). The pregnancy ratio for fresh and frozen nuclear transfer embryos has ranged from 0 to 54% (Fig. 3).

Table 5. Production of nuclear transfer cattle embryos by donor embryo cell number (1st and multiple generation procedures)

Donor embryo cell no.	Viable embryos	No. of transfers	No. pregnant	Pregnancy rate (%)
1–10	53	38	1	2·0
11–19	143	74	11	14·8
20–29	190	126	23	18·25
30–39	168	105	24	22·8
40–49	71	48	12	25·0
50–59	15	10	0	0·0
≥60	60	50	14	28·0
Total	700	451	85	18·8

Successful nuclear transfer transplantations and embryo development in farm animals have great implications. These achievements will provide the first opportunity to make large numbers of identical offspring for milk or beef production. Nuclear transfers will produce uniformity in calf crops and even have an impact on changing the sex ratio of a calf crop. Nuclear transplantation will have a great impact on transgenic production of farm animals since a potential transgenic embryo can be mass produced.

Transgenic animals

Technology for manipulating embryos of farm animals has developed primarily because of the techniques developed for the study of embryonic development of laboratory animals. Germ-line transformation of laboratory mice was first reported in 1980 (Gordon *et al.*, 1980). Reports from several other laboratories established that cloned foreign DNA was stably integrated into the genome and Mendelian germ-line transmission occurred following microinjection into the pronuclei of fertilized mouse embryos (Brinster *et al.*, 1981; Costantini & Lacy, 1981; Gordon & Ruddle, 1981; Wagner *et al.*, 1981). The extrapolation of this technology from laboratory animals to farm animals has not been as simple and straightforward as would be expected. The major drawback has been the high cost associated with obtaining a source of early-stage embryos from

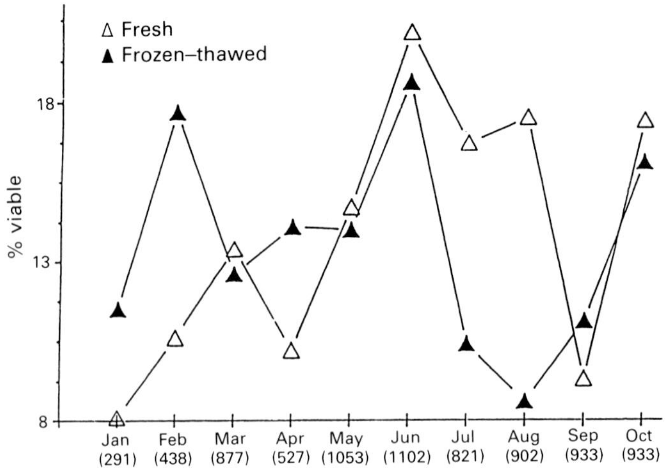

Fig. 1. Percentage production of viable cattle embryos after nuclear transfer. Blastomeres were physically separated from fresh or frozen–thawed embryos and fused (1–15 October only in October) to an enucleated oocyte. The resulting embryos were cultured and evaluated 6 days after fusion. Total no. of fusions given in parentheses.

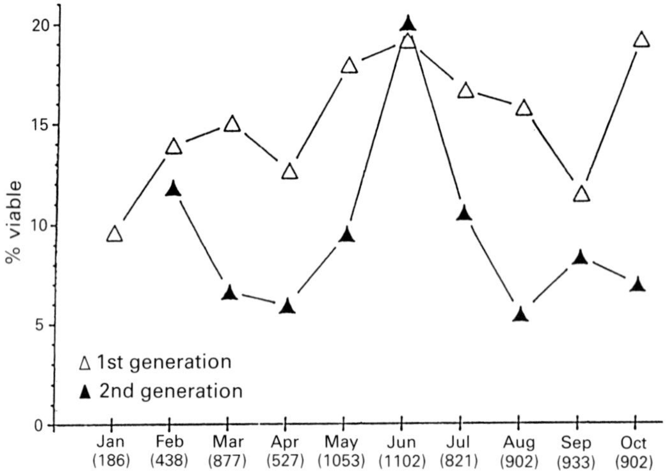

Fig. 2. Percentage production of viable cattle embryos after nuclear transfer. Donor embryos were collected and blastomeres were physically separated and fused to enucleated oocytes (1st generation). The resulting embryos were then used as donor embryos to produce a 2nd generation of embryos by separating blastomeres and fusing to enucleated oocytes. The efficiency was evaluated 6 days after fusion. Total no. of fusions given in parentheses; fusions in October on 1–15 October only.

farm animals and the high cost of maintenance of resulting offspring. The cytoplasm of ova from mice and rabbits is transparent, and the cytoplasm of pig, sheep and cattle ova is opaque, making the pronuclei invisible under light microscopy.

Techniques have been developed for visualizing and injecting pronuclei of sheep, pigs and cattle (Hammer *et al.*, 1985; Wall *et al.*, 1987; Biery *et al.*, 1988). The human growth hormone gene attached to the mouse metallothionein promotor (MT-hGH) has been injected into the pronuclei

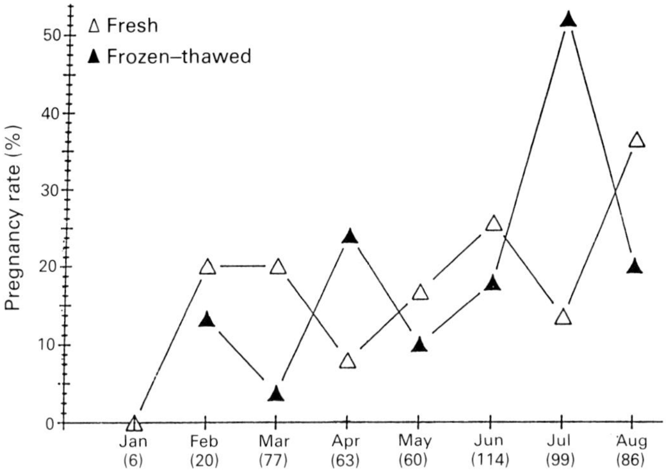

Fig. 3. Pregnancy rate in cows from embryos produced by nuclear transfer. The embryos (no. in parentheses) were transferred to a recipient cow after 6 days of culture of frozen and thawed for later transfer. Pregnancy rate was assessed at 90 days and at birth.

or nuclei of rabbit, pig and sheep ova (Hammer *et al.*, 1985, 1986). Integration of the gene into the DNA of each species was achieved and expression of the gene was observed in rabbits and pigs. The integration frequencies were 12·8% in rabbits, 10·4% in pigs and 1·3% in sheep. The integration frequency of a bacterial chloramphenicol acetyltransferase gene under the control of the Rous sarcoma virus promoter was 3·4% and expression was only observed in the placental tissue of one conceptus.

A system for large scale production of transgenic cattle is currently being evaluated at our laboratory. Bovine pronuclear embryos were collected and injected with a human oestrogen receptor gene linked to a skeletal actin promoter (ASK-HER). The injected embryos were cultured *in vitro* for 7 days before transfer to recipient cattle (Table 6).

The use of in-vitro matured and in-vitro fertilized oocytes is also under evaluation for production of transgenic cattle (Table 7).

Table 6. Production of transgenic cattle with ASK-HER

No. of ova collected	3902
No. of ova injected	1704 (44%)
No. of ova developed in culture	261 (15% of injected)
No. of pregnancies	79 (30% of transferred)
No. of transgenic cattle	1 known from 16 calves born

Tables 6 and 7 clearly demonstrate the efficiency of producing transgenic cattle to be very low but the zygotes produced by in-vitro maturation and fertilization of oocytes are very cost effective when compared to an in-vivo source of pronuclear embryos.

There are other potential methods of integrating genes into the genome of an embryo, such as the utilization of retroviral vectors (King *et al.*, 1985) and embryo-derived stem cells (Evans &

Table 7. Production of transgenic cattle from oocytes matured and fertilized *in vitro*

No. of ova matured	2408
No. of ova injected	858 (36%)
No. of ova developed in culture	49 (6% of injected)
No. of pregnancies	12 (24%)
No. of transgenic calves	Unknown

Kaufman, 1981). Lavitrano *et al.* (1989) reported that when spermatozoa incubated with PSV2CAT plasmid were used to fertilize mouse eggs *in vitro*, integration was observed in approximately 30% of 250 progeny. However, this finding needs to be repeated for mice and tested for other species.

The practicability of transgenic animals for improvement of farm animals cannot be assessed until questions of integration and expression are answered. While many of these questions will probably be answered with experiments on laboratory animals, many will only be answered with the actual production of transgenic farm animals. For example, how might over-expression of a gene like growth hormone affect the complex interplay that regulates growth rate, body composition, overall fitness, age, sexual maturity and reproductive capacity of an animal? Disease resistance genes may be better candidates for transgenic production because they may have fewer physiological repercussions.

The genes of choice

The evidence from large animal studies and theory indicates that single genes, in general, are unlikely to have a significant effect on commercial traits and success will depend on finding the exception, such as the Booroola gene (Land & Wilmut, 1987), which increases litter size in sheep.

Another example of single genes affecting polygenic traits is the double muscling gene of cattle. These genes have disadvantages as well as advantages. The double-muscle gene produces a higher lean yield but also increases calving difficulties. It may be possible to manipulate a number of genes or to manipulate genes that govern a whole cascade of enzymes. The elevated concentration of individual enzymes is unlikely to increase the flux through a complex pathway, but if all enzymes in a pathway are elevated the output would increase proportionally. Research with the pig shows the relevance of this knowledge. Selection for the level of four enzymes (glucose 6-phosphate dehydrogenase, 6-phosphogluconate dehydrogenase, NADP-linked malate dehydrogenase and NADP-linked isocitrate dehydrogenase) in the nicotinamide adenine dinucleotide phosphate (NADP) pathway led to marked changes in backfat (Muller, 1986). After 8 generations of selection, the phenotypes of the high and low lines differ by 3·6 standard deviations. The effectiveness of hormones such as growth hormone or growth hormone-releasing factor was demonstrated in larger transgenic mice (Palmiter *et al.*, 1982; Hammer *et al.*, 1985).

The expression of transgenes at unique physiological age or the expression of site-specific genes would be highly significant. If a growth hormone gene can be linked to a promotor which only expresses at birth and turns off at puberty, it may produce all the desired effects of early growth. Site specific expression of growth hormone in the mammary system may increase milk production without any other physiological effects. Transgenic sheep have been produced with a human anti-haemophilic factor IX gene linked to a sheep derived lactoglobulin promotor (Clark *et al.*, 1989). Two sheep expressed factor IX into their milk, demonstrating that expression can be site specific and that a novel gene can be integrated and expressed in a foreign genome.

Genetic markers

The contribution of recombinant DNA technology may lie in the identification and mapping of polymorphic loci in the genomes of livestock. The identification and mapping of restriction fragment length polymorphisms (RFLPs) have revolutionized human genetics by providing markers for a variety of inherited diseases. Only a few RFLPs in farm animals have been reported (Womack, 1988). If the DNA of domestic animals is as polymorphic as that of humans, one would expect approximately one polymorphic site in every 100 nucleotide pairs (Jeffreys, 1979). The widespread availability of DNA probes and the use of RFLPs for genomic analysis should increase the number of known DNA polymorphisms in farm animals.

The parasexual methods of somatic cell hybridization used extensively in human gene mapping can be applied to any mammalian species (Womack & Moll, 1986; Womack, 1987). Using cow–hamster hybrid cells, 37 bovine isoenzyme loci to 24 syntenic groups have been identified. Gene maps of sheep and pigs are being developed with the same methods at comparable rates.

The number of polymorphic markers required to saturate a mammalian genome depends on the desired density of mapped loci (Womack, 1987). If one assumes a genome size of 2500 cM for a given species, approximately 215 randomly distributed markers are necessary to generate 95% probability that any quantitative-trait locus will be within 20 cM of a polymorphic marker (Kashi *et al.*, 1986). Homology of the human, mouse and cattle maps is extensive (Womack & Moll, 1986); that is, groups of syntenic genes tend to be conserved in all 3 species. This knowledge will allow for the use of human and mouse mapping data to select probes for identification of polymorphisms at loci with predictable spacing along the bovine chromosomes.

I thank Ken Bondioli, Frank Barnes, Charles Looney and Mark Westhusin for their assistance in compiling information which they have generated; and Deborah Sheehan and Tina Weido for preparation of this manuscript.

References

Anderson, G.B. (1987) Identification of embryonic sex by detection of H-Y antigen. *Theriogenology* **27,** 81–97.

Biery, K.A., Bondioli, K.R. & DeMayo, F.J. (1988) Gene transfer by pronuclear injection in the bovine. *Theriogenology* **29,** 224, abstr.

Bondioli, K.R., Ellis, S.B., Prior, J.H., Williams, M.W. & Harpold, M.M. (1989) The use of male-specific chromosomal DNA fragments to determine the sex of bovine preimplantation embryos. *Theriogenology* **31,** 95–104.

Briggs, R. & King, T.J. (1952) Transplantation of living nuclei from blastula cells into enucleated frog eggs. *Proc. natn. Acad. Sci. USA* **38,** 455–463.

Brinster, R.L., Chen, H.Y. & Trumbauer, M.E. (1981) Somatic expression of herpes thymidine kinase in mice following injection of a fusion gene into eggs. *Cell* **27,** 223–231.

Clark, A.J., Bessos, H., Bishop, J.O., Brown, P., Harris, S., Lathe, R., McClenaghan, M., Prowse, C., Simons, J.P., Whitelaw, C.B.A. & Wilmut, I. (1989) Expression of human anti-hemophilic factor IX in the milk of transgenic sheep. *Bio/Technology* **7,** 487–492.

Costantini, F. & Lacy, E. (1981) Introduction of a rabbit β-globulin gene into the mouse germ line. *Nature, Lond.* **294,** 92–94.

Critser, E.S., Leibfried-Rutledge, M.L., Eyestone, W.H., Northey, D.L. & First, N.L. (1986) Acquisition of developmental competence during maturation *in vitro*. *Theriogenology* **25,** 150, abstr.

Elsden, R.P., Nelson, L.D. & Seidel, G.E. (1978) Superovulating cows with follicle stimulating hormone and pregnant mare's serum gonadotropin. *Theriogenology* **9,** 17–26.

Evans, M.J. & Kaufman, M.H. (1981) Establishment in culture of pluripotential cells from mouse embryos. *Nature, Lond.* **292,** 154–156.

Fukui, Y. & Ono, H. (1988) *In vitro* development to blastocyst of *in vitro* matured and fertilized bovine oocytes. *Vet. Rec.* **122,** 282.

Gordon, J.W. & Ruddle, F.H. (1981) Integration and stable germ line transmission of genes injected into mouse pronuclei. *Science, NY* **214,** 1244–1246.

Gordon, J.W., Scangos, G.A., Plotkin, D.J., Barbosa, J.A. & Ruddle, F.H. (1980) Genetic transplantation of mouse embryos by microinjection of purified DNA. *Proc. natn. Acad. Sci., USA* **77,** 77 380–77 384.

Hammer, R.E., Pursel, V.G., Rexroad, C.E., Jr, Wall, R.J., Bolt, D.J., Ebert, K.M., Palmiter, R.D. & Brinster, R.L. (1985) Production of transgenic rabbits, sheep and pigs by microinjection. *Nature, Lond.* **315,** 680–683.

Hammer, R.E., Pursel, V.G., Rexroad, C.E., Jr, Wall, R.J., Bolt, D.J., Palmiter, R.D. & Brinster, R.L. (1986) Genetic engineering of mammalian embryos. *J. Anim. Sci.* **63,** 269–278.

Jeffreys, A.J. (1979) DNA sequence variants in the G-gamma, A-gamma, delta and beta globin genes of man. *Cell* **18,** 1–10.

Johnson, L.A., Flook, J.P. & Hawk, H.W. (1989) Sex preselection in rabbits: live births from X and Y sperm separated by DNA and cell sorting. *Biol. Reprod.* **41,** 199–203.

Kashi, Y., Soller, M., Hallerman, E. & Beckman, J.S. (1986) Restriction fragment length polymorphisms in dairy cattle improvement. *Proc. 3rd World Congr. Genetics Applied to Livestock Production*, Lincoln, NE, Vol. 7, pp. 57–63.

King, W., Patel, M.D., Label, L.I., Goff, S.P. & Chi Nguyen-Huu, M. (1985) Insertion mutagenesis of embryonal carcinoma cells by retroviruses. *Science, NY* **228,** 554–558.

King, W.A. (1984) Sexing embryos by cytological methods. *Theriogenology* **21,** 7–17.

Land, R.B. & Wilmut, I. (1987) Gene Transfer and animal breeding. *Theriogenology* **27,** 169–179.

Lavitrano, M., Camaioni, A., Fazio, V.M., Dolci, S., Forace, M.G. & Spadafora, C. (1989) Sperm cells as vectors for introducing foreign DNA into eggs: genetic transformation of mice. *Cell* **57,** 717–723.

Lu, K.H., Gordon, I., Gallagher, M. & McGovern, H. (1987) Pregnancy established in cattle by transfer of embryo derived from *in vitro* fertilization of oocytes matured *in vitro*. *Vet. Rec.* **121,** 259–260.

McGrath, J. & Solter, D. (1984) Inability of mouse blastomere nuclei transferred to enucleated zygotes to support development *in vitro*. *Science, NY* **226,** 1317–1319.

Muller, E. (1986) Physiological and biochemical indicators of growth and composition. In *Exploiting New Technology in Animal Breeding*, pp. 132–139. Eds C. Smith, J. W. B. King & J. C. McKay. Oxford University Press, Oxford.

Palmiter, R.D., Brinster, R.L., Hammer, R.E., Trumbauer, M.E., Rosenfeld, M.G., Birnberg, N.C. & Evans, R.M. (1982) Dramatic growth of mice that develop from eggs microinjected with metallothionein-growth hormone fusion genes. *Nature, Lond.* **300,** 611–615.

Palmiter, R.D., Norstedt, G., Gelinas, R.E., Hammer, R.E. & Brinster, R.L. (1983) Metallothionein-human GH fusion genes stimulate growth of mice. *Science, NY* **222,** 809–814.

Prather, R.S., Barnes, F.L., Sims, M.M., Robl, J.M., Eyestone, W.H. & First, N.L. (1987) Nuclear transplantation of the bovine embryo: assessment of donor nuclei and recipient oocyte. *Biol. Reprod.* **37,** 859–866.

Wachtel, S.S. (1984) H-Y antigen in the study of sex determination and control of sex ratio. *Theriogenology* **21,** 18–28.

Wagner, T.E., Hoppe, P.C., Jollick, J.D., School, D.R., Hadinka, R.L. & Gault, J.B. (1981) Microinjection of a rabbit β-globulin gene into zyogtes and its subsequent expression in adult mice and their offspring. *Proc. natn. Acad. Sci., USA* **78,** 6376–6380.

Wall, R.J., Pursel, V.G., Hammer, R.E. & Brinster, R.L. (1987) Development of porcine ova that were centrifuged to permit visualization of pronuclei and nuclei. *Biol. Reprod.* **32,** 645–651.

Willadsen, S.M. (1979) A method for culture of micromanipulated sheep embryos and its use to produce monozygotic twins. *Nature, Lond.* **277,** 298–300.

Willadsen, S.M. (1986) Nuclear transfer in sheep embryos. *Nature, Lond.* **320,** 63–65.

Wilson, J.M. (1988) Superovulation: FSH update. *Proc. 7th Annual Conv. Am. Embryo Transfer Ass., Reno,* 31–43.

Womack, J.E. (1987) Genetic engineering in agriculture: animal genetics and development. *Trends in Genetics* **3,** 65–68.

Womack, J.E. (1988) Molecular cytogenetics of cattle: a genomic approach to disease resistance and productivity. *J. Dairy Sci.* **71,** 1116–1123.

Womack, J.E. & Moll, Y.D. (1986) Gene map of the cow: conservation of linkage with mouse and man. *J. Hered.* **77,** 2–7.

Xu, K.P., Greve, T., Callesen, H. & Hyttel, P. (1987) Pregnancy resulting from cattle oocytes matured and fertilized *in vitro*. *J. Reprod. Fert.* **81,** 501–504.

J. Reprod. Fert., Suppl. **41** (1990), 209–220

Printed in Great Britain

Abstracts of Posters

USDA regulatory activities concerning transgenic animals. G. P. Shibley*, A. Langston*, D. D. Jones† & D. B. Berkowitz‡, **USDA Animal and Plant Health Inspection Service (APHIS), Hyattsville, Md, USA; †USDA Office of Agricultural Biotechnology (OAB), Washington, DC, USA; and ‡USDA Food Safety Inspection Service (FSIS), Washington, DC, USA.*

The USDA, under existing authority, develops policies and procedures for regulating the introduction of transgenic animals into the environment and into the food chain. USDA/APHIS regulates importation, exportation, interstate movement, and environmental release (including field testing) of transgenic animals and veterinary biologicals under the Virus–Serum–Toxin Act and the animal quarantine laws. Regulations allow USDA to regulate importation, shipment and release of organisms and vectors used in biotechnology. APHIS regulates field tests of animals genetically engineered for infectious disease resistance.

USDA/FSIS regulates animals that enter the human food chain. When a gene coding for a substance (e.g. a growth hormone) is established in a food animal's germ line, regulations from USDA (APHIS and FSIS), Environmental Protection Agency (EPA), and Food and Drug Administration (FDA) are applicable, depending upon the gene product. We will discuss USDA (APHIS and FSIS) and FDA authority concerning release, shipment, and commercialization of transgenic animals.

A research model for constructing porcine genomic libraries. N. Li, Y. Cheng & C. Wu, *Beijing Agricultural University, Beijing 100094, China.*

In this paper we describe the detailed procedure for constructing pig genomic libraries with an EMBl-3 bacteriophage and a cosmid PHC-79, as well as a cosmid PBTI-10E. Many complete genomic libraries of Landrace pigs and mini-pigs (Xiang) have been constructed using the abovementioned vectors, because the two breeds of pigs are extremely different in body size, growth rate, meat composition and many other traits. The libraries can be used to study genetic polymorphism, gene family and mechanism of gene regulation of mammalian animals.

We have also reconstructed a cosmid PBTI-10 to generate a new cosmid, PBTI-10E, which is suitable for rapid and convenient cloning of large stretches of DNA of many animals. Furthermore, some modified techniques are presented for isolation of high molecular weight DNA and preparation of packaging proteins *in vitro*. In practice, application of all these techniques has given good results.

Porcine–murine heteromyelomas as fusion partners for the production of porcine monoclonal antibodies. Q.-e. Yang, M. Hollingshead, T. Kuhara, C. Hammerberg, S. Tonkonogy & E. V. De Buysscher, *College of Veterinary Medicine, North Carolina State University, Raleigh, NC 27606, USA.*

During the past years we have developed several murine–porcine heterohybridomas (murine P3 × porcine lymphoblastoid P-16) that produced porcine monoclonal antibodies (mAbs). Some of these heterohybrids lost antibody synthesizing capability, became 8-azaguanine resistant and their caryograms contained double minutes. We observed subsequently that these 8-azaguaine resistant, 2-mercaptoethanol independent, non-producing heteromyelomas functioned as excellent fusion partners with pig peripheral blood or thoracic duct lymphocytes. The resulting triple hybrids were shown to be stable in both growth and mAb production. Unfortunately, unlike in the rat, fusion of pig thoracic duct lymphocytes (TDL) with the heterohybrids resulted in only a few monoclonal IgA-producing triple hybrids. All pig TDLs were obtained after removal of the jejuno–ileal mesenteric lymphnodes. Flow cytometric analysis (FACS II, Becton Dickinson) of TDLs with anti-pig α-chain mAbs revealed no IgA-bearing TD lymphocytes. Cytoplasmic staining with fluorescent labelled antibodies revealed numbers of IgA-containing cells that were substantially lower than those reported for rat TDLs by other investigators. The role of the mesenteric lymph nodes in IgA-blast traffic in the pig is currently under investigation. Data will be presented on the histocompatibility make-up (murine/porcine Classes I and II), growth characteristics under the influence of various lymphokine preparations, and the rate of IgA synthesis by these heterohybrids.

Growth hormone restriction fragment length polymorphisms that segregate with 42-day live weight of mice. D. C. Winkelman, L. Querengesser & R. B. Hodgetts, *Department of Genetics, University of Alberta, Edmonton, Alberta, Canada, T6G 2E9.*

The known correlation between growth hormone concentrations and growth rate in a number of species prompted us to examine whether polymorphic restriction fragment alleles at the growth hormone locus in mice might be associated with differentiable rates of growth. An F_2 population of mice was generated from crosses between a line selected for high 42-day weight and an unselected control line. The original selected and control lines exhibited 42-day mean weights of $30{\cdot}6 \pm 3{\cdot}8$ and $20{\cdot}5 \pm 2{\cdot}6$ g, respectively. Since the two lines also differed with respect to the restriction fragments detected by hybridization to a rat growth hormone cDNA probe, an analysis of the F_2 generation was carried out to determine whether this polymorphism could be considered a quantitative trait locus for 42-day weight. The results of the analysis indicated that a polymorphic HindIII restriction fragment was correlated ($P < 0{\cdot}05$) with 42-day weight. However, the allele which was positively correlated with weight was the one which was fixed in the original control line, rather than the one from the selected line. While these findings suggest the possibility of using restriction fragment length polymorphisms for quantitative trait evaluation in livestock, they also emphasize the requirement for testing such potential quantitative trait loci in the appropriate genetic background.

Mapping the FOS and FES protooncogenes and the immunoglobulin lambda genes in cattle. T. C. Tobin & J. E. Womack, *Texas A & M University, College Station, TX 77843, USA.*

The FOS and FES protooncogenes and the IGL genes have been assigned to bovine syntenic groups. Panels of bovine–hamster hybrid somatic cells, which segregate cattle chromosomes, were screened for the presence or absence of bovine restriction fragments. Genes consistently present in the same hybrid clones were assumed to be on the same chromosome, or syntenic. Using this technique, FOS was assigned to bovine syntenic group U5, and FES was assigned to bovine syntenic group U4. These syntenic groups consist of human chromosomes 14 and 15 markers. Similarly, IGL was assigned to bovine syntenic group U23. This syntenic assignment was unexpected based on comparative mapping, as none of the U23 markers has been shown to be syntenic in other species. In additional experiments, the extent of variation of the germ-line configuration of the IGL genes is being determined. Further studies to determine the areas of genomic rearrangements between cattle and other mammals, and to characterize the immunoglobulin gene polymorphisms in cattle, are being performed, and will be of interest in comparative mammalian and evolutionary genetics, as well as having possible implications for breeding programmes.

Restriction fragment length polymorphisms as an aid in selection for quantitative traits in beef cattle. J. F. Baker & J. L. Rocha, *Texas A & M University, College Station, TX 77843, USA.*

One application for RFLPs would be their utilization as genetic markers for quantitative effects on traits of economic importance in livestock. Quantitative measurements on 900 animals from a 5-breed diallele study are being matched with the genotype constitution for 7 loci defined by restriction enzyme × probe combinations: bovine growth hormone-EcoRI, BGH-Taql, bovine prolactin-

Msp, pOsteonectin-EcoRI, chymosin-EcoRI, parathyroid hormone-Msp, and keratin 4-Msp. Preliminary results indicate that only two of these loci, BGH-Taql and PTH-Msp, display enough polymorphism (major allele frequencies of 0·84 and 0·6, with 5 and 4 allelic variants, respectively) for such an analysis. A linkage study with quantitative effects for these two loci, based on a half-sib family structure is being conducted. It was determined that, for the BGH-Taql locus, three breeds behave as inbred lines with incomplete fixation for different alleles: allele A frequency for Angus (1), Brahman (0·33) and Hereford (0·96). This led to an 'inbred-lines type of analysis' screening the Angus–Brahman and Hereford–Brahman F2s for the referred locus. Fewer polymorphic loci can only be utilized in an analysis for pleiotropic effects. There is also some evidence suggesting that one of the loci (Chy-EcoRI) may be under natural selection pressure. With regard to allelic frequencies, the Brahman breed usually differs from cattle of the more uniform Angus, Hereford, Holstein and Jersey breeds.

Methods for K-casein and β-lactoglobulin genotyping of bulls. S. Lien, S. Rogne, T. Steine, T. Langsrud*, G. Vegarudl & P. Alestrom, *Institute for Animal Science, Box 25, N-1432 As-NLH, Norway; and *Department of Dairy and Food Industries, Box 36, N-1432 As-NLH, Norway.*

Methods for K-casein and β-lactoglobulin genotyping in cattle have been developed. By analysis of DNA polymorphisms it is possible to discriminate between the alleles A and B for both protein types in bulls. A method has also been developed to analyse milk proteins by rapid isoeletric focussing on Pharmacia's PhastSystem. That the analysis of DNA polymorphisms correctly reflects the different alleles, was checked by determination of variants in milk from the same cows. By these new methods it is no longer necessary to test female progeny to decide the K-casein and β-lactoglobulin genotypes in breeding bulls. This means that the methods can be efficient tools used in breeding to increase the frequency of the most desirable milk protein variants.

Somatic cell mapping of the bovine somatostatin gene. A. B. Dietz & J. E. Womack, *Texas A & M University, College Station, TX 77843, USA.*

DNAs from cow–hamster and cow–mouse hybrid somatic cells segregating cow chromosomes have been analysed by Southern blotting and hybridization with a bovine somatostatin probe. The somatostatin gene (SST) segregated concordantly with superoxide dismutase-1 (SOD1) and α-crystallin (CRYA1) representing cattle syntenic group U10. Since homologues of these three genes have also been mapped in mice and humans, a comparison of the gene maps of the three mammalian species reveals some insight into the chromosomal rearrangements that have accompanied their evolution. We propose that the bovine chromosome carrying U10 is an ancestral chromosome, giving rise to human chromosome 21 and mouse chromosome 16, each requiring a minimum of one rearrangement.

Nuclear injection of bovine oocytes after in-vitro maturation. M. Gagné & M. Sirard, *Laboratoire Ontogénie & Reproduction and Department Zootechnie, Laval University Medical Centre, Québec, Canada G1V 4G2.*

In-vitro maturation of oocytes from domestic animals represents a large source of gametes for nuclear injection and should allow a rapid development of the technology. With these oocytes, pronuclear formation is controlled in time and fertilization rates are high. However, even if many offspring are obtained after in-vitro maturation, the development potential is still limited especially after successive manipulations. In Exp. 1, the effect of centrifugation (4 min, 15 000 **g**) and vortexing (2 min) were evaluated with bovine zygotes derived from the in-vitro system. In-vitro development to 2–8 cells was not affected (76% *vs* 72%) for control ($n = 211$) and treated ($n = 210$) zygotes at 18 h after insemination. In Exp. 2, the effect of pronuclear injection with CRF gene (2 μg/ml in 10 mM-Tris–0·2 mM-EDTA) at 20 h after insemination was evaluated by in-vivo development for 5 days in a rabbit oviduct. The embryos submitted only to centrifugation and vortexing resulted in a morula–blastocyst (> 32 cells) rate of 26% ($n = 185$) compared to the injected zygotes with only 5% ($n = 158$) achieving the same stage. For all experiments, the developmental rate to 32 cells after rabbit oviduct incubation is 23·4% ($n = 327$). We conclude that the procedure is traumatic by itself and the quality of oocytes obtained *in vitro* has to be raised if better results are required.

Effect on in-vitro fertilization of electroporation of bovine spermatozoa with a foreign gene. M. Gagné, F. Pothier & M. Sirard, *Laboratoire Ontogénie & Reproduction, and Department Zootechnie, Laval University Medical Centre, Québec, Canada G1V 4G2.*

The production of transgenic animals by pronuclear injection is still limited in domestic animals. The use of in-vitro fertilization can be helpful in many ways for the injection procedure, but also if spermatozoa are to become transgene carriers. Frozen–thawed bovine spermatozoa (10×10^6/ml) were submitted to electroporation (Bio-Rad) using 6 different conditions (voltage, 500, 1000 or 1500 V) and (capacitance, 1 or 25 *C*) in the presence (1 μg/ml) of the plasmid (pOCAT) previously radiolabelled with ^{32}P. For 5 replicates, the fractions of radiolabelled plasmid retained in spermatozoa after 2 washes (stable for at least 8 washes) were 15, 13, 9, 15, 14 and 10% for 500 V, 1000 V and 1500 V at 1 *C* and 500, 1000 and 1500 V at 25 *C*, respectively. Spermatozoa in contact with plasmid without electroporation retained only 6·5% of the label. In the same experiment, the electroporated spermatozoa were submitted to ionophore A23187 to evaluate the fraction of plasmid remaining after removal of the acrosome. The results were 7, 9, 9, 8, 11 and 9% for the same conditions and 5·8% for the spermatozoa in contact with pOCAT but not submitted to the electrical field. Using bovine oocytes matured *in vitro*, electroporation conditions were evaluated for fertilization rate. Using unlabelled plasmid in three replicates, fertilization rate was 83% ($n = 76$) and developmental rate to morula in rabbit oviduct was 24% ($n = 21$). These results are comparable with the controls. We are now evaluating integration rate after in-vitro fertilization.

Integration and stable germ-line transmission of human growth-hormone gene via microinjection into early medaka (*Oryzias latipes*) embryos. J. Lu, O. M. Andrisani, J. E. Dixon & C. L. Chrisman, *Purdue University, West Lafayette, IN 47907, USA.*

Gene constructs consisting of promoterless human growth hormone (hGH), or hGH fused with regulatory DNA from the mouse metallothionein (MT), viral thymidine kinase, rat cholecystokinin, or chicken β-actin (CBA) genes were microinjected, via the micropyle, into the cytoplasm of newly fertilized medaka eggs. Over 39% of the embryos reached hatching stage after DNA injection. Embryos with the CBA–hGH transgene hatched earlier than non-injected fish and had a higher survival rate than did the other gene construct groups. The growth performance of MT–hGH fish was significantly ($P < 0{\cdot}005$) greater than that of controls. The largest adult transgenic fish were 97% heavier than the mean weight of controls. Genomic DNA, extracted from whole juvenile fish or the caudal fin of adult fish, was analysed using slot-blot and Southern-blot hybridizations. Between 20 and 60% of injected fish contained entire hGH DNA. Over 50% of F_1 progeny derived from crosses between transgenic males and transgenic females tested positive for the transgene. Short generation interval, simple husbandry and egg collection, and ease of micromanipulation make the medaka an excellent model species for the study of gene control.

Lack of genetic transmission of avian leukosis proviral DNA in viraemic Japanese quail. D. W. Salter, A. Balander*, J. Bradac†, S. Hughes† & L. B. Crittenden, *Department of Microbiology & Public Health and *Department of Animal Science, Michigan State University, East Lansing, MI 48824, USA; †NCI-Fred. Can. Res. Fac., Frederick, MD 21701, USA; and ‡USA-ARS, PRL, East Lansing, MI 48823, USA*

A model system for avian germ-line research is needed to decrease the time and expense of in-vivo testing of avian retroviral vector constructs. Due to its short reproductive cycle and small size, the Japanese quail has previously served as an efficient avian model system for many genetic selection experiments. We therefore tried to reproduce our very successful chicken transgenic experiments (Salter *et al.*, 1986, 1987; Crittenden *et al.*, 1989; Salter & Crittenden, 1989) in Japanese quail. Subgroup A recombinant and wild-type replication-competent avian leukosis virus (ALV) containing exogenous or endogenous long terminal repeats (LTR) were injected into the yolk of fertile quail eggs just before incubation. In contrast to chickens, viraemic quail were obtained only with ALV containing exogenous LTR. Vertical (congenital and genetic) transmission of ALV was investigated by back-crossing viraemic females and males to male and female Japanese quail and testing the progeny for ALV antigens and proviral DNA. About 5% of the progeny of approximately 30 viraemic females were infected with ALV by, presumably, congenital transmission. None of the approximately 50 viraemic males infected at day of set or approximately 20 viraemic males infected *in ovo* through the viraemic females transmitted proviral DNA to their progeny. Therefore, Japanese quail will not serve as a model system for avian germ-line research using ALV vectors because, in contrast to viraemic chickens, viraemic quail failed to transmit ALV proviral DNA genetically.

Crittenden, L.B., Salter, D.W. & Federspiel, M.J. (1989) *Theor. appl. Genet.* **77,** 505–515.
Salter, D.W. & Crittenden, L.B. (1989) *Theor. appl. Genet.* **77,** 457–461.
Salter, D.W., Smith, E.J., Hughes, S.H., Wright, S.E., Fadly, A.M., Witter, R.L. & Crittenden, L.B. (1986) *Poult. Sci.* **65,** 1445–1458.
Salter, D.W., Smith, E.J., Hughes, S.H., Wright, S.E. & Crittenden, L.B. (1987) *Virology* **157,** 236–240.

Modification and characterization of helper cell lines for production of replication-defective avian retro-viruses; a comparison with a murine amphotropic helper cell line. P. J. Hippenmeyer & M. K. Highkin, *Monsanto Company, St Louis, MO 63198, USA.*

Foreign genes have been introduced into chick embryos by retrovirus-mediated transduction using a replication competent retrovirus. Improvements in the process require high titre stocks of replication-defective vectors generated from helper cell lines. Because a previously characterized helper cell line, D17C3, produces low levels of wild type virus, we attempted to engineer cell lines with improved characteristics. We have engineered generic helper lines from QT6 and D17 cells for production of reticuloendotheliosis virus (REV). The D17C3 line was re-evaluated using hygromycin as a selectable marker. A murine cell line, PA317, was evaluated for production of virus that infects chick cells. The majority of lines from the D17, D17C3 and QT6 experiments produced titres in the 10^2 to 10^3 c.f.u./ml range. Two lines produced 10^4 c.f.u./ml when assayed on QT6 cells. In the presence of polybrene, the titres increased approximately 50-fold. The mouse line produced 6- to 10-fold higher amounts of virus than did the QT6 and D17 lines and had a comparably higher titre on chick cells. The results indicated that the replication-defective virus stock from a QT6 producer line was relatively free of helper virus when the producer line was restricted to low passage.

Differential expression of ^{I}*alpha* and ^{II}*alpha* globin genes in goats. O. Bergersen & V. M. Fosse, *National Veterinary Institute, Department of Animal Genetics, P.O. Box 8156 Dep, 0033 Oslo 1, Norway.*

In Norwegian dairy goats, great variation has been reported in the ratio between ^{I}*alpha* and ^{II}*alpha* globin chains. Most goats show a ratio of approximately 3:1. Some animals have a reversed ratio of approximately 1:2. One goat has even been detected without any [I]*alpha* globin. The different *alpha* globin chains were quantified by use of an FPLC technique. Our results indicate that goats with a ratio of 1:2 were heterozygous for the 'reversed' type, while the animal without the expression of [I]*alpha* globin was homozygous for the 'reversed' type. One offspring from a normal sire and a homozygous 'reversed' dam also support this, since the offspring became heterozygous.

Genomic DNA samples from goats with a normal and a reversed ratio of *alpha* globin chain haplotypes were digested with 6 restriction enzymes and hybridized to an [II]*alpha* globin probe. The restriction patterns were similar in goats with either type of α-expression. This suggests that the different ratios of *alpha* globin expression could not be due to a deletion of [I]*alpha* and/or duplication or triplication of [II]*alpha* globin genes. The most likely explanation of the observed α-globin ratios is differential expression of the [I]*alpha* and [II]*alpha* globin genes in the different goats.

Negative effect of low molecular weight components of calf serum on progesterone production by bovine luteal cells in culture. J. D. Cavalcoli*, G. F. Ambroski† & R. A. Godke*, **Department of Animal Science, LAES, LSU Agricultural Center and* †*Department of Veterinary Microbiology and Parasitology, LSU School of Veterinary Medicine, Louisiana State University, Baton Rouge, LA 70803, USA.*

The objective of this experiment was to evaluate the cell-stimulatory activity of Solcoseryl (SOL) on bovine luteal cells in an in-vitro cell culture system. Solcoseryl (Solco Basle Ltd, Basle, Switzerland) is a calf blood haemodialysate ($M_r < 10\,000$). Bovine luteal tissue from a young post-pubertal beef female was dispersed using collagenase (Sigma, St. Louis, MO, USA) and mechanical dissociation. Ham's F-12 (Gibco, NY, USA) with insulin (5 µg/ml), transferrin (5 µg/ml), selenium (5 ng/ml) was the base medium (HF-12 + ITS) used across treatment groups. The 24-well culture plates (Gibco) were preincubated with HF-12 with 10% fetal calf serum before plating to facilitate cell attachment. Each culture well was seeded with 1×10^5 cells and 12 wells were randomly assigned to each treatment group. Treatment A was HF-12 + ITS medium (control), Treatment B was HF-12 + ITS medium with 10 ng LH/ml (equine pituitary LH: Sigma), Treatment C was HF-12 + ITS medium with 0·1% SOL and Treatment D was HF-12 + ITS supplemented with 10 ng LH/ml and 0·1% SOL. Plates were incubated at 37°C with 5% CO_2 in air (97% relative humidity). Media were harvested and replaced after 24 h of incubation and again at 48-h intervals for 15 days. All media samples were frozen for progesterone analysis by RIA.

Table 1. Progesterone concentrations (ng/ml) in media from cultured luteal cells

Days in culture	0	1	3	5	7	9	11	13	15
Treatment A	0·19	24·71	8·86	7·49	6·22	2·70	3·10	1·28	0·81
Treatment B	0·01	44·40	9·93	7·40	7·73	3·74	3·25	1·60	0·57
Treatment C	0·01	18·70	1·65	1·21	0·80	1·33	0·45	0·55	0·37
Treatment D	0·01	38·60	3·41	1·48	1·80	0·00	0·00	0·01	0·00

The results (Table 1) indicate that LH stimulated progesterone production while SOL decreased progesterone output ($P < 0{\cdot}05$) over the first 24 h and then continued over time in culture. Based on mean cell counts, components of SOL appeared to stimulate cell viability during Days 9–11 of culture; however, steroidogenesis was markedly reduced. In summary, the low molecular weight components of calf serum (Solcoseryl) elicited a negative response on progesterone production by this bovine luteal cell culture system.

Bovine alpha casein gene sequences direct high-level expression of active human urokinase in mouse milk. H. Meade & N. Lonberg*, *Biogen, Inc., Cambridge, MA 02142, USA; and *Memorial Sloan-Kettering Cancer Center, New York, NY 10021, USA.*

A possible means of producing heterologous proteins efficiently is in the milk of transgenic dairy animals. The first step in establishing this technology is to determine those milk-specific gene sequences capable of directing the expression of heterologous proteins into milk. We are utilizing transgenic mice as a model system to identify these sequences.

Cow milk contains a high level (10 mg/ml) of alpha S-1 casein, a milk-specific protein. We have cloned the bovine alpha casein gene and linked portions of this gene sequence to the gene encoding human urokinase. We then used the resulting construct to produce transgenic mice.

One transgenic line carries 2–3 copies of the casein–UK sequences. All females from this line produce high levels of active human urokinase in their milk. This trait has been stably inherited through 3 generations.

Using this model system we have shown that (1) the bovine alpha casein promoter is expressed in the mouse; (2) heterologous proteins can be expressed in the mammary gland; and (3) high levels of properly folded active protein can be secreted into the milk.

Expression of virally transduced mouse tyrosinase in tyrosinase-negative chick embryo melanocytes in culture. B. Whitaker*, T. Frew*, J. Greenhouse†, S. Hughes†, H. Yamamoto‡, T. Takeuchi‡ & J. Brumbaugh*, **Biological Sciences, University of Nebraska, Lincoln, NE 68588, USA; †Frederick Cancer Research Facility, Frederick, MD 21701, USA; and ‡Biological Institute, Tohoku University, Sendai 980, Japan.*

A putative mouse tyrosinase cDNA was inserted into a plasmid containing provirus of a replication competent Avian Leukosis Virus (ALV). The plasmid was transfected into proviral free (line 0) chick embryo fibroblasts and the viral supernatant harvested. This virus was used to infect cultured tyrosinase-negative (c^a/c^a) chick embryo melanocytes. These melanocytes are normally completely devoid of pigment. By 5 days after infection many cells had produced very dark discrete pigment granules. Mock-infected, Rous sarcoma virus-infected, and ALV-neo-infected cultures produced no pigment. This shows that the mouse cDNA is tyrosinase because it 'cured' the genetic defect in the chick melanocytes. It also shows that the mouse and chick tyrosinase genes are similar enough so that the mouse cDNA is translated, processed, packaged, and expressed in the chicken. The next step will be to infect fertile eggs and test for germ-line insertions. Transgenic chickens expressing tyrosinase will serve as models for transferring other genes. Tyrosinase expression can also serve as a marker for indicating gene transfer when transferred in tandem with another gene of choice.

A comparison of bGH expression in two cell lines directed by the CMV, SV40, RSV-LTR, or mouse metallothionein promoters. E. Fernandez, X. Chen & J. J. Kopchick, *Department of Zoology, Molecular and Cellular Biology Program, and the Edison Animal Biotechnology Center, Ohio University, Athens, OH 45701, USA.*

Expression of the bovine growth hormone (bGH) gene linked to different promoters has been studied in a mouse neuroblastoma cell line (NB41A3) and a human cell line (HELA). Chimaeric plasmids containing the bGH-gene and viral (CMV, SV40, RSV-LTR) or cellular [metallothionein (Met), phosphoenol pyruvate carboxykinase (PEPCK)]-transcriptional regulatory sequences (Trs), placed upstream from the bGH translational initiation codon, were used in transient expression studies. The results indicate that, in NB41A3 cells, only those cells transfected with plasmids containing either the CMV or RSV-LTR Trs express the hormone as determined by Western-blot analyses. In contrast, SV40-Trs directed the highest levels of expression in NB41A3 and HeLa cell lines, respectively. On the other hand, cells transfected with the Met–Trs-containing plasmid did not express the bGH gene under normal culture conditions (NB41A3 cells) or expressed the bGH gene at a very low level (HeLa cells). Nevertheless, expression of Met–bGH can be induced by increasing concentrations of $ZnCl_2$ in the culture fluid. Northern analysis of the total RNA shows that the bGH–mRNA is transcribed only when cells are transfected with plasmid containing CMV– or RSV–LTR (CMV–, SV40– and Met– for HeLa cells) Trs. In co-transfection experiments, in which bGH gene-containing plasmids were used with a plasmid containing the chloramphenicol acetyl transferase (CAT) gene linked to the CMV– or SV-40–Trs, different degrees of inhibition of CAT expression were seen in each cell line depending on the TRS which was linked to the bGH gene. These results indicate a possible competition between promoters for a putative cellular factor(s) involved in gene transcription.

In-vitro and in-vivo expression of bGH deletion mutants in exons IV and V. N. Y. Chen, S. J. McAndrew, L. DiCaprio, P. Wiehl, J. Yun, T. Wagner, S. Okada & J. J. Kopchick, *Department of Zoology, Molecular and Cellular Biology Program, and the Edison Animal Biotechnology Center, Ohio University, Athens, OH 45701, USA.*

The synthetic oligonucleotide, 5′-pd[CTAGTCTAGACTAG]-3′, which encodes unique translation termination codons in all three reading frames, was inserted at either a TthIII I site (exon IV) or SMA I site (exon V) of plasmid pbGH-10. This plasmid contains the bGH gene (5 exons and 4 introns) adjoining the mouse metallothionein I transcriptional regulatory sequences. Two derivatives of the parental plasmid, pbGH-4A and pbGH-5A, were constructed and used to study the expression of truncated bovine growth hormones *in vitro* and *in vivo*.

After transient transfection of these plasmid DNAs into cultured mouse L cells, expression but not secretion of the mutant growth hormones was demonstrated. The intracellular proteins encoded by the mutant DNAs were smaller in molecular weight (pbGH-4A = ~16 500, pbGH-5A = ~15 000) than wild type bGH (~22 000).

Both mutant DNAs were microinjected into the male pronucleus of fertilized mouse eggs. Transgenic mice generated in these experiments showed the same growth pattern as non-transgenic litter mates. Expression of mutant growth hormones in transgenic animals was assessed by Western and Northern blot analysis using mouse liver tissue and serum samples. In mice containing the pbGH-4A transgene, bGH-specific RNA and protein were found in liver samples. The apparent molecular masses of these polypeptides were similar to those found in transiently transfected mouse L cells. Surprisingly, the mutant pbGH-4A protein, not found in culture fluids derived from transfected mouse L cells, was found in the serum of transgenic mice. Neither bGH-specific RNA nor mutant growth hormone was detected in liver or serum samples derived from pbGH-5A transgenic mice.

Study on fluorescent test for diagnosing viability of frozen mouse and sheep embryos. Xiang-Dong Ma, *Institute of Animal Science, CAAS Beijing, China.* (Present address: Department of Immunology, Merck & Co., Inc., Rahway, NJ 07065, USA).

A modified DAPI (4′,6′-diamidino-2-phenylindole) fluorescent technique for evaluation of frozen–thawed embryos before transfer was studied. After thawing successively in 8 batches, 406 mouse embryos of late morula and early blastocyst stages were assigned Grades of A, B, C or D according to morphological criteria. After staining with DAPI, the embryos of each grade were reassigned according to fluorescent criteria into Grades I, II, III or IV. The embryos of all groups were then cultured under the same conditions. The culture results indicated that Grades I and II were more reliable than Grades A and B for embryonic survival rate, rate of expansion and hatching rate. The rates of recovery of live embryos were also significantly different (94·7% *vs* 68·7%). Similar results were obtained with sheep embryos. By means of in-vitro culture and transfer of some frozen embryos of mouse and sheep, it has been demonstrated that fluorescence treatment does not influence embryonic development either *in vitro* or *in vivo*.

Experiment on effects of laser on goat semen. Z. Liu, Y. J. An, C. J. Tan, X. Z. Ao & C. L. Liu, *Department of Animal Science, Inner Mongolia College of Agriculture and Animal Husbandry, Huhehaote, Inner Mongolia, 01000, China.*

To investigate the possibility of laser application in goat breeding, goat semen samples were irradiated with the following results. (1) Sperm motility, survival rate, respiration rate, fructolysis, $^{32}P_i$ and Ca^{2+} intake, Zn^{2+} release, and activity of GOT, AKP and LDH in seminal plasma were affected. (2) Sperm metabolism (respiration and fructolysis) was significantly affected and the permeability of the sperm membrane ($^{32}P_i$ and Ca^{2+} intake and Zn^{2+} release) was increased. (3) Activities of three kinds of enzymes, GOT, AKP and LDH, in seminal plasma were increased but not significantly, except for GOT. Therefore the damage to spermatozoa caused by the laser was not significant. The acrosomal abnormality rate was higher than in the control group due to acrosomal swelling of dead spermatozoa. (4) These experiments show that a suitable dose of laser irradiation may be useful in goat breeding. Normal kids have been obtained after insemination of does with laser-treated semen.

Improvements in freezing rabbit spermatozoa for biotechnology studies. Y. Q. Chen, C. Blanpain-Tobback, J. M. Li, X. Yang & R. H. Foote. *Cornell University, Ithaca, NY 14853, USA.*

The rabbit is useful for various biotechnology and genetic engineering experiments involving the use of zygotes or young embryos. Uniformity and repeatability of results can be affected by the uniformity of male gametes, a factor which can be controlled by successful cryopreservation of semen. In two breeding trials involving insemination of a total of 174 females no difference in fertility was observed between frozen and unfrozen semen, when semen from selected males and $1{\cdot}6 \times 10^6$ motile sperm were inseminated. However, the procedure used for freezing and thawing had considerable room for improvement as the proportion of unstained spermatozoa before freezing was 84%, which was reduced to 44% after freezing. To improve survival rate, 6 experiments using 5–6 males per experiment were factorially arranged to measure the effect of cooling, 'seeding' and freezing procedures. The egg-yolk acetamide extender used theoretically freezes at −4°C. With usual freezing in nitrogen vapour the extended semen supercooled to about −14°C. Mechanical seeding at −6 and −9°C increased the post-freeze–thaw motility of spermatozoa by 10–16% in different experiments ($P < 0{\cdot}01$). Subsequent freezing at −15°C/min was not significantly superior to −5 or −10°C/min ($P > 0{\cdot}05$). The best results (65% motility of frozen–thawed sperm) were obtained with seeding at −6°C and freezing at −15°C/min.

Localization of MHC genes involved in resistance to Marek's disease. M. G. J. Tilanus*§, G. A. A. Alberts†, E. Egberts*, A. J. Van Der Zijpp*, B. Hepkema‡ & J. J. Blankert*, **Agricultural University, Zodiac Institute, Wageningen, The Netherlands;* †*Euribrid BS, Boxmeer, The Netherlands; and* ‡*State University of Utrecht, The Netherlands.*

Associations of Marek's disease with the major histocompatibility complex have been described in inbred chicken lines. The so-called B-complex includes three defined loci: B-F (class I), B-L (class II) and an avian specific B-g (class IV), which is expressed on erythrocytes specifically. A high linkage disequilibrium between these loci has been suggested; however, it appears that in commercial pure lines recombination between the loci has been found more frequently. Precise characterization of the individual genes and alleles has been achieved by RFLP analysis using a B-G and a B-F cDNA probe and with antisera by haemagglutination and Western blotting. In one of the White

§Present address: University Hospital, 3508GA Utrecht, The Netherlands.

Leghorn lines a limited number of B-G extended haplotypes have been identified. Refined typing for the genes and alleles indicated the presence of recombinants between the individual loci. Usage of those recombinants in a challenge with Marek virus will allow the identification and localization of the MHC gene/allele responsible for resistance.

The bovine MHC and disease associations. Ø. Lie, D. I. Våge, I. Olsaker, M. Syed, F. Lingås, S. van der Beek, V. Fosse & M. J. Stear, *Department of Animal Genetics, Norwegian College of Veterinary Medicine/National Veterinary Institute, P.O. Box 8156, Dep. N-0033 Oslo 1, Norway.*

We have shown associations between bovine class I MHC antigens and clinical mastitis. We are examining the frequency of class I alleles defined serologically and class II alleles defined by restriction fragment length polymorphism (RFLP) analysis in two groups of cattle. Cattle in one group have a history of clinical mastitis; cattle in the other group have no recorded incidence of clinical mastitis. Animals are paired according to herd and lactation number. The RFLP analysis revealed 13 DQβ variants, and strong linkage disequilibrium with class I alleles. Preliminary analysis showed significant differences in frequencies between the two groups of cattle. We are currently using human oligonucleotides and the polymerase chain reaction (PCR) to produce bovine exon-specific DNA probes for RFLP and sequence comparisons.

Expression of extra class II and class IV genes of the major histocompatibility complex (MHC) in diploid and aneuploid chickens. S. E. Bloom, M. E. Delany, J. R. Putnam, W. E. Briles* & R. W. Briles*, *Department of Poultry and Avian Sciences, Cornell University, Ithaca, NY 14853, USA; and *Department of Biological Sciences, Northern Illinois University, De Kalb, IL 60115, USA.*

The MHC is one of the more biologically influential of the multigene families in animals and man. We have developed an approach for modulating MHC dosage in chickens and evaluating biological outcomes of such changes. One strain (FCT-15) of chickens segregates various MHC microchromosome copies so that offspring have 2, 3 or 4 copies per cell. A second strain (T-MFO-$R^8$15) segregates various class IV (B-G) genes over the range of 2–6 copies per cell. In the FCT-15 strain the expression of class II immune response antigens (Ia) on B-lymphocytes was related directly to the number of MHC copies per cell. Enhanced cell surface expression was associated with altered B-cell populations of embryonic and neonatal birds. Chickens containing a B-$G^{23,2}$ duplication (Briles R^8 recombinant) were crossed to FCT-MFO-15 chickens to obtain offspring having one R^8- and two B^{15}-encoding chromosomes. Such chickens were recovered and mated *inter se* to generate a B-G gene dosage series of 2–6 copies per chicken in the F_2 generation. Blood typing of 178 offspring revealed the co-expression of the B2, B23, and B15 haplotypes on erythrocytes of diploid and aneuploid chickens. These results support a gene dosage-dependent model for regulation of both class II and class IV MHC genes.

A strategy for cloning of lysozyme gene(s) in aquacultured salmonid fish. V. Fosse, M. Syed, B. Grinde, K. Røed & Ø. Lie, *Department of Animal Genetics, Norwegian College of Veterinary Medicine/National Veterinary Institute, P.O. Box 8146, Dep. N-0033 Oslo 1, Norway.*

In marine fishes non-specific defence mechanisms appear to play an important role in conferring disease resistance. We have purified and characterized lysozyme from the kidney of rainbow trout. The kidney was chosen because it contains the highest activity of the lysozyme in this species. At least two lysozyme species were identified. They have similar molecular weights (14 400) but different isoelectric points (9·5 and 9·65). Amino acid sequences, as well as activity profiles, indicate that rainbow trout lysozymes are of the c-type. We are currently engaged in isolating the lysozyme gene(s) from rainbow trout and Atlantic salmon by applying the PCR (polymerase chain reaction) technique. The primers used are based on amino acid sequence data from rainbow trout and DNA sequences from chicken lysozyme. The aim is to sequence the gene(s) and to identify adjacent regulatory elements. We also aim to characterize the factors that influence lysozyme gene expression by their interaction with the regulatory elements.

Introduction of large DNA molecules (> 100 kb) into the germ line of transomic mice. J. Bennett & J. Gearhart, *Johns Hopkins University School of Medicine, Baltimore, MD 21205, USA.*

The ability to produce mice with cloned, large DNA molecules (> 100 kb) integrated in their genomes would open up new avenues of investigation involving the regulation of gene expression in mammals. Until now, the largest gene construct successfully microinjected and integrated into transgenic mice was 48·5 kb (Costantini & Lacy, 1981). Here we report the development of procedures for producing transomic mice, i.e. mice with cloned, unique sequences of DNA > 100 kb integrated into the germ line. Following pulsed field-gel electrophoresis of yeast artificial chromosomes (YACs), bands containing YACs with HeLa cell DNA inserts of 350 kb (and greater) were cut from the low melting point agarose, treated with agarase in the presence of the tetravalent cation spermine and microinjected into mouse pronuclei. Pilot studies demonstrated that neither the isolation nor microinjection steps sheared the DNA. Transomic mice have been identified by hybridization to human Alu hrDNA probes and to yeast telomeric sequences. Possible applications of this technology in examining the role and regulation of gene expression at critical stages of development are discussed.

Costantini, F. & Lacy, E. (1981) *Nature, Lond.* **294,** 92–94.

Differential effects of growth-hormone gene constructs on adrenal development in transgenic mice. J. G. M. Shire*†, A. Moyerhofer* & A. Bartke*, †*The Jackson Laboratory, Bar Harbor, ME 04609, USA; and *Southern Illinois University, Carbondale, IL 62901, USA.*

The adrenals have been investigated in mice hemizygous for each of 4 transgenes. Three contained the metallothionein promoter linked to human growth hormone (hGH) genes A or B, or to the bovine growth hormone gene. The fourth had the bovine growth hormone gene linked to the PEPCK promoter. The original transgenic mice were made by T. E. Wagner and J. S. Yun at Ohio University, Athens, OH. The two human GH genes caused disproportionate increases in adrenal weight in males, whilst the adrenals of the bovine GH transgenic were only enlarged in allometric proportion (Shea *et al.*, 1987) to the overall increases in body weight. Adrenals were very large (up to 20 mg) in some transgenic females, but their analysis was complicated by developmental changes affecting the juxtamedullary X-zone. This zone was very large in bovine GH transgenics and was undergoing the 'fatty' involution characteristic of normal females. Transgenic females from both the hGH-A and hGH-B stocks showed lipophilic 'brown' degeneration, only found in certain old or oestrogen-treated females. hGH-A and hGH-B males also showed 'brown' degeneration, unlike normal males and the bovine GH transgenics. Only the hGH-A protein has prolactin-like effects in rodents (Selden *et al.*, 1988), but the adrenal effects of the B-gene suggest it may also be prolactin like. Some effects may be indirectly changing hypothalamic neurotransmitters (Bartke *et al.*, 1988).

Funded by Wellcome Trust (J.G.M.S.) and NIH grants HD20001 and HD 09042 (A.B.).

Bartke, A., Steger, R.W., Hodges, S.L., Parkening, T.A., Collins, T.J., Yun, J.S. & Wagner, T.E. (1988) *J. exp. Zool.* **248,** 121–124.
Selden, R.C., Wagner, T.E., Blethen, S., Yun, J.S., Rowe, M.E. & Goodman, H.M. (1988) *Proc. natn. Acad. Sci. USA* **85,** 8241–8245.
Shea, B.T., Hammer, R.E. & Brinster, R.L. (1987) *Endocrinology* **121,** 1924–1930.

Expression of human protein C in the liver of transgenic mice. J. Vitale, G. Monastersky, P. Bourdon, N. Capalucci, E. Cohen, B. Roberts, P. DiTullio, R. Wydro, G. Moore & K. Gordon, *Integrated Genetics, One Mountain Road, Framingham, MA 01710, USA.*

We have produced transgenic mice which contain a liver-specific fusion gene. It was our goal to use transgenic mice as a means of establishing cell lines for the production of human therapeutics. By targeting genes to endogenous sites of production, we plan to produce therapeutic proteins in a biologically active form. In our experiments we used the liver-specific promoter/enhancer for mouse albumin gene fused to the cDNA for human protein C. Protein C is a therapeutic with clinical significance in the prevention of blood clots. Several lines of transgenic mice were generated. All had human protein C in their blood, although the levels varied. Expression levels remained constant in subsequent generations. The secreted protein C comigrated with native human protein C when analysed by Western blotting. Organ cultures established from perfused livers expressed protein C.

Expression of the human growth-hormone transgene in heterozygous and homozygous mice. J. S. Yun, Y. Li, D. C. Wight, R. P. Portanova* & T. E. Wagner, *The Edison Animal Biotechnology Center, and *College of Osteopathic Medicine, Ohio University, Athens, OH 45701, USA.*

Transgenic mice carrying the mouse metallothionein-I/human growth hormone (mMT-I/hGH) fusion gene exhibit high levels of serum human growth hormone (hGH), enhanced somatic growth, and female sterility. Over the past several years, we have transmitted the transgene via the male germ line, by mating transgenic males with non-transgenic females, thereby producing several generations of heterozygous progeny containing only a single (paternal) copy of the transgene. Recently, we succeeded in producing homozygous transgenic mice by mating male transgenics with female non-transgenics which were the recipients of ovaries transplanted from female transgenic mice. In the heterozygous and homozygous transgenic animals: (1) serum hGH levels are highest in the young, gradually decrease with age and become stable in adults, (2) hGH levels are higher (1·5-fold) in males than in females, and (3) hGH levels are increased (about 10-fold) by Zn induction. However, in comparison to heterozygous animals of the same sex, the homozygous transgenic mice attain a greater body weight and have higher (about 2-fold) levels of hGH-mRNA and serum hGH under basal conditions and in response to Zn induction. In summary, the expression of the transgene is qualitatively similar in the heterozygous and homozygous transgenic mice, but the level of transgene activity is greater in the homozygous animal. We interpret this finding to indicate that, in the homozygous animal, both copies (maternal and paternal) of the transgene are active and expressed independently.

A 796 bp promoter sequence from the human neuropeptide Y gene directs tissue-specific gene expression in transgenic mice. G. C. Waldbieser, C. D. Minth, J. E. Dixon & C. L. Chrisman, *Purdue University, West Lafayette, IN 47907, USA.*

Neuropeptide Y (NPY) is the most abundant neuropeptide found in the central nervous system, and is a potent stimulator of food intake in mammals. To study the regions of the NPY gene which confer tissue-specific activity, we have produced several gene constructs containing the chloramphenicol acetyltransferase (CAT) gene fused with various lengths of promoter DNA from the human NPY gene. Approximately 900 copies of an NPY-CAT transgene containing 796 bp of NPY promoter were microinjected into each (C57BL/6J $\times$ SJL/J) F_2 embryo. Of the 81 embryos transferred to pseudopregnant ICR females, 16 young were born and 2 tested positive for gene incorporation. We have established transgenic lines from both founder animals, and both lines appear to contain a single integrated site. The transgene is integrated on the X chromosome in one line, while the other line carries the transgene on an autosome yet to be determined. Extracts from several tissues have been prepared from mice of both lines and analysed for CAT enzyme activity. To date, we have found significant levels of CAT activity only in brain tissue.

Attempts to improve efficiency of embryo transfer in transgenic mouse experiments. B. Pintado, R. J. Wall*, V. G. Pursel* & S. Martin, *Department of Produccion Animal, CIT-I.N.I.A, Carretera de La Coruna Km 5 900, 28040 Madrid, Spain; and *U.S. Department of Agriculture, Agricultural Research Service, Reproduction Laboratory, Beltsville, MD 20705, USA.*

Normally, 20–25% of microinjected mouse eggs, transferred to recipient mothers, result in live born young. We theorized that the number of eggs surviving to term might be increased by improving efficiency of embryo transfer. In the first experiment, 10 µl adrenaline were applied topically to the ovarian bursa at concentrations of 0, 0·1, 0·3 and 1 µg/µl 30 sec before exposing the oviducal ostium; 509 microinjected eggs were transferred to 24 recipients/treatment. The haemostatic property of adrenaline, which improved visualization of the infundibulum, was apparent at 0·1 µg/µl. Recipients were killed on Day 14 or 15 of gestation, and implantation sites of live and resorbing fetuses were counted. Mean pregnancy rate was 70% (range 54–83%); no. live fetuses/pregnant recipient was 7·4, and embryos resulting in live fetuses was 24%. There was a tendency toward higher pregnancy rate with 1 µg/µl and larger litter size with the dose of adrenaline of 0·1 µg/µl. In the second experiment, microinjected eggs were either transferred all into one oviduct or half into each oviduct; 930 eggs were transferred to 40 recipients. For monolateral and bilateral transfers, pregnancy rate was 80% and 75%, number of live fetuses/pregnant recipient, 7·7 and 9·1, and transferred embryos resulting in live fetuses, 30% and 34%, respectively. A trend towards larger litters was observed for bilateral transfers. The application of adrenaline to the bursa reduces bleeding and does not alter embryo survival.

Male hypogonadism and altered steroidogenesis in transgenic mice carrying a hGR/c-myc fusion gene. F. Pothier, M. V. Govindan, G. Pelletier & A. Bélanger, *MRC Group in Molecular Endocrinology, Laval University Medical Centre, Quebec, Canada G1V 4G2.*

Our group has isolated the promoter region of the human glucocorticoid receptor gene (hGR) from the λ EMBL-3 human genomic library. Gene transfer studies with hGR-CAT chimaeric plasmids into CV-1 cells showed that the hormonal regulatory sequences of hGR are contained within a 4·0 kb EcoRI-XbaI fragment. With a view to investigating the regulation of the hGR *in vivo*, we have produced transgenic mice by microinjection of the 4·0 kb fragment of the hGR promoter linked to c-myc as reporter gene. By Southern blot analysis, 5 animals (3 males and 2 females) were found to contain the injected sequences. Surprisingly, all the transgenic males (except 2 mosaic founders) derived from different founders were sterile and had atrophic testes and prostate glands. Histological analysis of the testes indicated the presence of normal Leydig cells with under-developed seminiferous tubules. No c-myc mRNA has been detected in tissues of these transgenic mice. Steroid analyses of the testes show a 20-fold increase of 17-hydroxypregnenolone, 17-hydoxy-progesterone and a 10-fold increase of progesterone and cortisol. Our results suggest that the observed effects are related to the sole presence of the hGR promoter competing for the same factors regulating the endogenous mouse GR. Although these results have to be extensively investigated, we think that this approach can constitute a first step towards a 'genetic castration'. If proven, it would be tempting to use competing DNA fragments to derive strains of farm animals in which only males would be sterile.

Selection of transgenic preimplantation embryos by PCR. T. Ninomiya, M. Hoshi, A. Mizuno, M. Nagao & A. Yuki, *Research Institute of Life Science, Snow Brand Milk Products, Co., Ltd, 519 Shimoishibashi, Ishibashi-machi, Shimotsuga-gun, Tochigi 329-05, Japan.*

The polymerase chain reaction (PCR) was tested to select transgenic embryos before they were transferred to recipients. The selection system involves bisection of morulae, selection of the half-morulae containing target sequences within 7 h, and transfer of the sister half-morulae. PCR analysis of morulae derived from transgenic mice showed that 36 of 41 implanted embryos from the PCR-positive morulae were transgenic, and that 27 of 28 implanted embryos from the PCR-negative morulae were wild type. The system was tested on fertilized mouse eggs into which pSV2-gpt-gE1A DNA had been injected. The injected DNA was detected in 30 of 84 morulae derived from the microinjected eggs. None of the 7 implanted embryos developed from the PCR-negative morulae had detectable amounts of transgenes, and 1 of the 2 successfully implanted embryos from the PCR-positive morulae was transgenic. This system was shown to be useful primarily for selecting transgene-negative preimplantation embryos before transfer and could be applied to generate transgenic livestock.

LIST OF AUTHORS CONTRIBUTING

INDEXES

LIST OF AUTHORS CITED

Entries in **bold type** indicate citations in the reference sections.

SUBJECTS